Geotechnical
Materials in
Construction

Other McGraw-Hill Books of Interest

BEALL AND MASONRY • *Design and Detailing*

BRANTLEY AND BRANTLEY • *Building Materials Technology*

BREYER • *Design of Wood Structures*

BROCKENBROUGH AND MERRITT • *Structural Steel Designer's Handbook*

BROWN • *Foundation Behavior and Repair*

BROWN • *Practical Foundation Engineering Handbook*

FAHERTY AND WILLIAMSON • *Wood Engineering and Construction Handbook*

GAYLORD AND GAYLORD • *Structural Engineering Handbook*

MERRITT AND RICKETTS • *Building Design and Construction Handbook*

MERRITT • *Standard Handbook for Civil Engineers*

NEWMAN • *Design and Construction of Wood Framed Buildings*

NEWMAN • *Standard Handbook of Structural Details for Building Construction*

SHARP • *Behavior and Design of Aluminum Structures*

To order or receive additional information on these or any other McGraw-Hill titles, please call 1-800-822-8158 in the United States. In other countries, contact your local McGraw-Hill representative.

Geotechnical Materials in Construction

Marian P. Rollings
Vicksburg, Mississippi

Raymond S. Rollings, Jr.
Vicksburg, Mississippi

McGraw-Hill

New York   San Francisco   Washington, D.C.   Auckland   Bogotá
Caracas   Lisbon   London   Madrid   Mexico City   Milan
Montreal   New Delhi   San Juan   Singapore
Sydney   Tokyo   Toronto

Library of Congress Cataloging-in-Publication Data

Rollings, Marian P.
    Geotechnical materials in construction / Marian P. Rollings and
Raymond S. Rollings.
        p.   cm.
    Includes bibliographical references and index.
    ISBN 0-07-053665-1 (hardcover)
    1. Engineering geology—materials.   2. Soil stabilization—
Materials.   3. Geosynthetics.   I. Rollings, Raymond S.   II. Title.
TA705.R587   1996
624. 1'51—dc20   95-20882
                                                                      CIP

McGraw-Hill

A Division of The **McGraw·Hill** Companies

1 2 3 4 5 6 7 8 9 0   DOC/DOC   9 0 0 9 8 7 6 5

ISBN 0-07-053665-1

*The sponsoring editor for this book was Larry S. Hager, the editing
supervisor was Fred Bernardi, and the production supervisor was
Pamela A. Pelton. It was set in Century Schoolbook by Dina John of
McGraw-Hill's Professional Book Group composition unit.*

Printed and bound by R. R. Donnelley & Sons Company.

McGraw-Hill books are available at special quantity discounts to use
as premiums and sales promotions, or for use in corporate training programs. For more information, please write to the Director of Special
Sales, McGraw-Hill, 11 West 19th Street, New York, NY 10011; or contact your local bookstore.

This book is printed on acid-free paper.

To our parents and our children

Contents

Part 2   Basic Geotechnical Construction Techniques

Part 3   Specialized Construction Techniques

Construction Series Preface

Construction is America's largest manufacturing industry. Ahead of automotive and chemicals, construction represents 14 percent of GNP. Yet construction is unique in that it is the only manufacturing industry where the factory goes out to the point of sale. Every end-product has a life of its own and is different from all others, although component parts may be mass produced or modular.

Because of this uniqueness, the construction industry needs and deserves a literature of its own, beyond reworked civil engineering texts and trade publication articles.

Whether management methods, business briefings, or field technology, it will be covered professionally and progressively in these volumes. The working contractor aspires to deliver to the owner a superior product ahead of schedule, under budget, and out of court. This series, written by constructors and for constructors, is dedicated to that goal.

M. D. Morris, P.E.
Series Editor

Preface

Today we have many texts on various aspects of geotechnical engineering. These provide a wealth of information on soil mechanics, foundation engineering, specialized practices, and similar topics, and yet despite this wealth of information, we continue to encounter problems with soils during construction. Problems in the field may develop from decisions made during the design stage, concerning project specifications and plans, or during the construction process. These problems are most commonly of a basic, fundamental nature rather than due to more esoteric causes. We have undertaken production of this book to try to address some of the day-to-day problems encountered in working with geotechnical materials. As such, this book is not a theoretical treatment of geotechnical engineering, but instead it offers a pragmatic approach to problems commonly encountered in working with geotechnical materials, including the nature of geotechnical materials, testing, compaction, stabilization, and site improvement. We also have included chapters covering several specialty topic areas. Some of these are specific geotechnical problems such as expansive clays, collapsible soils, and freezing effects that cause distinctive problems in design, construction, and service. We also have included a chapter on geotextiles and one on impoundments, landfills, and liners. These last two chapters cover subjects that have become a major part of geotechnical work in the modern construction market.

The objective of this book is to provide solutions and guidance to overcome common construction problems with geotechnical materials. The intended audience includes engineers and architects who must design projects with these materials, provide plans and specification for the work, and then provide on-site construction engineering that must deal with these materials. We have purposely avoided a theoretical treatment of the issues, so prior education in mathematics, engineering, and science is not required to make use of this book's material. Consequently, this book also will be useful to contractors and quality-control/quality-assurance personnel working with geotechnical materials. Most projects today require multidiscipline teams, and we hope this book will prove useful to other specialists such as engineering geologists or environmental engineers who also must work with geotechnical materials as part of these teams. The student studying engineering, architecture, or geology may find this book useful as a guide to actual field problems and their solution and as a supplement to his or her more theoretical academic studies.

The impetus to prepare this text came from our long-time friend Mr. M. D. Morris of McGraw-Hill, to whom we are thankful. Admittedly, there were periods over the years that were spent preparing the manuscript that we wondered about the friendship that prevailed on us to undertake this task. Similarly, we appreciate the assistance of Mr. L. Hager, senior editor with the McGraw-Hill Professional Book Group. We are particularly indebted to Professor George Sowers for his thorough review of the manuscript and his many helpful comments. A number of colleagues graciously agreed to review and comment on specific portions of the manuscript, and we would like to thank them for their efforts. These include

Mr. Ross Bentsen, The Asphalt Institute, Lexington, Kentucky

Mr. David Coleman, Soil Testing Engineers, Jackson, Mississippi

Mr. Joe Fluet, JEF Associates, Boynton Beach, Florida

Dr. Starr Kohn, Soil Materials Engineers, Plymouth, Michigan

Professor Christopher Mathewson, Texas A&M University, College Station, Texas

Professor John Metcalf, Louisiana State University, Baton Rouge, Louisiana

Mr. Ray Steinle, Rust Engineering, Denver, Colorado

Any errors or mistakes in the text are solely the authors' responsibility, however.

Drawings for the text were prepared by Mr. Dennis Mathews of CADD Systems, Vicksburg, Mississippi. Photographic reproductions were done by L & L Photo of Vicksburg, Mississippi. Their assistance is greatly appreciated.

Acknowledgments

We gratefully acknowledge permission to reproduce copyrighted material from the following sources where noted in the text: American Concrete Institute, American Society of Civil Engineers, American Society for Testing and Materials, Blackie Academic and Professional,

Boston Society of Civil Engineers, Chapman and Hall Inc., CRC Press, Inc., Elsevier Scientific Publishing Company, Institute of Civil Engineers, Institution of Highway Engineers, International Fabrics Association International, Japanese Society of Soil Mechanics and Foundation Engineering, John Wiley and Sons, Journal of Soil Science, McGraw-Hill Book Company, Organization for Economic Cooperation and Development, Pentech Press Limited, Portland Cement Association, The Asphalt Institute, The Association of American Geographers, and the Transportation Research Board. We also gratefully acknowledge use of material from government sources as noted in the text, including the U.S. Environmental Protection Agency, Federal Aviation Administration, Federal Highway Administration, U.S. Air Force, U.S. Army, and U.S. Army Corps of Engineers.

In addition, we would like to acknowledge permission to use photographs from the following sources and individuals as noted in the text: Dynapac Mfg., Inc., Seaman-Maxon, Inc., Wacker Corporation, Mr. Less Perrin, Ft. Worth, Texas, and Dr. Randy Ahlrich, Vicksburg, Mississippi.

Ray and Marian Rollings

1

Introduction

The real purpose of books is to trap the mind
into doing its own thinking.   C. MORLEY

1.1 Objective

Geotechnical materials—soils and rocks—are at the heart of all civil engineering efforts. All the weight of the lofty spires of our soaring cathedrals or of the bridge girders boldly spanning a chasm must ultimately be brought to bear back on soils or bedrock. Failure to realize this simple fact and to provide sound foundation support has caused many problems for engineers, architects, and contractors over the years. Geotechnical materials perform many mundane tasks in engineering such as providing backfill, but they also can be processed to provide important engineering functions such as impermeability for a landfill liner or structural strength for an airfield pavement base course. As the primary constituent of asphalt concrete and portland cement concrete, they provide us with some of our most economical and high-quality construction materials. It is virtually impossible to visualize any civil engineering effort that is not affected in some manner by geotechnical materials. Consequently, it behooves anyone involved in engineering design and construction to be familiar with the idiosyncrasies of geotechnical materials and to be aware of practical techniques for construction with these materials.

Today we have many excellent books on classic geotechnical topics such as soil mechanics or foundations. In addition, there are many fine references on specialized topics dealing with various aspects of geotechnical engineering, e.g., excavation, dewatering, drilled piers, expansive soils, etc., and for the specialist, we have in-depth theoretical treatises on topics such as advanced soil mechanics and probability applications. However, construction problems in the field are generally due to a failure to grasp the fundamental principles of dealing with geotechnical materials rather than to theoretical factors. This text concentrates on examining some of these basic techniques and principles of building with geotechnical materials.

We have attempted to draw together the fundamental information for dealing with geotechnical materials in field construction without producing another academic text on soil mechanics or delving into material that is adequately covered elsewhere. We have drawn heavily on our personal experience to identify the topics included in this text, and in some ways, this book is a collection of things we wished we had learned in school before running headlong into the problem in the field. This book is written for the practicing engineer, architect, or contractor. We have purposely eschewed theoretical discussions and development of topics that are more appropriately found in academic texts. Students may find this book useful as a supplement and balance to their more theoretical academic studies. Quality-control and quality-assurance personnel also will find much of the material here directly applicable to their endeavors.

We have attempted to limit our material in the text to that necessary to understand geotechnical materials and to overcome field construction problems. These problems may arise from decisions made either during the design stage, during preparation of the plans and specifications, or during construction. Consequently, we feel that the engineer or architect designing a project and preparing the designs and specifications must be as familiar with the fundamental principles of dealing with geotechnical materials as the field construction engineer or contractor.

This book also treats aggregates, whether used for production of concrete, for a structural base course, or for some other purpose, as geotechnical materials. This is a departure from conventional geotechnical texts with which we are familiar, but we have found that often it is the geotechnical engineer in an organization that is tapped to prepare specifications for these materials, evaluate their suitability, determine the adequacy of their placement, and so forth. While we do not discuss the specifics of asphalt and portland cement concrete production, we do cover aggregate production, characteristics, and use in various portions of this text.

The nine technical chapters in the book are broadly divided into three parts:

Part 1   Geotechnical Materials and Testing

Part 2   Basic Geotechnical Construction Techniques

Part 3   Specialized Construction Techniques

The three chapters in Part 1 provide a basic overview of geotechnical materials. Chapter 2 covers fundamental information on formation, engineering characteristics, and classification of soils and aggregates to aid the reader who is unfamiliar with geotechnical materials or as a refresher for the more experienced reader. Chapter 3 provides a brief description of the impact of geologic conditions on geotechnical materials, while Chap. 4 covers materials testing. These are crucial items. Geology determines the nature of the materials with which we work, and testing is used to measure material properties employed in engineering analysis or to determine compliance with specifications. The three chapters in Part 2 cover compaction, stabilization, and site improvement. Excavation and moisture control are basic construction techniques encountered in geotechnical work, but they are already covered in depth in other references (e.g., Church 1981, Powers 1981). Part 3 includes one chapter covering a potpourri of special topics that fail to fit neatly into any specific category. Geotextiles have become a basic tool in geotechnical construction and are covered in Chap. 9. The final chapter provides an overview on impoundments, landfills, and liners. These last topics are increasingly important in geotechnical engineering as the amount of environment-related work increases.

1.2 Theory Versus Practical Experience

We once previously expressed our personal views on civil engineering as follows (Rollings 1991):

> Civil engineering is a learned profession requiring specialized education and training and demanding skills not possessed by the lay public....Although engineers apply the principles of science to design and build structures of various kinds for use by the public, it is the production of some structure or the changing of some natural condition for the benefit and use of mankind that separates the engineer from other highly educated and trained groups such as scientists....Engineering is learned by the practice of the profession and not by education alone. It is an art that develops a concept, conceives a plan or design to bring the concept to fruition, and then builds the concept with full consideration of function, safety, economy, and aesthetics. It is not simply a string of calculations, nor is it a collection of sterile plans and specifications. The miracle of engineering is the production of some concrete product for the benefit of people.

Science and theory are the handmaidens of civil engineering but do not alone constitute the profession of civil engineering. The art of civil engineering requires applying these theoretical tools with practical knowledge of materials, construction, field conditions, and human nature to build a useful and safe structure. Satisfactory results can only be achieved when both theory and practical knowledge are used together synergistically. To use one without the other often leads to unsatisfactory and occasionally catastrophic results.

Early civil engineering construction in the United States was primarily a craft learned by doing and self-study, and practical experience was the requisite precursor to practicing civil engineering. As U.S. universities began turning out trained and scientifically educated engineers—the U.S. Military Academy first (1802), followed by Rensselaer Polytechnic Institute (1835), Union College (1845), and Harvard and Yale (1846)—the dependence on empiricism waned, and scientific principles began to come to the forefront. Although science and mathematics can provide elegant solutions for the engineer and added insight into problems, the uncompromising world of geology, materials, construction, and complex loading ensures that there remains a great deal of art beyond science alone in actual civil engineering construction.

As early as 1893, cautions against unwarranted reliance on simplified analytical approaches were being put forth (Anderson 1893):

> There is a tendency among the young and inexperienced to put blind faith in formulas, forgetting that most of them are based upon premises which are not accurately reproduced in practice, and which, in any case, are frequently unable to take into account collateral disturbances, which only observation and experience can foresee, and common sense provide against.

When the natural complexity and uncertainties involved with geotechnical work are factored into the problem, the situation worsens. The late Karl Terzaghi (1961) provided the following commentary on the difficulties of applying purely analytical and rigorous scientific approaches to the art of geotechnical engineering when he observed:

> Many problems of structural engineering can be solved solely on the basis of information contained in textbooks, and the designer can start using this information as soon as he has formulated his problem. By contrast, in applied soil mechanics, a large amount of original brain work has to be performed before the procedures described in the textbooks can safely be used. If the engineer in charge of earthwork design does not have the required geological training, imagination and common sense, his knowledge of soil mechanics may do more harm than good. Instead of using soil mechanics he will abuse it.

Another experienced foundation engineer made similar observations while commenting on causes for failures in the field (Immerman 1963):

> The principles of this science [soil mechanics] may not be successfully applied by the mere substitution of figures for symbols in some formula, nor may results be obtained in the same manner as that in which a beam size is obtained from a steel handbook, having previously determined the section modulus. The successful application of this fine sci-

ence [soil mechanics] requires the combination of mature judgement with a realization of the possible shortcomings in the formula being applied or a realization that the formula presupposes conditions difficult of attainment in the field. The science has not failed—the tool has just been improperly handled.

The need to maintain a blend of sound theory with a practical grasp of design, construction, and materials behavior is crucial in civil engineering works and particularly for those using geotechnical materials. Failure to maintain this balance results in costly and expensive failures. The late Gregory Tschebotarioff was a widely experienced engineer and educator who made the following observation on the prevalent cause of failures in civil engineering works (Tschebotarioff 1973):

> About 80 percent of all soil and foundation failures the writer had to study over the past quarter century were primarily caused by one or more of the following: either the structural engineers involved had no proper understanding of the relevant soil-engineering problems; or the soil engineers did not appreciate the structural problems; or the engineering administrators had no real grasp of either field.[1]

After reviewing over 500 failures with which he had been involved over his career, another eminent engineer and educator, George Sowers (1993), similarly observed

> Unfortunately, 88 percent of the failures have a human cause: not understanding contemporary technology or not using contemporary technology when they understand it. The 88 percent of problems related to human deficiencies challenge our profession, because most of those could have been prevented.

Clearly, engineering must maintain a balance of the scientific and the practical because these are the tools that can combine to produce economical and safe engineered products. This poses a particularly challenging problem for educating and training new engineers because there are two different aspects to the art of engineering that must be mastered.

Modern engineering universities are largely research institutions. This has developed for a variety of reasons, including financial necessity and cultural pressures within the scientific academic community. The engineering practitioner often has responded with sharp criticism of the increasingly scientific bent of engineering education and the declining prestige of practice-oriented training within the engineering student's curriculum. Porter (1991) described the current situation as follows:

---

[1]G. P. Tschebotarioff, *Foundations, Retaining and Earth Structures,* McGraw-Hill, New York, 1973. Reprinted with permission of McGraw-Hill, Inc.

The student is typically prepared to competently meet technological challenges, but not prepared in the practical skills and knowledge necessary to immediately compete in many civil engineering work environments....Employers tend to hire those who can do the work required. The resulting trend is that engineers, and particularly civil engineers, are losing their status as team leaders and managers and becoming technical mules for engineering services managed by nonengineers. This trend is reducing the quality of engineering services, because technological impacts are neglected or mismanaged by teams deficient in competent leadership with engineering expertise.

There are some chilling parallels between Tschebotarioff's (1973) observations on engineering management's contributions to failures and those of Porter (1991) on current management of engineering works, although almost two decades separate their writings.

Current U.S. professional engineering licensing laws recognize the need for both theoretical knowledge and experience. They typically require that professional engineers demonstrate some minimal scientific knowledge (initial fundamentals examination for engineer-in-training status) and practical design knowledge (later practice-oriented examination). However, these examinations also must be supplemented with documented evidence of at least 4 years of engineering experience under a licensed engineer before an applicant is granted a professional license to practice engineering.

As a profession, engineers historically have recognized the need to have both academic scientific training and practical knowledge and experience. The academic community can and does provide the scientific and theoretical training needed to practice engineering. The debate centers on the degree to which academia can and should provide practice-oriented training. Realistically, the engineering academic staff at a university must compete for positions, tenure, and pay raises within a system that rewards research, theoretical excellence, and scientific publication. This is a university-wide culture that is outside the control of the engineering faculty; hence their personal interests must coincide with these university goals if they are to be personally successful at the university. It would be nice to add additional practice-oriented courses on design and construction practicalities taught by experienced engineers to the students' required course work, but the existing curriculum is already so overloaded that it is increasingly difficult to finish an undergraduate civil engineering degree in 4 years. To do more would require a 5-year or more professional degree in civil engineering, as is done for law and medicine. This has been resolutely resisted by both academia and practicing professionals.

Academia does not appear equipped under current practices to consistently provide the practical knowledge to supplement the scientific training needed to practice civil engineering. At best, academia can include added emphasis on design practices, case studies, and especially warnings on the limitations of theoretical approaches within their existing course work. However, the onus of providing a new engineer's practical engineering must today, of necessity, fall on the practicing professionals.

Practical experience can be gained in one of two ways: from personal experience or from someone else's experience. The first method is the best teacher but also the most painful. Learning from someone else's experience can be very valuable, and we all share a professional obligation to share our experiences with others. Case studies and descriptions of failures and problems such as found in the American Society of Civil Engineers' *Journal of the Performance of Constructed Facilities* or similar publications can provide the reader valuable insight into what went wrong on other projects and how to avoid similar problems. The engineering practitioner, however, must make the time to prepare these contributions for publication, and the reader must make the effort to study and learn from these mistakes of others. Such material will then also be readily available to academic faculty for inclusion in their classroom lectures where appropriate.

By concentrating on the construction and behavioral aspects of geotechnical materials, we hope that this text will provide practice-oriented guidance. Hopefully, it will serve as a useful supplement to more common theoretical geotechnical texts. Ideally, the reader will find help in solving (or even better, avoiding) his or her geotechnical construction problems.

1.3 Responsibilities, Risk, and Quality Control

At one time, engineers were intimately involved in all aspects of an engineering project. The famous bridge engineer, John Roebling, started his contributions to U.S. engineering by manufacturing wire rope as a substitute for the clumsy hemp rope in use on American canals. With his excellent German technical education, supplemented with intellectually broadening studies under and friendship with the famous German humanist George Wilhelm Friedrich Hegel (Kelley 1991), Roebling soon became the leading U.S. bridge engineer and builder of the last century. His untimely death prevented him from completing his crowning achievement, the Brooklyn Bridge, but he left the project in the capable hands of his engineer son, Washington Roebling, and daughter-in-law, Emily Roebling (who arguably came to function as the first U.S. woman engineer). This remarkable engineering family was involved in production of wire rope, state-of-the-art design in suspension bridges, marketing of their wire rope and of their bridge concepts, arranging financing and public support, introducing technical innovations such as pneumatic caissons, and overseeing the day-to-day construction and inspection of their works.

Today we no longer have such a streamlined approach and have produced many different specialists who labor on such projects. The owner, being typically a corporate entity or government agency, has a procurement or engineering staff that will contract with an engineer or architect for design services. The engineer or architect, in turn, often subcontracts work to various specialists and consultants (e.g., geotechnical, structural, stormwater control, environmental impact, traffic planning, etc.) and then reassembles this work into a package of plans and specifications with any necessary supporting documentation suitable for review by the owner and for any needed permit applications. After revisions mandated by the owner and possibly regulatory agencies, the plans and specifications are ready to send out to contractors for bid.

The owner may use a simple low-bid process open to anyone licensed and bonded to do the work, or the bids may be solicited from only the owner's list of approved and prequalified contractors. The first approach is typically required for public-agency procurement. The contractor usually has 4 to 6 weeks to prepare his or her bid, and the lowest bidder gets the contract to do the work in accordance with the plans and specifications. This contractor then will subcontract out work to other specialty contractors. Inspection of the contractor's work may be done by the owner's staff, the original designer, an independent consultant, or the contractor (under the oxymoron *contractor quality control*), or it may not be done at all. Intermixed with this process are material and equipment manufacturers who are marketing to owners, engineers, architects, and contractors to get their products specified, approved for use under the specification, or accepted as a substitute for that required in the specification. There are numerous variants to this scenario (e.g., construction manager approaches, turnkey construction, design-build, etc.), but suffice it to say that our modern engineering construction process is complex with many overlapping areas of responsibility. The design engineer or architect is also often far removed from what is happening in the field.

Starting somewhere in the 1950s, the number of lawsuits involving engineering and construction began increasing rapidly, and litigation considerations are today a part of every design and construction project. This highly litigious climate has fractured participants into separate and often hostile and suspicious camps centered on the owner, engineer/architect, and contractor. Some steps have been taken to reduce the incidence of litigation and its heavy financial cost (e.g., contractual agreements to use alternate dispute resolution). However, the current construction climate does not encourage team efforts, and most parties to the construction effort conscientiously try to avoid as much responsibility, risk, and liability as possible.

Probably the single biggest difference in geotechnical construction work and the remainder of the construction industry is having to deal with the uncertainty of what lies below the surface. Any subsurface drilling program samples only a minuscule fraction of the total underlying materials that may affect the project. Only when excavation or loading reveals the true nature of underlying materials can we be certain of their behavior, and even then nature may still surprise us. Too often the site investigation may be done very superficially (e.g., "lets put down a couple of borings and see what we've got out there"), or it is let to the lowest bidder, who may not have sufficient experience or local knowledge to conduct the work adequately. The field information may or may not be made available to the contractor preparing the bid. If it is made available, there is often a specific exculpatory clause releasing the owner from any responsibility for the subsurface conditions and directing the contractor to carry out his or her own subsurface studies. This is generally unreasonable because the time the owner allows for bid preparation is usually too short for an adequate site investigation in time to influence the bid preparation.

A general legal principle holds that without specific contract provisions to the contrary or unless the owner misrepresents or fails to disclose known subsurface conditions, the contractor bears the risk of unforeseen subsurface conditions (Postner 1991), but there is considerable variation in interpretation within different legal jurisdictions (Jarvis 1963). The owner will pay for shifting the risk onto the contractor, and this will be reflected in higher bid prices and increased likelihood of claims during construction. It will be more economical in the long run for the owner to shoulder this risk for subsurface conditions.

Acceptance of this responsibility typically will require paying for an adequate initial site investigation by competent personnel, providing all available information to the bidders, including a reasonable changed conditions clause in the contract, and paying for the designer to monitor the geotechnical work to identify any departures from the design assumptions. Dealing with uncertainty in construction is discussed extensively by the American Society of Civil Engineers (1963 and 1991) and by Sowers (1971).

No single document is as important as the specifications for construction of a satisfactory project. These specifications should be (1) technically accurate and adequate, (2) definitive and clear, (3) fair and equitable, (4) easy to use, and (5) legally enforceable (Dunham and Young 1971). They also must be constructable in the real world.

The specifications prepared by the design engineer or architect will largely determine if interactions and construction on a project will proceed smoothly or with difficulty.

Simons (1992) identifies three obstacles to achieving satisfactory construction once a proper set of specifications is prepared:

1. *The low-bid owner* attempts to gain a final product worth more than he or she is willing to pay.

2. *The change-order contractor* bids low to get the job and then floods the owner with claims for changes.

3. *The ivory-tower designer* treats the specifications as sacrosanct, refuses to visit the site to identify changes from his or her design assumptions, and/or is unable to accept real-world construction limitations on the practicality of carrying out a design.

The low-bid owner looks constantly for ways to reduce expenditures and may shortchange the initial site investigation or quality of materials used in the work. He or she may fail to call in consultants where needed, not allow or pay for the original designer to monitor construction, and limit or drop requirements for testing. In the end, you will generally get no more than you pay for, and this type of owner will invariably get poor work.

No perfect set of specifications has ever been prepared, and the change-order contractor can often fabricate alternate interpretations of specification clauses to allow a claim (whether legitimate or not). Such contractors are not the norm, happily. If encountered, they should be held as rigidly as possible to the specifications (resulting in very unpleasant working conditions) and barred from future jobs where possible.

There are always legitimate changes in a job, however, and the contractor should be fairly compensated for these. The ivory-tower designer generally has little or very narrow experience, which prevents him or her from adjusting designs and specifications for real-world conditions or changes in site conditions that depart from the original design assumptions. One also should bear in mind that the job was bid on the specifications, and failure to require compliance with the specifications without valid technical reason unbalances the bidding process and discriminates unfairly against the other bidders who did not get the job but prepared their bid to meet the published specifications. There are also times that the requirements of the specification or design concept are critical, and they must be enforced scrupulously, although the importance may not be apparent to the construction and field forces. Again, we see that engineering is an art of balance that uses science rather than a rigid theoretical discipline.

The literature on failures in engineered structures commonly calls for increased inspection to help avoid recurrence of such calamities. While there are different philosophies on how inspection may be accomplished, we firmly believe that it is crucial that the original designer have significant inspection responsibilities for which he or she is compensated (Rollings 1991). Only the designer is aware of all the nuances and intent in the original design, and it is the designer who is best equipped to identify changes or events that impact these original assumptions and to determine satisfactory solutions to departures from these assumptions. To accomplish this, the designer will need a team of trained and experienced inspectors. Unfortunately, these conditions are often not met in actual construction.

Many aspects of human nature come into play to complicate good interaction between design and construction. The late Karl Terzaghi (1958) described some of this internal conflict as follows:

> On the other hand, in the realm of earthwork and foundation engineering the absence of continuous and well-organized contacts between the design department and the men in charge of the supervision of the construction operations is always objectionable and can even be disastrous....it [the interaction between design and construction personnel] often depends on whether or not the design and construction departments are on friendly terms with each other. More often than not the two departments despise each other sincerely, because their members have different backgrounds and mentalities. The construction men blame the design personnel for paying no attention to the construction angle of their projects, but they are blissfully unaware of their own shortcomings. The design engineers claim that the construction men have no conception of the reasoning behind their design, but they forget that the same end in design can be achieved by various means, some of which can be easily realized in the field, whereas others may be almost impracticable....In any event, the construction men have no incentive to find out whether or not the design assumptions are in accordance with what they experience in the field during construction and serious discrepancies may pass unnoticed. If conditions are encountered which require local modifications of the original design, the construction engineer may make these changes in accordance with his own judgment, which he believes is sound, although it may be very poor. Important changes of this kind have been made on the job without indicating the change on the field set of construction drawings....The contractor cannot be expected to be interested, or even aware of, the reasoning behind the design. His sole aim is to perform the work covered by the contract at a minimum expense. (Occasional discrete departures from the specifications reduce the cost quite considerably.) The inspectors, too, may be inclined to consider uncomfortable items in the specifications as superfluous refinements, conceived in the hothouse atmosphere of the design department....Therefore, a consultant can never be sure how a structure was built unless he maintains continuous contact with the construction operations.

1.4   Conventions

Like much of the United States, the authors of this book are divided on the desirability of switching to SI units of measure: One wishes to switch and one sees the switch as costly and unnecessary. Since the United States still shows no serious signs of changing, this book generally uses U.S. customary units. Exceptions to this rule are made when reporting results from other researchers who used SI units and for some laboratory results and applications where SI units are common (e.g., permeability in centimeters per second). A table of conversions for U.S. customary units to SI units is given in App. A.

Laboratory testing is an essential part of geotechnical engineering and testing. Information on specific tests in this book is illustrative and is intended to help the reader understand principles behind the testing and limitations of the tests. This is not a laboratory manual on testing with details on how to conduct the tests. We firmly believe that every time a laboratory test is run, the current approved test procedure should be consulted before testing commences. Memory is a fickle instrument upon which to rely for laboratory testing. We have included a listing of useful ASTM standards that may prove helpful in geotechnical work in App. B.

Space limits the amount of information that may be presented on any single topic, so numerous references have been included to assist the reader in finding more information on specific topics. References are placed alphabetically at the end of each chapter and are noted in the text using the Harvard system of reference with author name and publication date.

1.5   References

American Society of Civil Engineers. 1963. "Who Pays for the Unexpected in Construction," *Journal of the Construction Division,* 89 (CO 2): 23–58.

American Society of Civil Engineers. 1991. "Who Pays for the Unexpected in Construction," in *Preparing for Construction in the 21st Century,* American Society of Civil Engineers, New York, pp. 148–215.

Anderson, W. 1893. "The Interdependence of Abstract Science and Engineering," in *Minutes of the Proceedings of the Institute of Civil Engineers,* vol. 114 (1892–1893, part IV), London, p. 255.

Church, H. K. 1981. *Excavation Handbook,* McGraw-Hill, New York.

Dunham, C. W., and R. D. Young. 1971. *Contracts, Specifications, and Law for Engineers,* 2d ed., McGraw-Hill, New York.

Immerman, H. 1963. "Unexpected in Subsurface Construction," *Journal of the Construction Division,* 89 (CO 2): 40–43.

Jarvis, R. 1963. "Legal Aspects of the Unexpected in Construction," *Journal of the Construction Division,* 89 (CO 2): 43–52.

Kelley, M. N. 1991. "Second Roebling Lecture," in *Preparing for Construction in the 21st Century,* American Society of Civil Engineers, New York, pp. 1–6.

Porter, J. C. 1991. "One Perception of Engineering Academia," *Journal of Professional Issues in Engineering Education and Practice,* 117 (3): 214–227 (discussion and closure in vol. 119, no. 1).

Postner, W. J. 1991. "Who Pays for the Unexpected in Construction?—The Lawyer's Point of View," in *Preparing for Construction in the 21st Century,* American Society of Civil Engineers, New York, pp. 181–186.

Powers, J. P. 1981. *Construction Dewatering, a Guide to Theory and Practice,* Wiley, New York.

Rollings, R. S. 1991. "Responsibility and Liability of Inspection," in *Preparing for Construction in the 21st Century,* American Society of Civil Engineers, New York, pp. 531–536.

Simons, B. 1992. "Improper Uses of Construction Specifications," in *Materials: Performance and Prevention of Deficiencies and Failures,* American Society of Civil Engineers, New York, pp. 316–324.

Sowers, G. 1993. "Human Factors in Civil and Geotechnical Engineering Failures," *Journal of Geotechnical Engineering,* 119 (2): 238–256.

Sowers, G. 1971. "Changed Soil and Rock Conditions in Construction," *Journal of the Construction Division,* 97 (CO 2): 257–269.

Terzaghi, K. 1961. "Past and Future of Applied Soil Mechanics," *Journal of the Boston Society of Civil Engineers,* reprinted in *Contributions to Soil Mechanics, 1954–1961,* Boston Society of Civil Engineers, Boston, Mass., pp. 400–429.

Terzaghi, K. 1958. "Consultants, Clients, and Contractors," *Journal of the Boston Society of Civil Engineers,* reprinted in *Contributions to Soil Mechanics, 1954–1961,* Boston Society of Civil Engineers, Boston, Mass., pp. 289–292.

Tschebotarioff, G. P. 1973. *Foundations, Retaining and Earth Structures,* 2d ed., McGraw-Hill, New York.

1

Geotechnical Materials and Testing

A man's judgment is no better than his information.
UNKNOWN

2

Soils and Aggregates

...the soil was dirt cheap—a versatile, strong, durable material if handled correctly, a pile of mud if not. Even a child knows this instinctively and empirically; some modern engineers have forgotten it.　　　　SOWERS, 1981

2.1 Introduction

To the engineer, soil is the unconsolidated material of the earth's crust that can be excavated without blasting. Although such a definition might not be acceptable to other disciplines, such as agronomy or geology, it is adequate for civil engineering practice. Natural soils exist in a wide variety of forms and provide the foundation support and much of the construction material for many engineered structures. Natural aggregate deposits and quarried rock also provide large volumes of construction material. They also constitute about 95 percent by weight of asphalt concrete and 75 percent of portland cement concrete.

Because of the complexity of soils and rocks and their importance in engineering work, a number of specialists may prove useful in planning and executing work with these materials. These might include geotechnical engineers (civil engineers specializing in work with soils), engineering geologists, petrologists (geologists specializing in the study of rocks), geomorphologists (scientists who study landforms), and pedologists (scientists who study surficial soils, particularly for agricultural purposes). This chapter can serve only as a basic introduction to the broad topic area of soils and aggregates and will concentrate on identifying how soils form, how they are classified for engineering work, characteristics of natural soils, and what characteristics of soils and aggregates affect their behavior in construction.

2.2 Geologic Factors

The earth is in a constant cycle of change, with erosion, deposition, weathering, uplift, and other factors altering and changing the surface features and the materials that comprise these features. As described in Table 2.1, the earth's rocks are identified as igneous, sedimentary, or metamorphic depending on their process of formation. Intact rock formations may have a direct impact on civil engineering construction, as in tunneling, or may be the source of much of our construction material. These rocks are also the parent material of all soils. Geology provides a framework for studying the formation and distribution of geotechnical materials. Two useful references to applications of geology in civil engineering are Legget and Karrow (1983) and Mathewson (1981).

2.2.1 Weathering

Soils develop from the mechanical and chemical weathering of rock. Mechanical processes such as abrasion or temperature variation dislodge fragments from the parent rock, which is *physical weathering.* *Chemical weathering,* on the other hand, alters some rock minerals and selectively removes others, thereby transforming the rock into a chemically different material. The physical residue of weathering may remain in place, in which case it is a *residual soil,* or the residue may be moved by water, wind, or other natural forces to a new location to form a *transported soil.* The mechanics of transportation influence the characteristics of the transported soil. Both residual and transported soils initially developed from weathering of rock are further altered in place by continued weathering to develop a surficial layered series of horizons. These processes of soil development determine the varying characteristics of soils and their engineering properties.

Mechanical weathering. Abrasion by natural forces such as ice, water, and wind is a basic mechanical weathering process that dislodges

TABLE 2.1 Classification of Rocks

Classification	Formation	Examples
Sedimentary	Rock formed by sedimentation or cementing at or near earth's surface	Limestone, sandstone, shale, mudstones
Igneous	Rock formed by solidification of molten rock	Granite, gabbro, basalt
Metamorphic	Transformation of in-situ rock by high temperature and pressure	Gneiss, marble, slate

rock fragments in a wide variety of sizes. Glaciers have covered much of the northern hemisphere and parts of the southern and have been a major form of mechanical weathering. Flowing water also has the ability to dislodge material, as evidenced by alluvial fans in arid regions or the deeply eroded American badlands. Wind can be another effective weathering agent, particularly in arid regions and if accompanied by blowing sand.

Exfoliation along planes within the rock structure or cracking related to temperature variation is another form of mechanical weathering. As a rock is alternately heated and cooled, different thermal expansion and contraction rates of minerals within the rock may lead to splitting and cracking. Water that collects within these cracks expands when it freezes and may lead to further cracking. Roots from vegetation can then invade these cracks, leading to additional wedging forces that may accelerate the breakup of the rock.

Chemical weathering. Water is the basic agent of chemical weathering. Water collects carbon dioxide from the air when it falls as rain and may collect other salts and organic acids as it flows across ground. In this form it can physically and chemically alter rock minerals. For instance, feldspar minerals will weather chemically into clay minerals, quartz particles, and soluble carbonates of various kinds. Other minerals such as gypsum or halite (salt) are soluble in water and can be removed completely in a wet climate. Pure limestone ($CaCO_3$) is soluble in water containing carbon dioxide (carbonic acid). Pure limestone seldom exists in nature, however. Instead, the calcium carbonate portion is removed by solution, and a clayey or cherty, clayey soil from the insoluble limestone constituents is left behind. A number of complex chemical processes, including hydrolysis, chelation, cation exchange, oxidation, and carbonation, play a role in chemical weathering of the parent material to form soils (Mitchell 1993).

Mechanical weathering and chemical weathering seldom act alone. Often the greatest effect of mechanical weathering is to break down the parent rock so that a larger volume of material is then exposed to chemical attack.

2.2.2 Factors affecting weathering

Climate. Climate is a major factor that affects both mechanical and chemical weathering. Figure 2.1 shows the relative effectiveness of different climates on weathering. Hot, humid climates are conducive to chemical weathering, whereas cold or arid regions are dominated by mechanical weathering. Optimal conditions for mechanical weath-

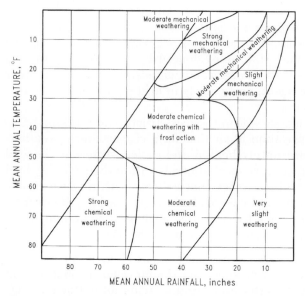

Figure 2.1   Climatic impact on weathering (*Peltier 1950*).

ering occur with a combination of moderate rainfall and cool temperatures that develops the maximum potential for disintegration under cyclic freezing and thawing.

Topography.   Topography controls both the rate of erosion and the length of time available for weathering. For example, steep slopes limit the time that water can chemically attack material but allow the water to rapidly erode soil particles and rock fragments. Mountain talus slopes are composed of piles of relatively unweathered rock fragments and are examples of intense physical weathering in areas with steep terrain.

Composition.   Rock composition determines the rate, type, and final products of weathering. Weak clay shales will be more susceptible to mechanical abrasion than stronger, more competent granites. Limestones and similar carbonate rocks can be attacked rapidly by solution. The specific minerals of the rock will weather at different rates to produce different kinds of new minerals. Figure 2.2 is Bowen's reaction series of mineral stability. *Mineral stability,* or the resistance of the mineral to decomposition, increases as one proceeds down the series (i.e., augite is more stable than olivine, hornblende is more stable than augite, etc.). The specific chemical constituents of the minerals also show variable weathering characteristics. Calcium, sodium, magnesium, and potassium are lost first. This is followed by silica and then alumina and iron last.

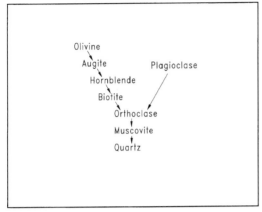

Figure 2.2 Bowen's reaction series.

Age. Older rocks are usually exposed to weathering for a longer time than young rocks, so their chemical weathering and mechanical weathering are more advanced under otherwise identical conditions. Consequently, rocks in the arid climate of Australia (oldest continent) would be more heavily weathered than those of a similarly arid region in the relatively young regions of California.

2.2.3 Development of the clay fraction

The fine fraction of a soil and its plasticity have a major impact on the engineering characteristics of the soil. The specific clay minerals that develop depend on the parent material, climate, topography, vegetation, and time. Once formed, these may be further altered by continued weathering, sedimentation in salt water, or other geologic processes. Table 2.2 shows the effect that different specific clay minerals have on the Atterberg limits. The Atterberg limits will be discussed in further detail later, but for this discussion it will suffice to recognize that Atterberg limits reflect a clay mineral's reactivity with water. Thus higher plastic limit, plasticity index, and liquid limit correlate with more plastic behavior and poorer engineering characteristics in the soil. The clay mineral carries an electric charge that attracts both water and ions, termed *exchangeable ions,* to neutralize the charge. As also seen in Table 2.2, the identity of this exchangeable ion, as well as the specific clay mineral, has a major influence on the Atterberg limits. Real soils will contain a potpourri of clay minerals (e.g., smectite and illite) and altered and unaltered particles of the original parent rock minerals (e.g., feldspar and quartz). Many very complex processes operate to produce soils and aggregates with which

TABLE 2.2   Atterberg Limits of Selected Clay Minerals

Clay mineral	Exchangeable ion	Liquid limit (%)	Plastic limit (%)	Plasticity index (%)
Montmorillonite	Na	710	54	656
	K	660	98	562
	Ca	510	81	429
	Mg	410	60	350
	Fe	290	75	215
	Fe*	140	73	67
Illite	Na	120	53	67
	K	120	60	60
	Ca	100	45	55
	Mg	95	46	49
	Fe	110	49	61
	Fe*	79	46	33
Kaolinite	Na	53	32	21
	K	49	29	20
	Ca	38	27	11
	Mg	54	31	23
	Fe	59	37	22
	Fe*	56	35	21
Attapulgite	H	270	150	120

*After five cycles of wetting and drying.
SOURCE: *Soil Mechanics*, W. T. Lambe and R. V. Whitman, Copyright © 1969. Reprinted by permission of John Wiley and Sons.

we will build our structures. Their inherent complexity and variability should be recognized and treated with respect.

The complexity of a soil's development can cause major construction headaches. For example, a silty gravel developed from in-situ weathering of an igneous rock, gabbro, in the Sultanate of Oman had low plasticity and was a reasonably good construction material. Because of the general scarcity of construction material in the local area, this soil promised to provide a good subbase and fill material, and large volumes were used in roads, airfield pavements, and as general fill. However, pockets of this soil with only 10 percent fine silt- and clay-sized particles would cause the plasticity index to vary from the normal value of about 5 percent up to 35 percent. Obviously, these pockets of soil contained a much more active clay mineral, reflecting their past parent material and weathering history.

The pockets of highly plastic soil could not be identified visually, and it was impractical to conduct sufficient detailed testing to find all these scattered pockets. Therefore, the contractor conducted sufficient testing to try to minimize the number of these pockets in selected barrow areas, but he recognized that some of the material would be inadvertently incorporated within the structures he was building. By closely observing the pavements under construction, the contractor found that the weak areas could be identified as trucks hauled base course aggregates, concrete, etc., over the subbase materials. These weak areas could then be removed and replaced with better materials before the overlying paving was placed. The contractor recognized that the material with which he had to build was very complex and varied. Recognizing this, he took a pragmatic approach of testing to avoid as much trouble as possible but then used construction traffic to essentially proof test the materials and find the weak materials he knew were missed in the original testing.

2.3   Soil Characteristics

2.3.1   Particle size

Soils are often described by the particle size of their individual components, as defined in Table 2.3. The distribution of soil particle sizes is determined as described in ASTM C 136 by shaking the soil through a nest of sieves for particles larger than 0.075 mm (No. 200 sieve). Standard U.S. sieve designations and their corresponding mesh size are shown in Table 2.4. The distribution of fine silt- and clay-sized particles smaller than 0.075 mm is determined by a sedimentation test using a hydrometer and following procedures described in ASTM C 117 and D 422 for soils and aggregates, respectively. Trying to separate silts and clays on the basis of size alone is misleading, since some clay minerals are larger than the 0.002 mm used to designate clay-sized particles and some nonclay minerals can

TABLE 2.3   Definition of Soil Components

Soil component	U.S. standard sieve		Size (mm)	
	Passing	Retained on	Maximum	Minimum
Cobbles	—	3 in	—	75
Gravel	3 in	No. 4	75	4.75
Coarse gravel	3 in	$3/4$ in	75	19
Fine gravel	$3/4$ in	No. 4	19	4.75
Sand	No. 4	No. 200	4.75	0.075
Coarse sand	No. 4	No. 10	4.75	2.00
Medium sand	No. 10	No. 40	2.00	0.425
Fine sand	No. 40	No. 200	0.425	0.075
Fines	No. 200	—	0.075	—
Silt	—	—	0.075	0.005
Clay	—	—	0.005	—

Note: Definitions of the different soil components by particle size vary between different organizations. For example, the maximum size particle for clays is often taken as 0.002 mm rather than the 0.005 mm used in ASTM D 422 and shown here.

TABLE 2.4   ASTM Standard Sieve Sizes

U.S. customary designation*	Nominal sieve opening (in)	Standard SI designation†	Permissible variation in average sieve opening
3 in	‡	75 mm	± 2.2 mm
1½ in	1.5	37.5 mm	± 1.1 mm
1 in	1.0	25 mm	± 0.8 mm
¾ in	0.750	19 mm	± 0.6 mm
½ in	0.500	12.5 mm	± 0.39 mm
⅜ in	0.375	9.5 mm	± 0.30 mm
¼ in	0.250	6.3 mm	± 0.20 mm
No. 4	0.223	4.75 mm	± 0.15 mm
No. 8	0.0937	2.36 mm	± 0.080 mm
No. 12	0.0661	1.70 mm	± 0.060 mm
No. 16	0.0469	1.18 mm	± 0.045 mm
No. 20	0.0331	850 μm	± 35 μm
No. 40	0.0165	425 μm	± 19 μm
No. 60	0.0098	250 μm	± 12 μm
No. 80	0.0070	180 μm	± 9 μm
No. 100	0.0059	150 μm	± 8 μm
No. 140	0.0041	106 μm	± 6 μm
No. 200	0.0029	75 μm	± 5 μm

*Complete listing of sieve sizes is given in ASTM E 11.
†The SI designations in column 3 are considered the international standard and conform to recommendations of the International Standards Organization, Geneva, Switzerland.
‡The nominal opening sizes in column 2 are only approximately equal to SI designated size.
SOURCE: Copyright ASTM. Reprinted with permission.

be smaller than 0.002 mm. Consequently, it is most correct to refer to silt-sized or clay-sized particles when the identification is based on grain size alone.

Fine silt and clay particles often will adhere to the surface of the larger soil grains, or they may form hardened clods or agglomerated particles in the soil. If a soil is simply shaken dry through sieves, the calculated fine content will be incorrect. This is a serious potential error because the quantity of fines in a soil or, particularly, an aggregate often controls its performance. To obtain a correct measurement of the fines in a soil or aggregate, the soil will have to be broken up and washed through the No. 200 sieve to ensure that all the hardened fines and coatings are removed. Failure to use washed gradations to evaluate soils and aggregates can lead to totally misleading results.

The results of sieve and hydrometer tests for several soils are plotted in Fig. 2.3. This curve is referred to as a *grain-size curve* and provides useful insight into the composition of the soil. At a glance, one can tell if the soil is predominantly fine-grained clay or sand. A soil such as the silty gravel in Fig. 2.3 that has a smooth, gradual curve

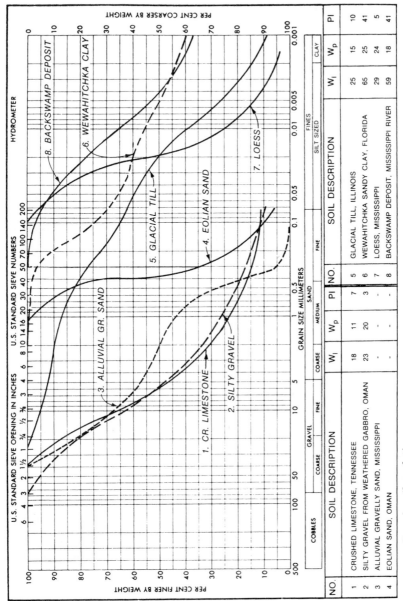

Figure 2.3 Sample soil gradation curves.

NO.	SOIL DESCRIPTION	W_l	W_p	PI
1	CRUSHED LIMESTONE, TENNESSEE	18	11	7
2	SILTY GRAVEL FROM WEATHERED GABBRO, OMAN	23	20	3
3	ALLUVIAL GRAVELLY SAND, MISSISSIPPI	-	-	-
4	EOLIAN SAND, OMAN	-	-	-

NO.	SOIL DESCRIPTION	W_l	W_p	PI
5	GLACIAL TILL, ILLINOIS	25	15	10
6	WEWAHITCHKA SANDY CLAY, FLORIDA	65	25	41
7	LOESS, MISSISSIPPI	29	24	5
8	BACKSWAMP DEPOSIT, MISSISSIPPI RIVER	59	18	41

25

showing the presence of all particle sizes from minimum to maximum is called *well graded.* The gravelly sand in Fig. 2.3 appears somewhat similar, but the characteristic hump indicates that the soil is *gap-graded,* or is missing some particle sizes. In this example, the gravelly sand is relatively short on coarse and medium sand sizes between the No. 4 and 20 sieves. The eolian sand from Oman shows that almost all its particles are fine sands between the No. 30 and 100 sieves. Such a soil is termed *poorly* or *uniformly graded.* The grading of soils and aggregates will later be seen to have important engineering significance.

Several parameters can be determined from a soil's grain-size curve. Three commonly used numerical values are D_{10}, D_{30}, and D_{60}. These are defined to be the particle diameters at which 10, 30, and 60 percent of the soil, respectively, is finer. For example, in Fig. 2.3, the D_{10} of the eolian sand is 0.1 mm; i.e., 10 percent of the soil is finer than this particle size and 90 percent is larger. These values are used to compute the uniformity coefficient C_u and the coefficient of curvature C_c. These coefficients are defined as

$$C_u = \frac{D_{60}}{D_{10}}$$

$$C_c = \frac{(D_{30})^2}{(D_{10})(D_{60})}$$

These coefficients offer a crude quantitative measure of how well graded or how poorly graded a grain-size curve may be. If the uniformity coefficient is greater than 4 for gravelly soils or 6 for sandy soils, the soil is well graded. The coefficient of curvature for a well-graded soil should be near unity. Coefficients of curvature between 1 and 3 are considered representative of well-graded material.

A dry soil exists as an aggregate of individual, discrete particles. If water is added to this dry soil, it will progressively behave first as a plastic material and then as a liquid as the quantity of water increases. Soils rich in clay minerals that exhibit the ability to be molded at some water content are called *plastic* or *cohesive,* whereas those such as clean sands that do not are called *nonplastic* or *cohesionless.* The plastic behavior of a soil is a function of the quantity of silt- and clay-sized particles it contains and the specific mineralogy of these particles.

Illustrative Example 2.1   Determine the coefficient of curvature and uniformity coefficient for the silty gravel, alluvial gravelly sand, and eolian sand in Fig. 2.3, and classify their gradation characteristics.

	Soil types		
	Silty gravel	Alluvial gravelly sand	Eolian sand
Particle size (mm)			
D_{60}	9.00	6.40	0.40
D_{30}	1.38	0.53	0.27
D_{10}	0.08	0.34	0.10
Percentage soil component (%)			
Gravel	52	44	0
Sand	38	56	94
Fines	10	0	6
Coefficients			
Uniformity C_u	112.5	18.8	4.0
Curvature C_c	2.64	0.13	1.8

solution: The silty gravel with its uniformity coefficient greater than 4 (gravel) and coefficient of curvature between 1 and 3 meets the requirements for a well-graded soil. The alluvial gravelly sand is poorly graded because its coefficient of curvature falls below the range of 1 to 3. The eolian sand uniformity coefficient is too low (must be greater than 6 for sands), so it too is a poorly graded soil.

2.3.2 Plastic behavior

Plastic behavior is measured by the soil's Atterberg limits, named for the Swedish soil scientist who first proposed their use. The *plastic limit* w_p or PL is defined as the soil moisture content at which the soil just begins to crumble when rolled into threads $\frac{1}{8}$ in in diameter. The *liquid limit* w_l or LL is defined as the soil moisture content at which a standard groove cut in a pat of soil will close over the length of $\frac{1}{2}$ in when the cup containing the soil is dropped 25 times from a height of 1 cm onto a hard rubber pad. These definitions are arbitrary, and the standard test procedures described in ASTM D 4318 must be followed exactly to obtain reproducible and meaningful results. The difference between the liquid and plastic limits is the *plasticity index PI*. Conceptually, the plastic limit is the moisture content at which a soil begins to exhibit plastic behavior, the liquid limit is the moisture content at which the soil behaves as a very viscous liquid rather than as a solid, and the plasticity index is the range of moisture content over which the soil behaves plastically. The *shrinkage limit* is the moisture content at which further decreases in moisture content do not cause further shrinkage. This limit is seldom used in routine engineering work in the United States but is more common in some overseas areas. Atterberg limits are only run on remolded soils and on the fraction of the soil that passes the No. 40 sieve.

Skempton (1953) observed that for a specific soil deposit, a plot of the percent of the soil particles smaller than 2 μm as the abscissa and the plasticity index as the ordinate formed an approximately straight line. He defined a clay's activity or A value to be equal to the slope of this line, or

$$A = \frac{PI}{(\% \text{ of soil finer than 2 μm})}$$

Typical activity values for specific clay minerals are shown in Table 2.5, and in a rough sense, the A value is a measure of the plasticity per clay-sized particle. As the A value of a soil increases, the potential impact of the clay fraction on the soil's engineering properties increases. Both the type of clay mineral and the amount of the clay mineral present influence the soil behavior. These two factors can be interrelated by use of Skempton's activity or A value.

Another common index for soil is called the *liquidity index LI*. This is defined as

$$LI = \frac{w - w_p}{w_l - w_p} = \frac{w - w_p}{PI}$$

where LI = liquidity index
w = in-situ or natural moisture content that exists in the field
w_l = liquid limit
w_p = plastic limit
PI = plasticity index

A liquidity index of less than 0 indicates a dry soil, an index of 0 identifies a soil at its plastic limit, and an index above 1 shows that the soil's water content is above its liquid limit. Consequently, the soil

TABLE 2.5   Activity Values for Some Clay Minerals

A value	Clay mineral
≤0.50	Kaolinite, halloysite ($2H_2O$)
Around 1.0	Illite, attapulgite, allophane, halloysite ($4H_2O$)
>1.50	Montmorillonite

SOURCES: From Mitchell, 1993, and Carter and Bentley, 1991.

with a liquidity index of 1 or higher should be expected to behave as a thick slurry if remolded, it may lose strength under impact stresses or if vibrated, and it will be unstable and may flow under its own weight if an excavation is attempted. Natural soils often exist in the field at or above their plastic limit, so liquidity indices of 0 to 1 are common.

Illustrative Example 2.2 Determine the activity value for clay soils with the given characteristics.

Soil	Percent of total sample finer than		Percent of minus No. 40 material finer than	PI	A
	No. 40	0.002 mm	0.002 mm		
Loess, MS	100	5	5	5	1.00
Glacial till, NH	81	14.5	18	10	0.56
Wewahitchka clay, FL	98	44	45	41	0.91
Backswamp deposit, LA	100	42	42	41	0.98
Eagle Ford clay, TX	92	35	38	30	0.79
Goodfellow clay, TX	99	34	34	35	1.03
Seguin clay, TX	100	47	47	28	0.60
Buckley clay, CO	88	23	26	16	0.62
Ellsworth clay, SD	86	41	48	22	0.46

solution: The first four soils are from Fig. 2.3, and the required gradation and plasticity data can be determined from there. The other five clays are from Texas, Colorado, and South Dakota. Note that soils such as the glacial till or Wewahitchka clay had material larger than the No. 40 sieve. Therefore, the percent finer than 0.002 mm taken from the grain-size curves in Fig. 2.3 had to be adjusted to determine the percent finer than 0.002 mm for only the fraction passing the No. 40 sieve.

2.3.3 Weight-volume relationships

Soil exists as a three-phase system consisting of solid mineral matter, water, and gas. The gas is normally air, but in problems dealing with decomposition of organic material (marshy areas, landfills, etc.), it may contain products such as methane. Figure 2.4 illustrates this conceptual division of soil into solid, liquid, and gaseous components. A number of important geotechnical measurements spring from this concept, and some basic relationships are defined in Table 2.6.

The moisture content of a soil or aggregate for engineering is defined to be the weight of the water in a sample divided by the weight of the solids (see Table 2.6). This is a very simple but useful test. Soils dried to constant weight at constant temperature will give different results depending on the value of the temperature used for drying (Lambe 1951), so test methods and specifications generally

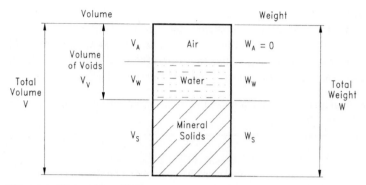

Figure 2.4 Three-phase diagram.

require drying at a standard test temperature of 105 or 110°C. Also, some other fields such as geology commonly define moisture content to be W_W/W_T, and some instruments report moisture content as V_W/V_T. The engineering definition will be used throughout this book.

Several unit weights or densities are defined in Table 2.6, but the dry unit weight or density is particularly useful in engineering work. The amount of moisture in a material may change depending on drying conditions, rain, etc., but dry unit weight determined from the weight of the solid phase and the total volume will remain unchanged as long as the soil volume does not change. Since the dry unit weight of a soil or aggregate is a constant measure, it is used to control compaction density requirements in the field.

TABLE 2.6 Basic Weight-Volume Relationships

Property	Symbol	Definition
Total volume	V	—
Volume of solids	V_S	—
Volume of water	V_W	—
Volume of air	V_A	—
Volume of voids	V_V	$= V_W + V_A$
Void ratio	e	$= V_V/V_S$
Saturation (%)	S_r	$= V_W/V_V$
Total weight	W	$= W_W + W_S$
Weight of solids	W_S	—
Weight of water	W_W	—
Moisture content (%)	w	$= W_W/W_S$
Mass unit weight (density)	γ	$= W/V$
Dry unit weight	γ_d	$= W_S/V$
Unit weight of solids	γ_s	$= W_S/V_S$
Unit weight of water	γ_w	$= W_W/V_W$ (typically 62.4 lb/ft^3)
Specific gravity of solids	G_s	$= \gamma_s/\gamma_w$

Several important void relationships are also defined in Table 2.6. The *porosity* of a material is defined as the ratio of the volume of voids (volume of air plus volume of water) to the total volume. The porosity is a measure of void space in a sample and will be related to the ability of a material to allow flow and absorb moisture. If the volume of a sample changes, then its porosity will change due to possible changes in both the volume of voids and the total volume. The volume of solids is generally a constant, so the void ratio or ratio of the volume of voids to the volume of solids is a convenient parameter for problems where volume changes are occurring. In such cases, only the volume of voids is changing, and this parameter is commonly used in geotechnical settlement calculations where a constant measure of volume of void change is needed. The *saturation* of a soil or aggregate is simply the ratio of the volume of water to the volume of voids so that at 100 percent saturation, the volume of water is equal to the volume of the voids, and the volume of air is zero. Soils below the water table are usually 97 to 100 percent saturated, and some fine-grained soils above the water table also can be similarly saturated from capillary action in the soil.

There are a number of interrelationships between the various parameters, and some of these are summarized in Table 2.7. Several example problems follow to illustrate the use of the relationships in Table 2.6 and 2.7. The basic strategy is to simply list

TABLE 2.7   Common Interrelationships for Weight-Volume Parameters

$$e = \frac{G\gamma_w V}{W_S - 1}$$

$$n = \frac{1 - W_S}{G\gamma_w V}$$

$$n = \frac{e}{1 + e}$$

$$S = \frac{W_W}{\gamma_w V_V}$$

$$Se = G_s w$$

$$\gamma_d = \frac{100\gamma}{100 + w}$$

what is known (e.g., moisture content, wet density, specific gravity), draw a three-phase diagram such as Fig. 2.4, and list on it what is known (e.g., W_s, W_w). Often an assumption such as a total volume of 1 ft³ is needed to allow the relative calculations, and then the relationships in Table 2.6 are used to determine the needed unknown quantities.

Illustrative Example 2.3   A sand cone field density test finds the density of a soil to be 125.0 lb/ft³, and an oven-dried sample of the soil determines the soil moisture content to be 18.3 percent. The specific gravity of the solids is 2.65. Find the saturation of the soil.

solution

$$w = W_W/W_S = 0.183$$
$$W_W = 0.183 W_S$$
$$W = W_S + W_W = W_S + 0.183 W_S = 1.183 W_S$$
$$\gamma = W/V = 125.0 \text{ lb/ft}^3$$

Assume that V is 1 ft³, and the preceding becomes, by substitution,

$$1.183 W_S = 125.0 \text{ lb}$$
$$W_S = 105.7 \text{ lb}$$
$$W_W = W - W_S = 19.3 \text{ lb}$$
$$\gamma_w = W_W/V_W = 62.4 \text{ lb/ft}^3$$

Substituting for W_W and rearranging terms gives

$$V_W = (19.3 \text{ lb})/(62.4 \text{ lb/ft}^3) = 0.309 \text{ ft}^3$$
$$G_s = \gamma_s/\gamma_w = 2.65$$
$$\gamma_s = 2.65 \times 62.4 \text{ lb/ft}^3 = 165.4 \text{ lb/ft}^3$$

This is the weight of 1 ft³ of the solids with no voids. Now the volume of 105.7 lb of the solids can be found as

$$V_S = W_S/\gamma_s = 0.639 \text{ ft}^3$$
$$V_V = V - V_S = 1.000 \text{ ft}^3 - 0.639 \text{ ft}^3 = 0.361$$

Note that this includes the volume of both the air and water. Saturation can now be found as

$$S_r = V_W/V_V = 85.6\%$$

Entering the various weights and volumes as they are calculated on a diagram such as Fig. 2.4 is a convenient method of keeping track of the results as the problem progresses.

Illustrative Example 2.4   Determine the dry density of a saturated soil that has a moisture content of 5 percent and a specific gravity of solids of 2.65.

solution   As in the preceding example, the moisture content and specific gravity can be used to determine

$$W = 1.05W_S$$
$$\gamma_s = 165.4 \text{ lb/ft}^3$$

Assume that V_S is 1.0 ft³, and the definition of γ_s gives

$$W_S = 165.4 \text{ lb}$$
$$W = 1.05W_S = 173.7 \text{ lb}$$
$$W_W = 0.05W_S = 8.27 \text{ lb}$$
$$V_W = W_W/\gamma_w = 0.132 \text{ ft}^3$$

The volume of air in a saturated soil is 0, so the total volume is

$$V = V_S + V_W = 1.132 \text{ ft}^3$$

And the soil density is

$$\gamma = W/V = 153.4 \text{ lb/ft}^3$$
$$\gamma_d = W_S/V = 146.1 \text{ lb/ft}^3$$

2.4   Soil Classification

An engineer needs a systematic, repeatable method for classifying soils. Three common systems—the U.S. Department of Agriculture (USDA) Textural System, the Unified Soil Classification System (often referred to as the USCS), and the American Association of State Highway and Transportation Officials (AASHTO, formerly AASHO) System—will be reviewed here. The Federal Aviation Administration (FAA) Classification System is no longer in common use, but descriptions of the method can be found in some older textbooks (e.g., Yoder and Witczak 1975).

The USDA Textural, Unified, and AASHTO systems will be described in the following sections. It is important to note the subtle differences between these systems because they are not consistent in their use of terms such as *clay* or their identification of what constitutes a fine-grained soil. Such inconsistencies can cause confusion when the engineering performance of a soil is estimated from its classification.

2.4.1   USDA Textural Soil Classification System

The USDA classification system (USDA 1975) defines *sands* to be particles between 2.000 and 0.050 mm, silts to be particles between 0.050 and 0.002 mm, and clays to be smaller than 0.002 mm. The soil texture is classified only on the basis of its relative content of these par-

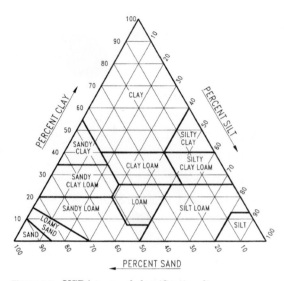

Figure 2.5 USDA textural classification diagram.

ticle sizes. The triangular textural classification chart in Fig. 2.5 is used to determine the appropriate soil textural classification. For example, a soil with 15 percent sand, 55 percent silt, and 30 percent clay would classify as a silty clay loam. If a soil contains a significant portion of particle sizes larger than 2 mm, the textural classification from Fig. 2.5 is preceded with the descriptor *gravelly* for fragments up to 3 in in diameter, *cobbly* for fragments between 3 and 10 in, and *stony* or *bouldery* for fragments larger than 10 in.

Note in Table 2.3 that the definition of sand and silt sizes used by the USDA differs from the definitions used by others. Also, this system requires a hydrometer test in addition to the sieve gradations so that the relative percentages of silt and clay can be determined. Because this system relies on particle size alone and neglects the impact of plasticity characteristics on soil properties, it has serious limitations for engineering and construction. However, the system is widely used by soil scientists, and they often have extensive information on surficial soils that can be adapted to construction.

Illustrative Example 2.5 Determine the USDA soil classification for the soils in Fig. 2.3.

solution The eolian sand, alluvial gravelly sand, crushed limestone, and silty gravel cannot be classified by this system because no hydrometer analysis was done, so the relative silt and clay content are unknown. The following can be found from Fig. 2.3.

	Glacial till	Loess	Wewahitchka clay	Backswamp deposit
Percent finer than				
2.000 mm	90	100	100	100
0.050 mm	60	90	65	91
0.002 mm	20	5	44	42
Percent of				
Sand	33	10	35	9
Silt	44	85	21	49
Clay	23	5	44	42

Textural classification is only based on the material smaller than 2 mm, so the 10 percent of the glacial till larger than 2 mm is not included in calculating the relative percent of sand, silt, and clay. From Fig. 2.5 the soils are classified as

Glacial till	Loam
Loess	Silt
Wewahitchka clay	Clay
Backswamp deposit	Silty clay

2.4.2   Unified Soil Classification System

The USCS was originally developed by Casagrande (1948) for classifying soils for military airfield construction. It divides soils into coarse- and fine-grained soils depending on whether 50 percent or more of the soil is retained on the No. 200 sieve or if 50 percent or more passes the No. 200 sieve. The coarse-grained soils are further defined as gravel, designated by G, if 50 percent or more of the coarse soil fraction larger than the No. 200 sieve is retained on the No. 4 sieve. If more than 50 percent of the coarse fraction passes the No. 4 sieve, the soil is a sand, designated by S. Clean coarse-grained soils with less than 5 percent passing the No. 200 sieve are further designated as well graded (W) or poorly graded (P) depending on the value of the coefficients C_c and C_u defined earlier. If a gravel has a C_c between 1 and 3 and a C_u equal to or greater than 4, it is well graded; otherwise, it is poorly graded. For sands, C_c also must be between 1 and 3 and C_u must be greater than or equal to 6 to be designated well graded. The USCS classification for clean coarse-grained soils consists of two symbols. The first symbol describes the soil's composition; the second describes the grading. Hence a GW is a well-graded gravel, and an SP is a poorly graded sand.

Fine-grained soils are divided into silts, designated by M, or clays, designated by C, on the basis of their plasticity characteristics. Soils that plot above the A line on the plasticity chart in Fig. 2.6 are clays, and those plotting below the line are silts. The fine-grained soils are

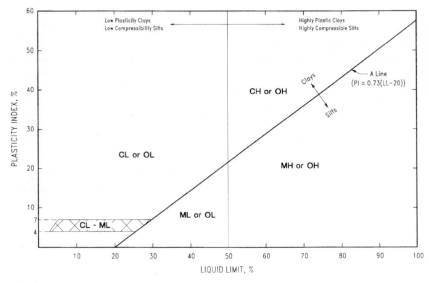

Figure 2.6 Plasticity chart.

further designated as high plasticity (for a clay) or compressibility (for a silt) with the letter H if their liquid limit is 50 or more or low plasticity or compressibility (L) if their liquid limit is below 50. The USCS classification of fine-grained soils consists of two symbols. The first identifies it as a clay (C) or a silt (M), and the second identifies it as high (H) or low (L) plasticity (for a clay) or compressibility (for a silt). Consequently, a clay of high plasticity, or "fat clay," is designated as CH, and a clay of low plasticity, or "lean clay," would classify as CL.

Coarse-grained soils with more than 5 percent passing the No. 200 sieve are designated with combined symbols reflecting the character of both the fine and coarse fractions. For coarse-grained soils with 12 percent or more passing the No. 200 sieve, the first symbol reflects the coarse fraction's composition as before. The second symbol is either a C or M depending on whether the liquid limit and plasticity index plot above or below the *A* line in Fig. 2.6. Therefore, a GC is a soil with over 50 percent larger than the No. 200 sieve (i.e., a coarse-grained soil), and over 50 percent of the coarse fraction is larger than the No. 4 sieve (i.e., a gravel). The results of the Atterberg limits run on the portion of the soil passing the No. 40 sieve plot above the *A* line, so the soil would be designated as a clayey gravel with the classification of GC. If a coarse-grained soil has between 5 and 12 percent passing the No. 200 sieve, it is designated with a dual symbol such as GW-GM to reflect both the gradation and plasticity characteristics of the soil.

Organic soils contain decaying vegetable matter and are often associated with difficult engineering problems. If the soil is dark brown or black and composed predominately of vegetable matter in various stages of decomposition, it is classified as a peat and designated as Pt. Such materials often have a distinct organic odor and may be spongy with a fibrous texture. If the ratio of the liquid limit after oven drying to the liquid limit without oven drying is less than 0.75, the soil is considered organic. Such soils are designated with the first classification symbol as O. The second symbol of the organic soil designation is either L or H depending on whether the liquid limit is less than 50 or greater than or equal to 50. The USCS provides for a verbal description of organic clay or organic silt depending on whether the soil plots above or below the A line, but there is no differentiation between organic silts and clays in the OL or OH symbols.

The USCS uses gradation characteristics to classify clean coarse-grained soils and differentiates between clay and silt on the basis of their plasticity characteristics alone. Intermediate soils with significant portions of both fine and coarse particles are identified with mixed or dual symbols. The USCS system is summarized in Table 2.8.

Illustrative Example 2.6 Determine the USCS classification for the soils in Fig. 2.3.

solution First, determine if the soils are coarse- or fine-grained.

	Percent finer than		
Soil	No. 4	No. 200	Classify as
Crushed limestone	45	11	Coarse
Silty gravel	48	10	Coarse
Eolian sand	100	6	Coarse
Glacial till	92	66	Fine
Wewahitchka clay	100	70	Fine
Loess	100	96	Fine
Backswamp deposit	100	95	Fine

Next, the relative gravel or sand proportions and the gradation curve characteristics of each coarse-grained soil must be determined. Illustrative Example 2.1 shows how to calculate the C_u and C_c coefficients.

	Percent					
Soil	Gravel	Sand	Fines	C_u	C_c	Grading
Crushed limestone	55	34	11	4.2	9.5	P
Silty gravel	52	38	10	112.5	2.6	W
Gravelly sand	44	56	0	18.8	0.1	P
Eolian sand	0	94	6	4.0	1.8	P

The crushed limestone and silty gravel classify as gravels, while the other two are sands. The crushed limestone, silty gravel, and eolian sand have between 5 and 12 percent fines, so they will require dual symbols. The crushed limestone

TABLE 2.8 Unified Soil Classification System

Criteria for assigning soil symbols and descriptions				Soil classification	
				Symbol	Description
Coarse-grained soils, ≥ 50% retained on No. 200 sieve	Gravels, ≥50% coarse fraction retained on No. 4	< 5%* fines	$C_u \geq 4$ and $1 \leq C_c \leq 3$	GW	Well-graded gravel
		< 5%* fines	$C_u < 4$ and/or $1 > C_c > 3$	GP	Poorly graded gravel
		>12%* fines	Fines ML/MH	GM	Silty gravel
		>12%* fines	Fines CL/CH	GC	Clayey gravel
	Sands, ≥50% coarse fraction passes No. 4	<5%* fines	$C_u \geq 6$ and $1 \leq C_c \leq 3$	SW	Well-graded sand
		<5%* fines	$C_u < 6$ and/or $1 > C_c > 3$	SP	Poorly graded sand
		>12%* fines	Fines ML/MH	SM	Silty sand
		>12%* fines	Fines CL/CH	SC	Clayey sand
Fine-grained soils, ≥ 50% passing No. 200 sieve	Silts and clays, $LL \geq 50$	Inorganic	Above A line†	CL	Lean clay
		Inorganic	Below A line†	ML	Low-compressibility silt
		Organic	$\dfrac{\text{Oven-dried } LL}{\text{Original } LL} < 0.75$	OL	Organic silt or clay
	Silts and clays, $LL \geq 5$	Inorganic	Above A line†	CH	Fat clay
		Inorganic	Below A line†	MH	High-compressibility silt
		Inorganic	$\dfrac{\text{Oven-dried } LL}{\text{Original } LL} < 0.75$	OH	Organic clay or silt
highly organic soils			Dark, odorous organic matter	Pt	Peat

*5 to 12 percent, use dual symbol.
†Hatched zone, use dual symbol.

liquid limit and plasticity index from Fig. 2.3 plot as a CL-ML symbol on Fig. 2.6. ASTM D 2487 states that this somewhat ambiguous case will classify as a -GC for gravels with between 5 and 12 percent fines. The silty gravel plots as an M, as does the nonplastic eolian sand. The four coarse-grained soils can now be classified as

Crushed limestone	GP-GC
Silty gravel	GW-GM
Gravelly sand	SP
Eolian sand	SP-SM

The fine-grained soils' classifications are determined from their plasticity characteristics and Fig. 2.6 as

Soil	Atterberg limits		Classification
	LL	PI	
Glacial till	25	10	CL
Wewahitchka clay	65	41	CH
Loess	29	5	ML
Backswamp deposit	59	41	CH

2.4.3 AASHTO Classification System

The AASHTO Classification System (AASHTO M145 or ASTM D 3282) divides soils into eight groups, A-1 through A-8. Soils with 35 percent or less passing the No. 200 sieve are considered coarse-grained soils and comprise groups A-1 through A-3. This definition of a coarse-grained soil is significantly different from that used in the USCS discussed previously. Soils with more than 35 percent passing the No. 200 sieve are considered fine-grained soils and are classed A-4 through A-7 depending on their liquid limit and plasticity index. Group A-8 includes highly organic soils such as peat. Soils are further subdivided into subgroups such as A-1-a or A-7-5 on the basis of gradation and plasticity characteristics. Tables 2.9 and 2.10 show the specific requirements to classify soils into groups and subgroups. To use this table, start on the left side with group A-1 and progressively check each group's and subgroup's requirements. The first group or subgroup that satisfies a specific soil's characteristics is the proper AASHTO classification. AASHTO studies found that soils with more than 35 percent passing the No. 200 sieve, liquid limits above 40, and plasticity indices above 10 are potentially troublesome as pavement subgrades. The AASHTO classification quantifies this concept with a numeric group index added to the classification. The *group index* is calculated as

$$GI = (F-35)[0.2 + 0.005(LL - 40)] + 0.01(F - 15)\,(PI - 10)$$

TABLE 2.9 AASHTO Classification for Major Groups

General classification	Granular materials[a]			Fine-grained materials[b]			
Group classification	A-1	A-3[c]	A-2	A-4	A-5	A-6	A-7[f]
Sieve analysis, % passing							
2.00 mm (No. 10)	—	—	—	—	—	—	—
0.425 mm (No. 40)	50 max	51 min	—	—	—	—	—
0.075 mm (No. 200)	25 max	10 max	35 max	36 min	36 min	36 min	36 min
Atterberg limits, on material passing No. 40							
Liquid limit	—	—		40 max	41 min	40 max	41 min
Plasticity index	6 max	NP[e]	[d]	10 max	10 max	11 min	11 min
General rating as subgrade	Excellent to good			Fair to poor			

[a]35 percent or less passing No. 200.
[b]More than 35 percent passing No. 200.
[c]Placement of A-2 to right of A-3 is for left-to-right elimination process used in classifying soil and does not reflect relative quality.
[d]See Table 2.9 for values.
[e]Nonplastic.
[f]Group A-8 is highly organic soil and peat that generally provides unsatisfactory support as a subgrade.

TABLE 2.10 AASHTO Classification for Subgroups

General classification	Granular materials						Fine-grained	
Group classification	A-1		A-2				A-7	
Subgroup classification*	A-1-a	A-1-b	A-2-4	A-2-5	A-2-6	A-2-7	A-7-5†	A-7-6‡
Sieve analysis, % passing								
2.00 mm (No. 10)	50 max	—	—	—	—	—	—	—
0.425 mm (No. 40)	30 max	50 max	—	—	—	—	—	—
0.075 mm (No. 200)	15 max	25 max	35 max	35 max	35 max	35 max	36 min	36 min
Atterberg limits, soil passing No. 40								
Liquid limit	—	—	40 max	41 min	40 max	41 min	41 min	41 min
Plasticity index	6 max	6 max	10 max	10 max	11 min	11 min	11 min	11 min
General rating as subgrade	Excellent to good						Fair to poor	

*Groups A-3 through A-6 have no subgroups.
†Plasticity index of subgroup A-7-5 is equal to or less than $LL - 30$.
‡Plasticity index of subgroup A-7-6 is greater than $LL - 30$.

where GI = group index
  F = percent of soil passing the 3-in sieve finer than the No. 200 sieve expressed as a whole number
  LL = liquid limit
  PI = plasticity index

The inverse of the group index may be considered an indicator of the support value of the soil (i.e., 0 is good, 20 or more is very poor). Any negative group indices are reported as 0. For subgroups A-2-6 and A-2-7, only the PI portion of the equation is used to calculate the group index. The soil's group index is reported as a whole number in parentheses behind the classification, as in A-7-5 (12).

Illustrative Example 2.7 Find the AASHTO classification for the soils in Fig. 2.3.

solution The data from Fig. 2.3 needed to classify these soils using Tables 2.8 and 2.9 are tabulated below with their final classification.

Soil	Percent passing			Atterberg limits		Classification
	No. 10	No. 40	No. 200	LL	PI	
Crushed limestone	30	16	11	18	7	A-2-4 (0)
Silty gravel	34	19	10	23	3	A-1-a (0)
Gravelly sand	50	19	0	NP	NP	A-1-a (0)
Eolian sand	100	76	6	NP	NP	A-3 (0)
Glacial till	90	81	62	25	10	A-4 (0)
Wewahitchka clay	100	98	70	65	41	A-7-6 (22)
Loess	100	100	96	29	5	A-4 (0)
Backswamp deposit	100	100	95	59	41	A-7-6 (36)

The limestone failed to meet the A-1 group because its PI was too high. The group index for the glacial till is negative, so it is reported as 0. The group indices for the Wewahitchka clay and the backswamp deposit are only calculated with the PI portion of the group index equation.

2.4.4 Comparison of classification systems

Regardless of which classification system is used, each has certain advantages and certain limitations, and each was developed for different purposes. The USDA's agriculture-oriented system relies purely on particle size distribution and provides no guidance on the plasticity characteristics so important to engineers. The USDA has done extensive soil mapping throughout the United States, so USDA classification of surficial soils is readily available. Happily, recent soil survey reports from the USDA usually include extensive correlations with other classification systems and with the local soil's engineering classification properties. Because these soil surveys are so readily available, they provide an excellent resource for orientation to local

conditions and for preliminary planning. The two engineering classification systems, AASHTO and USCS, address both particle size distribution and plasticity characteristics. The USCS divides soils into 15
categories, with provision for some dual-symbol classifications for
intermediate conditions. AASHTO uses eight major soil groups with
eight further divisions for three of the groups. Each classification system has a different historical background, so one tends to find the
highway industry using the AASHTO system and the geotechnical
and airfield community using the USCS. A 1974 review of the AASH-
TO, FAA, and Unified classification systems found that the Unified
system was more useful for predicting remolded soil behavior than
the (then current) AASHO or FAA systems (Yoder 1974).

Some care must be exercised when moving between the different
classification systems because their basic definitions of important
parameters, such as what constitutes a fine- or coarse-grained soil,
differ. Table 2.11 compares how 610 soils taken from a depth of 6 to
12 in at sites in the South (20 states), Central (7 states), Western
Mountain (3 states), North Central (4 states), and Northeast (6
states) classified using the USDA and Unified systems. There is considerable overlap between the categories in the different systems. The
differences between the Unified and AASHTO classification of fine-
grained soils can be particularly confusing. Figure 2.7 provides a comparison of how the two systems classify fine-grained soils based on
their Atterberg limits. Considerable caution is needed when trying to
make direct comparisons of these two soil classification systems.

2.5   Aggregates

Aggregates are fundamental geotechnical materials that are crucial
to the U.S. construction industry. Over 6000 companies produce over
2 billion tons of aggregates annually, with sales of $8.8 billion to the
industry (Huhta 1991). This production includes both sand and gravel
from natural deposits and crushed stone from quarries. The production of crushed stone has exceeded that of sand and gravel in the
United States since 1975, and the relative importance of crushed
stone to the construction industry has grown steadily.

2.5.1   Sources of aggregates

Sand, gravel, and other construction fill materials may be excavated
from relatively dry natural deposits by power shovels, scrappers,
front-end loaders, clamshells, and similar construction equipment, or
if the deposit is below water, dredges or cranes with clamshells can
effectively recover the material. Clean sand and gravel are particularly valuable to the construction industry and are used in portland

TABLE 2.11 Frequency of USDA Classifications Occurring in Unified Soil Classification Categories

	Percent of USDA soil type classifying in USCS category													
USDA Textural Classification	Coarse-grained						Fine-grained					Organic		
	GM	GC	SP-SM	SM	SM-SC	SC	ML	CL-ML	MH	CL	CH	OL	OH	Pt
Sand	—	—	47	50	3	—	—	—	—	—	—	—	—	—
Loamy sand	—	—	—	98	—	2	—	—	—	—	—	—	—	—
Sandy loam	<1	—	—	41	8	14	19	4	—	9	—	2	2	—
Sandy clay loam	—	—	—	11	3	25	7	—	—	54	—	—	—	—
Sandy clay	—	—	—	—	—	50	—	—	—	50	—	—	—	—
Loam	<1	<1	—	—	—	—	22	11	<1	52	3	3	6	—
Silty loam	—	—	—	—	—	—	36	10	3	45	2	2	2	—
Silt	—	—	—	—	—	—	91	—	—	9	—	—	—	—
Clay loam	—	—	—	—	—	—	5	—	3	75	15	—	2	—
Silty clay loam	—	—	—	—	—	—	1	1	1	67	27	—	3	—
Silty clay	—	—	—	—	—	—	5	—	—	24	71	—	—	—
Clay	—	—	—	—	—	—	3	—	10	13	66	—	8	—
Peat	—	—	—	—	—	—	—	—	—	—	—	—	—	100

SOURCE: Based on data from USAE Waterways Experiment Station, 1963.

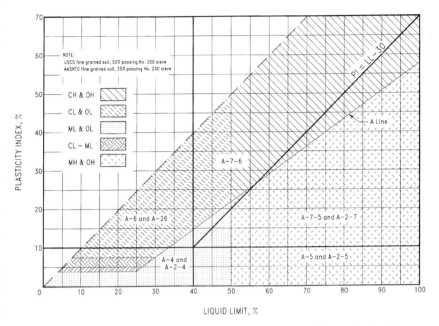

Figure 2.7 Comparison of USCS and AASHTO classifications of fine-grained soils.

cement concrete, asphalt concrete, pavement bases, and fill. Other natural deposits such as clays, silts, fine sands, or materials containing appreciable quantities of fine or plastic material have limited value in construction as high-quality aggregate material for uses such as concrete, pavement base courses, or railroad ballast. Some waste products such as blast furnace slag and a variety of lightweight artificial aggregates such as expanded clay are also used in construction.

Once the sand and gravel aggregate is recovered, it normally undergoes further processing. This may include screening to control particle size, crushing to improve particle characteristics, and washing to remove vegetation, light particles, clay, and similar undesirable material. Table 2.12 lists some common methods of improving aggregate properties (often also referred to as *benefication*).

Crushed stone is playing an increasingly important role in construction. Limestone is the source of about 70 percent of the crushed stone in the United States, with granite providing another 15 percent and traprock (dark, fine-grained igneous rocks), dolomite, sandstone, marble, quartzite, and miscellaneous other materials contributing progressively lesser amounts (Bureau of Mines 1987). The quality of rocks within any given classification varies and must be investigated individually. However, Table 2.13 provides some general qualitative assessments of the engineering characteristics of some common rock types.

TABLE 2.12 Sample Methods of Aggregate Benefication

Method	Principle	Effect
Screening	Material processed over vibrating screens (can be combined with spray washing on screens)	Divide by particle size
Crushing	Material run through a crusher	Angular shape
Log washers	Material is run through an inclined unit with dual rotating shafts mounted with paddles. Continuous flow of water carries fine material out of low end of the unit while cleaned aggregate is discharged at the upper end.	Removes deleterious material (e.g., clay) present in the aggregate or as a coating
Sand classifying unit	Continuous flow of water containing sand is fed into horizontal unit. Coarse sand settles first; finer sands later; fine contaminants are carried out of far end by the water flow.	Divides sand into fractions based on particle size
Screw classifier	Water and sand are fed into the low end of an inclined unit with a rotating screw auger. Sand is moved up the unit and out of the unit by the screws. Waste water at the low end carries off fines and lightweight contaminants.	Removes lightweight and fine contaminants
Rotary scrubber	Water and aggregate fed into a revolving, inclined drum equipped with lifting angles. Aggregate tumbles on self on way to discharge.	Capable of removing large quantities of soluble contaminants
Jig benefication	Mechanical or air pulses agitate water, allowing material to sink to bottom of unit in layers of different density.	Separate on basis of specific gravity
Heavy media separator	Aggregate fed into medium of given specific gravity. Denser particles sink; lighter particles do not.	Precise separation on basis of specific gravity of medium

TABLE 2.13 Summary of Typical Engineering Properties of Rocks

Rock type	Mechanical strength	Durability	Chemical stability	Surface	Impurities	Crushed shape
Igneous						
Granite, syenite, and diorite	Good	Good	Good	Good	Possible	Good
Felsite	Good	Good	Questionable	Fair	Possible	Fair
Basalt, diabase, and gabbro	Good	Good	Good	Good	Seldom	Fair
Peridotite	Good	Fair	Questionable	Good	Possible	Good
Sedimentary						
Limestone and dolomite	Good	Fair	Good	Good	Possible	Good
Sandstone	Fair	Fair	Good	Good	Seldom	Good
Chert	Good	Poor	Poor	Fair	Likely	Poor
Conglomerate and breccia	Fair	Fair	Good	Good	Seldom	Fair
Shale	Poor	Poor	—	Good	Possible	Fair to poor
Metamorphic						
Gneiss and schist	Good	Good	Good	Good	Seldom	Good to poor
Quartzite	Good	Good	Good	Good	Seldom	Fair
Marble	Fair	Good	Good	Good	Possible	Good
Serpentine	Fair	Fair	Good	Fair to Poor	Possible	Fair
Amphibolite	Good	Good	Good	Good	Seldom	Fair
Slate	Good	Good	Good	Poor	Seldom	Poor

SOURCE: J. J. Waddell and J. A. Dobrowolski. *Concrete Construction Handbook*, 1993, reproduced with permission of McGraw-Hill, Inc.

Rock from a quarry or gravel is crushed to produce the final desired aggregate. Typically, oversized material is removed on a grizzly (an inclined vibrating machine with bars that allow properly sized material to pass and catch and divert oversized material). The properly sized rock then is crushed in a primary crusher and one or more additional secondary crushers. These crushers use compression, shear, impact, attrition, or some combination of forces to break the rock into smaller pieces. Rapid reduction in particle size by applying large forces in a crusher can produce excess fines. Consequently, reduction usually occurs in stages using two or more crushers to progressively reduce the particle size produced at each stage. Intermediate screening after each stage of crushing separates the material into acceptably sized particles for use and into particles requiring additional crushing.

Figure 2.8 illustrates the conceptual operation of typical crushing units. Jaw crushers are common primary crushers and are able to handle very hard rock. The jaw faces may be smooth, or if the rock tends to break into a slabby shape, they may be corrugated to encourage better breakage. Roll crushers use one or more rolls with smooth or corrugated surfaces for further particle size reduction after previous crushing. Impact compactors, including the hammer mill, tend to produce good cubical particles and are suitable as secondary or tertiary crushing units. The cage mill throws the rock feed from the center outward, where impact with successive rows of spinning pins breaks the rock into smaller particles. This crusher is often effective in removing softer material in the rock matrix. Cone crushers use a rapid wobbling movement of the cone to compress and shear the feed and are common secondary or tertiary crushers. The gyratory crusher is somewhat similar in concept to the cone crusher but has a longer conical crushing surface eccentrically rotating within the mantle. It also has a larger feed opening and a slower operating speed and can be used either as a primary or a secondary crusher. It is more effective as a primary crusher for slabby rock than is the jaw crusher. Additional washing, screening, and/or other processing also may be needed at various stages to remove undesirable fines or deleterious materials. Rock contaminants such as salt, chlorides, and sulfates that are common problems in arid regions are inherently weak and tend to concentrate in the finer fractions during processing and crushing. Consequently, considerable reductions in contaminants are possible by progressively separating and removing the finer material at various stages of processing (Fookes and Higgenbottom 1980).

The selection of the specific crushers depends on the final product desired and rock characteristics such as strength, toughness, abrasiveness, crystal structure, degree of weathering, natural bedding, fracture, and foliation. For instance, gabbro and basalt are often very strong construction aggregates but are expensive to crush compared

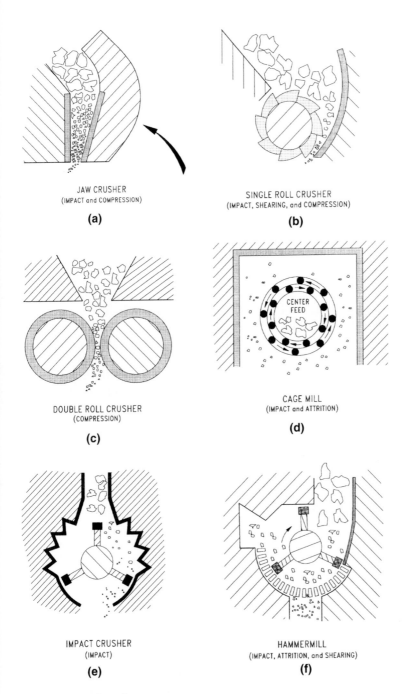

JAW CRUSHER
(IMPACT and COMPRESSION)

(a)

SINGLE ROLL CRUSHER
(IMPACT, SHEARING, and COMPRESSION)

(b)

DOUBLE ROLL CRUSHER
(COMPRESSION)

(c)

CENTER
FEED

CAGE MILL
(IMPACT and ATTRITION)

(d)

IMPACT CRUSHER
(IMPACT)

(e)

HAMMERMILL
(IMPACT, ATTRITION, and SHEARING)

(f)

Figure 2.8 Typical crushers.

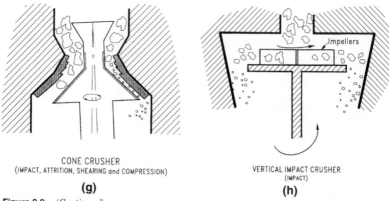

CONE CRUSHER
(IMPACT, ATTRITION, SHEARING and COMPRESSION)
(g)

VERTICAL IMPACT CRUSHER
(IMPACT)
(h)

Figure 2.8   (*Continued*)

with weaker rocks such as limestone. If the basalt has weathered, however, it may be weak and contain an excess of clay. Sandstone crushing characteristics depend primarily on the strength of the cementing material, and rocks such as schist or shale that have distinct foliation or bedding planes break into poorly shaped, elongated particles. Natural rock is too variable to allow any general rules to be made concerning its suitability for crushing. Each potential source should be individually investigated and evaluated by a competent engineering geologist and by pilot crushing operations to assess the characteristics of the final aggregate product.

2.5.2 Strength

The strength of aggregates is a function of a variety of factors, including inherent strength of the particle and particle shape, density, and gradation. Aggregates must be strong enough to withstand handling during excavation, screening, transporting, and placement, and they also must be sufficiently strong to withstand loads while they are in use.

Crushed aggregate particles achieve higher strengths than rounded particles because of superior interlock of the crushed particles. The improved rutting resistance of aggregates with crushed particle shapes rather than rounded shapes has been recognized widely in aggregate base courses under pavements (Haynes and Yoder 1963, Barksdale 1973, Chisolm and Townsend 1976, Ahlrich and Rollings 1994). In aggregates for portland cement concrete, crushing improves strength and flexural strength more than compressive strength. Crushed particles also make a harsher, less workable concrete mixture.

Asphalt concrete relies heavily on the aggregate characteristics for strength of the mix, and crushed particles significantly improve the strength of the asphalt concrete mix (Ahlrich 1991a, Ahlrich and Rollings 1994). The importance of crushed particles in the fine (sand-sized) portion of the asphalt mix is often overlooked. During construction of an asphalt concrete airfield taxiway in the western United States, the natural sand content was increased from the specified maximum of 15 percent of the aggregate to over 30 percent. In less than 1 year, the pavement was badly rutted and cracked because of the lowered stability of the asphalt concrete with the high proportion of rounded natural sand particles (Rollings and Rollings 1992). Whether the aggregate is to be used as fill, pavement base course, portland cement concrete aggregate, or asphalt concrete aggregate, crushing is a consistent method of improving its strength.

Ideally, aggregate particles should be cubical. Flat and elongated particles tend to cause problems with compaction, particle breakage, loss of strength, and segregation. Limits on flat and elongated particles (ASTM D 4791) are sometimes incorporated into specifications. Suggested upper limits on flat and elongated particles vary from 10 percent (Barksdale 1972) to 30 percent (Corps of Engineers 1981) for base course aggregates, from 8 percent (Federal Aviation Administration 1989) to 20 percent (Corps of Engineers 1984) for asphalt concrete aggregates, and up to 20 percent (Department of the Army 1987) for portland cement concrete. Qualitatively, flat and elongated particles can cause problems, but there is no clear correlation between these specified limits on the particles and field performance.

Generally, a well-graded material with a complete distribution of particle sizes will achieve higher density than will a uniformly or gap-graded material. This increased density from tighter interpacking of particles generally will lead to higher strength also. The maximum density of a graded soil or aggregate generally has been determined by the following relationship, which has been referred to as the *power grading law, Talbot equation,* or *Fuller curve* (Barksdale 1991, Department of Commerce 1962, Nijboer 1948, Yoder and Witczak 1975):

$$p = 100 \times \left(\frac{d}{D_{max}} \right)^n$$

where p = percent of material smaller than d (i.e., percent passing d)
d = grain size in question (use consistent units for all sizes in equation)
D_{max} = maximum size aggregate used in material
n = a power

Depending on specific particle characteristics such as angularity and shape, the maximum density for the aggregate generally exists when n is equal to 0.45 to 0.50. Illustrative Example 2.8 shows an example calculation to determine the gradation to achieve maximum density using this equation.

Specifications for materials such as pavement base course aggregate gradations that need to achieve high density and strength are not selected to meet the maximum density grading alone, however. Local agency experience and tradition play a significant role in selecting actual gradation specification requirements.

The percentage of material passing the No. 200 sieve (fines) is also a crucial parameter governing the behavior of aggregates. The optimal percentage of fines to achieve high density is higher than the optimum for high strength (Yoder and Woods 1946). Triaxial testing of base course aggregates found that depending on the specific gradation and state of stress, a critical fine content exists between about 5 and 15 percent (Ferguson 1972). Above this critical value, deformations under load increase very rapidly. Laboratory tests by Barksdale (1972) indicated that at a constant fines content, coarser gradations tended to rut less than finer ones.

The plasticity characteristics of the fines in an aggregate are also critical for determining their behavior. If the aggregates are to be used in asphalt or portland cement concrete, the fines generally should be nonplastic. For soils and aggregates to be used as fill and base or subbase courses, the plasticity index is an indicator of the material's potential strength loss in the presence of moisture. Figure 2.9 shows the change in California bearing ratio (CBR) values measured on soaked and unsoaked subgrade soil and base course aggregate samples from 10 airports in the southeastern and southwestern United States. The CBR test is described in detail in Chap. 4 and is simply a crude index of soil strength as measured by penetration of a 3-in^2 piston into the soil. As the plasticity index increases in Fig. 2.9, the soils and aggregates consistently show a pronounced decrease in CBR. A plasticity index of 6 percent has been a traditionally accepted dividing point between plastic and nonplastic behavior (Ahlvin 1991), and many specifications set the maximum allowable plasticity index between 4 and 6 for materials to be used in the base and subbase of pavements.

Illustrative Example 2.8   Determine the optimal grading for an aggregate with a 1-in maximum size particle using the power law and n equal to 0.45.

solution

Sieve size	Size (mm)	Percent passing
1 in	25	100
½ in	12.5	73.2
No. 4	4.75	47.3
No. 8	2.36	34.6
No. 20	0.85	21.8
No. 40	0.425	16.0
No. 100	0.15	10.0
No. 200	0.075	7.3

Sample calculation for No. 4 sieve:

$$p = 100 \times (4.75/25)^{0.45} = 47.3 \text{ percent}$$

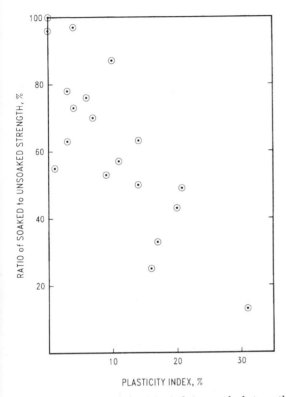

Figure 2.9 Influence of plasticity index on soaked strength.

2.5.3 Durability

Aggregates must be durable under the conditions to which they will be exposed. This may include being exposed to wetting and drying, freezing and thawing, abrasion, and impact during construction and while in service. In one instance, a well-graded crushed limestone aggregate with 1½-in maximum-sized aggregate, classified as a GW under the USCS, was used as a 6-in-thick base course for an airfield pavement in central Missouri. The same crushed limestone but with a 3-in maximum particle size and graded as a GP was used for the 17-in-thick subbase. Reported strength for this material in terms of CBR was 80, which would represent a moderate-quality base course for airfield pavements. The original construction was in 1959–1960, and by 1981, the base and subbase materials had deteriorated to a clayey, sandy gravel containing sufficient clay to allow capillary rise of water, and the CBR had dropped to 60. In 1993, dynamic deflection testing on the pavement surface suggested that the CBR had continued to deteriorate to a level in the range of 24 to 30. The effects of in-situ weathering have essentially destroyed much of this material's load-carrying capacity.

Commonly specified aggregate durability tests such as the sulfate soundness test (ASTM C 88), cyclic freezing and thawing test (AASHTO T 103), and LA abrasion test (ASTM C 131 and 535) are discussed in Chap. 4. Unfortunately, no single test correlates well with the durability of aggregates in the field when exposed to natural weathering processes or abrasive action (e.g., Marek 1991, Rollings 1988, Dolar-Mantuani 1978, Meininger 1978, Reidenour et al 1976, West et al 1970). Consequently, laboratory durability tests are probably best considered as screening tests. Field service records of an aggregate's performance under similar exposure conditions and field investigations of the natural soils formed by this rock's weathering are probably the best indicators of future performance.

2.5.4 Aggregates for asphalt and portland cement concrete

Much of the high-quality aggregate used in construction goes into asphalt concrete (asphalt cement and aggregates) and portland cement concrete (portland cement, aggregates, and sometimes various admixtures and pozzolans). Aggregates used in these applications may encounter some special problems. For asphalt concrete, these include stripping and development of tender mixes. For portland cement concrete, the problems include deleterious materials, D-cracking, and alkali-aggregate reaction. More detailed information on aggregates for use in asphalt and portland cement concrete may be found in American Concrete Institute (1994), Roberts et al (1991), Barksdale (1991), Scherocman (1991), Department of the Army (1985), Neville (1981), American Society of Testing and Materials (1978).

Stripping in asphalt concrete. Stripping in asphalt concrete occurs when moisture replaces the asphalt film coating on the aggregate. The loss of the asphalt cement requires complete rehabilitation of the pavement and can occur in very short periods of time. Research work (e.g., Petersen 1984, Curtis et al 1989, 1992) has made strides in this area, but stripping remains only partially understood. In fact, use of the term *stripping* is probably a misnomer because several mechanisms are likely to contribute to water damage to asphalt concrete and are usually lumped together as stripping. Moisture damage to asphalt concrete is probably one or some combination of the following mechanisms (Dunning 1989):

1. Displacement of the asphalt cement by water on the aggregate surface (true stripping)

2. Emulsion of the asphalt cement

3. Pore pressure development in the asphalt concrete

There are a variety of tests to examine stripping potential [see Lottman (1978), Roberts et al (1991), Al-Swailmi and Terrel (1992) for discussions of conventional testing procedures and new procedures from the Strategic Highway Research Program]. However, there is not a universally accepted approach yet. Most of these tests subject the asphalt concrete sample to some level of water saturation by immersion, boiling, or vacuum pressure and then determine some loss of asphalt cement coating on the aggregate or some change in asphalt concrete engineering properties (Marshall stability, unconfined compressive strength, indirect tensile strength, etc.). The laboratory procedures have been criticized as too severe and as not representative of actual in-service pavements, and thereby, they may needlessly indicate that some aggregates are stripping susceptible when they are not (Crawford 1989). Because of the severity of the stripping problem and the uncertainty surrounding its causes and ways to evaluate it, some agencies routinely require antistripping agents such as lime or one of the commercially available proprietary antistripping compounds in all their asphalt concrete.

Several steps may be taken to help protect against stripping. Stripping requires water or water vapor at high concentrations, so proper drainage is needed. This should be clearly addressed in design. Too often we tend to build pavements in a bathtub configuration that traps water in the structure, leading to a number of unpleasant consequences, with increased stripping possibly being one of them.

Second, proper selection of asphalt concrete aggregates will determine how much stripping potential exists. Acidic aggregates (e.g., siliceous aggregates) tend to be hydrophilic ("water loving"), whereas basic aggregates (e.g., limestone composed of $CaCO_3$) tend to be hydrophobic ("water hating"). An aggregate's propensity to asphalt stripping has been suspected to be a function of its hydrophobic or hydrophilic tendency, with the hydrophilic aggregates giving up asphalt for water and resulting in stripping. In the presence of water, acidic aggregates containing silica such as quartz have a negative surface charge, whereas basic aggregates such as limestone or dolomite have a positive charge. Aggregates can be classified as a function of their surface charge (e.g., Mertens and Borgfeldt 1965), which may provide a guide to selection of aggregate types to reduce stripping potential. Higher positive surface charges generally will be associated with greater stripping resistance.

Tender asphalt concrete mixes. Another problem with asphalt concrete construction that is strongly related to the aggregates used in the mix is the existence of so-called tender mixes that displace and crack under the rollers during construction. This unstable behavior may be caused by a variety of factors such as rolling at too high a temperature or excess asphalt cement in the mix. However, usually the cause of tender mixes is the aggregate in the mix. Crushed materials are usually short on the desired quantity of No. 50 and 100 sieve–sized sand particles for asphalt concrete. Commonly, a number of crushed and uncrushed materials are blended for the aggregate of an asphalt concrete mix, but as noted in Sec. 2.5.2, an excess of uncrushed material will result in a loss of strength in the mix. Natural sand often is used to make up the deficiency in the No. 50 and 100 sieve–sized crushed sand fraction. If care is not used, this results in too much natural sand in the mix and also develops a hump in the gradation curve in the region of the No. 16 to 50 sieves. This hump has been found to be characteristic of many mixes showing tenderness under the rollers. The stability of an asphalt mix is strongly influenced by the particle shape of the sand fraction, and an excess quantity of natural sand is also characteristic of tender mixes.

An extreme example of a problem gradation is shown in Fig. 2.10. Although this mix was generally within the specification band, a natural sand content in excess of 30 percent and the hump at the No. 16 sieve size resulted in a tender mix and one that was unstable under aircraft traffic. To overcome these deficiencies, the aggregate gradation has to be adjusted to decrease and, if possible, eliminate the hump, and crushed sand needs to replace some of the natural sand.

Deleterious materials in portland cement concrete. Most specifications for concrete aggregates include limits on deleterious materials. In the widely used ASTM C 33, Standard Specification for Concrete Aggregates, these limits cover clay lumps and friable particles, chert with a specific gravity of less than 2.40, coal and lignite, and the amount of material passing the No. 200 sieve. The specific content of these deleterious materials allowed in the aggregate varies depending on whether it is fine or coarse aggregate that contains the contaminant, the use of the concrete (e.g., exposed architectural concrete, footings, pavements, etc.), and the climate severity.

Clay lumps and soft particles break down easily, cause localized pitting or popouts, and, if present in sufficient quantity, can reduce strength and abrasion resistance. Low-specific-gravity chert and other lightweight particles are often weak and vulnerable to freezing and thawing damage, which results in unsightly popouts. Coal and lignite particles are lightweight with the potential problems just noted and also may cause staining. If these organic materials are finely divided and present in sufficient quantities, they can affect the setting of the portland cement concrete. Fine material passing the No.

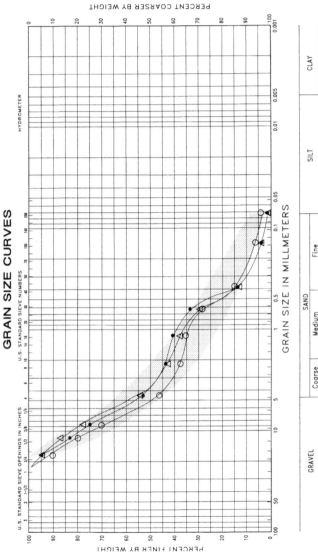

Figure 2.10 Example of aggregate grading hump causing asphalt concrete tenderness during construction (*from data reported by Ahlrich 1991b*).

| | Specific Gravity | | |
	Natural Sand, %	Retained on No. 4	Passing No. 4
● Sample 1	37.2	2.67	2.54
○ Sample 2	30.7	2.63	2.63
△ Sample 3	34.2	2.65	2.63

200 sieve and particularly clay particles will increase the water demand of the concrete to maintain workability and are associated with reduced workability, increased shrinkage and cracking, lowered strength, and reduced abrasion resistance. An excess of clay fines also can result in poor skid resistance in pavements (Grau 1979). The problems associated with fine material passing the No. 200 sieve are primarily related to the clay particles. Consequently, many specifications allow an increase in the amount of fine material if the fines are generated by crushing. When deleterious materials are present, usually they can be controlled by selective quarrying and processing or by some combination of benefication methods such as those listed in Table 2.12. Standard ASTM tests for determining the content of these deleterious materials in an aggregate are available and generally prove to be an effective method of avoiding some of the potential concrete problems associated with these materials.

D-cracking in portland cement concrete.   Certain sedimentary aggregates and a few metamorphic or partially metamorphosed sedimentary rocks have pore structures that are vulnerable to damage when the aggregate is critically saturated and frozen (Stark 1976, Schwartz 1987). The aggregate is cracked by the forces generated by freezing and ice formation, and the cracks propagate into the surrounding mortar. The damage appears as a series of fine cracks paralleling slab joints, as shown in Fig. 2.11. This cracking is known as *durability* or *D-cracking* and is particularly prevalent in pavements. The damage initially begins on the underside of slabs and adjacent to joints where the moisture content of slabs placed on ground is highest. It is a progressive form of deterioration that will eventually reduce the concrete to rubble.

There is no simple test to determine an aggregate's susceptibility to D-cracking (Schwartz 1987). At present, the relatively expensive and time-consuming rapid freezing and thawing test (ASTM C 666) of concrete made with the aggregate under question is the most reliable method of identifying D-cracking–susceptible aggregate. Other tests generally have failed to correlate with field performance of aggregates (Schwartz 1987), but they may prove useful for screening aggregates in a local area before more in-depth testing is undertaken.

If possible, D-cracking–susceptible aggregate should be avoided. Reducing the maximum aggregate size used in the concrete is the most effective method of minimizing D-cracking damage if such aggregates must be used. However, it is not always effective, and the amount of size reduction needed for protection varies from source to source (Schwartz 1987). Factors such as fine aggregate, mix design, cement, reinforcing, slab size, traffic, etc., have no bearing on development of D-cracking (Schwartz 1987). Vapor barriers and drainage

Figure 2.11   D-cracking in an Indiana pavement with asphalt concrete corner patch.

for the underside of the pavement slab successfully slowed the rate of development of D-cracking in some cases, but in other tests they were ineffective and in no case did they prevent D-cracking (Stark 1970, Stark and Klieger 1973, Klieger et al 1974, American Concrete Pavement Association 1988). Blending durable aggregate with D-cracking–susceptible aggregate may reduce the severity of the deterioration to manageable proportions, but it is not a solution to the problem (Thompson et al 1980). Reduction in the proportion of coarse aggregate with a corresponding increase in the fine aggregate proportion of the concrete mixture is also done. This, however, can lead to other problems. Rollings and Wong (1992) found that the use of a 70 percent fine aggregate to 30 percent coarse aggregate blend to avoid D-cracking problems with the local limestone coarse aggregate in Kansas resulted in poor workability, construction problems, and a nondurable concrete paste. Although the problem of D-cracking was avoided with this aggregate proportion, it caused other problems as severe as the one it tried to solve. Heating aggregates to dry them in the expectation that it will take longer to critically saturate them in service has not worked to avoid D-cracking (Schwartz 1987). However, pretreatment of the aggregate to coat the surface or impregnate the aggregate pores seems to have some potential, but more work is needed in this area (Schwartz 1987).

Alkali-aggregate reaction in portland cement concrete.   The alkalis in
portland cement react with the minerals in the concrete aggregate.
While not all reactions are detrimental, reactions with certain specific
forms of silica form a gel that absorbs available free water and
expands. These expansive forces then cause volume distortions of the
material, cracking, loss of strength, and further progressive deterioration. Expansive chemical reactions between portland cement alkalis
and certain relatively rare carbonate aggregates also occur, but they
are much less common (American Concrete Institute 1989).

Figure 2.12 shows characteristic surface cracking from this alkali-
silica reaction that occurred at an airfield pavement near
Albuquerque, New Mexico. Tremendous expansive forces are generated by these reactions that cause apparent growth of the concrete,
which can cause upheavals in asphalt shoulder materials and tilt
light poles and inlets to drainage systems. Two pins that were originally adjacent to one another at Pease AFB, New Hampshire, were
separated almost a foot over the course of 10 years from differential
concrete expansion that occurred from alkali-aggregate reaction. The
two pins were located across from one another at a slip joint between
two very large parking aprons. This slip joint allowed each apron to
move parallel to the joint independently from the other apron.

Either fine or coarse aggregate may be involved in alkali-aggregate
reactions. The example from New Mexico has reactive silica in the

Figure 2.12   Alkali-silica reaction cracking in a New Mexico pavement.

coarse aggregates, whereas the reactive material in the New Hampshire example was opal (a form of reactive silica) that was in the concrete fine aggregate. The reactions may appear in months, or it may take decades. The symptoms of alkali-silica reactions, i.e., cracking and movements causing distress to adjacent structures and shoulders, were noticed approximately 6 years after construction in the New Hampshire example, but no mention of them was made in a pavement inspection conducted 4 years after construction.

Water is a necessary ingredient in the reaction to allow the gel to swell and cause the cracking and volume changes characteristic of these reactions. However, adequate moisture is found generally in pavements and dams even in desert regions and at least seasonally in bridge decks and columns in dry climates (Stark 1991). Consequently, most concrete structures can provide sufficient moisture to allow these reactions to occur.

There are three approaches to handling the alkali-aggregate reaction problem:

1. Avoid reactive aggregates.
2. Use low-alkali cements.
3. Counter the reaction with a pozzolanic additive.

There is a three-pronged approach to evaluating aggregates for potential alkali-silica reactivity. First, a trained petrographer can examine the aggregate for evidence of reactive aggregates, as outlined in ASTM C 295. However, even an experienced petrographer can encounter problems judging the reactivity of some borderline materials that may be reactive in some cases and not in others (Diamond 1978). Next, a chemical test (ASTM C 289) can be run, but it uses reaction conditions that differ from field conditions and cannot be considered as a predictor of the reaction in the field. However, it allows a quick screening of reaction potential within a few days and is especially useful when coupled with petrographic examination. Finally, mortar bar tests (ASTM C 227) can be run using the aggregate in question, and the resulting degree of expansion can be measured. Unfortunately, although the tests are considered reasonably reliable, they require 3 to 6 months to complete. A variant of this test (ASTM C 342) has been used in Oklahoma, Kansas, Nebraska, and Iowa because of the slow development of alkali reactions in these areas, possibly reflecting the local climate and concrete mixture practices (Diamond 1978). While avoidance of reactive aggregates is the best approach when feasible, there are some limitations on our ability to quickly and effectively determine if an aggregate will prove to be reactive or not.

If reactive aggregates or potentially reactive aggregates must be used for concrete, the danger of developing destructive reactions can be reduced or eliminated by reducing the alkali content of the cement or by using some pozzolanic materials to counter the reactions. Low-alkali cements are becoming increasingly scarce and expensive and are not universally available. Some pozzolanic materials such as fly ash, blast-furnace slag cements, or blended cements (ASTM C 595) can counter or mitigate the destructive alkali-silica reactions (Diamond 1978). The effectiveness of different materials varies, however, and a modified mortar bar test (ASTM C 227) provides a measure of the effectiveness of the additive in a period of 14 days.

2.5.5   Aggregate handling

A great deal of money is spent processing aggregates to meet specified requirements such as gradation, plasticity, etc. However, mishandling of aggregates results in segregation and contamination of the aggregates that often defeat the original processing. Any time an aggregate is allowed to fall from conveyors into coned stockpiles, trucks dump aggregates down a slope, or equipment such as a grader or dozer pushes aggregates over extended distances, coarser aggregate particles tend to separate from the other particles, resulting in segregated areas with excess coarse particles and other areas with excess fines.

During placement of a crushed-limestone road base course in the Mideast, adjustable side wings on an aggregate spreader box were opened as wide as possible to increase placement productivity. As the dozer pushed this spreader box forward, the larger aggregate particles tended to roll to the outside and accumulated on the outside edges of the aggregate placement. Figure 2.13 shows the resulting aggregate segregation with accumulated large aggregate particles at the juncture between two placement lanes. This segregated material had to be removed and replaced. The placement width of the spreader box also was decreased by closing the adjustable wings to avoid recurrence of this problem.

Figure 2.14 shows several methods of handling aggregate that generally result in contamination and segregation. Under average conditions, about 3 percent of a stockpile dumped directly on the ground is lost to contamination (American Concrete Paving Association 1975). Consequently, often it will be cost-effective to provide a stable, paved storage platform for stockpiles. Generally, any action that drops material into a cone or down a slope or any extended pushing or blading of the aggregate will result in unwanted segregation. Placement of aggregate in a single cone causes the greatest segregation of any storage technique, but it remains common in practice (Transportation

Figure 2.13 Example of segregated base course aggregate.

Research Board 1967). The effect of segregation in a coned stockpile can be countered to some extent by recovering aggregate with a front-end loader operating perpendicular to the flow of aggregate during placement and by spreading recovery operations across the entire face of the stockpile.

If aggregates are stockpiled in different size groups and then reblended at a plant for base course aggregate or portland cement concrete or asphalt concrete, segregation will be minimized. Continuously graded materials tend to segregate in stockpiles more than uniformly graded materials. Figure 2.15 shows several improved methods of storing aggregates. Basically, the aggregates should be dumped in single truck loads and left alone, or they should be spread in uniform horizontal layers to minimize segregation.

2.5.6 Economics and aggregates

Soils and aggregates used in construction must be as economical as possible to reduce overall project costs. Consequently, locally available materials are used to the extent possible, and the high costs of transporting materials over long distances and extensive benefication processes should be used only when the cost of the improved quality can be justified by performance in the final product. However, the selection of any material to be used in construction must be based on

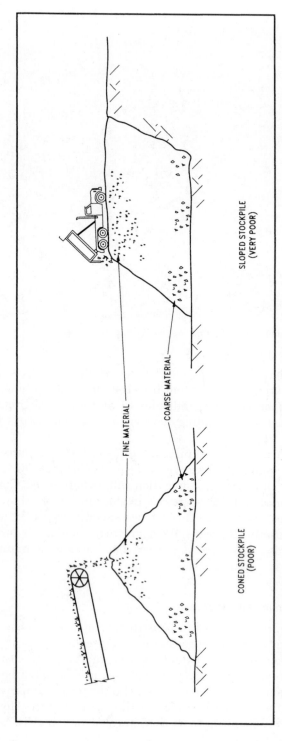

Figure 2.14 Aggregate handling techniques resulting in segregation and contamination.

FINE MATERIAL

COARSE MATERIAL

CONED STOCKPILE
(POOR)

SLOPED STOCKPILE
(VERY POOR)

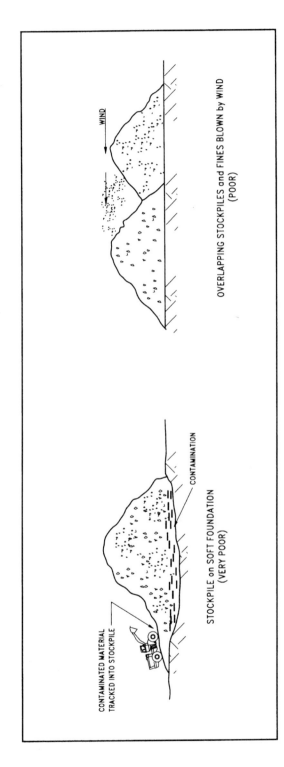

Figure 2.14 (*Continued*)

65

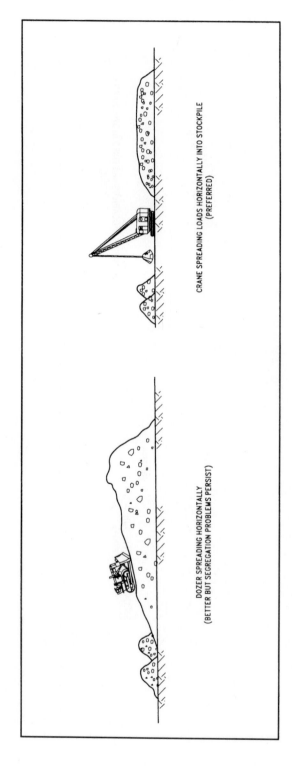

Figure 2.15 Preferred aggregate handling techniques.

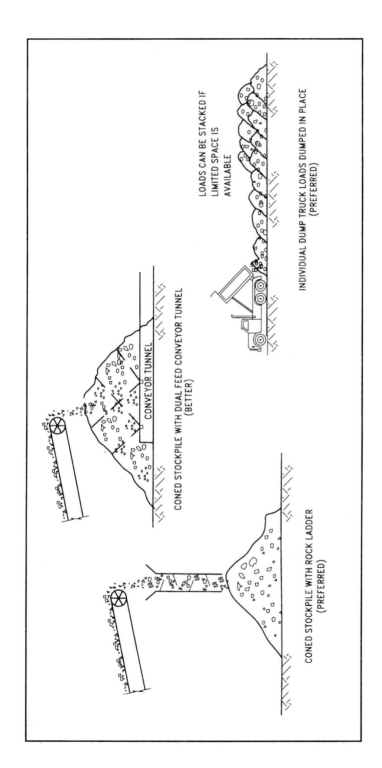

LOADS CAN BE STACKED IF LIMITED SPACE IS AVAILABLE

INDIVIDUAL DUMP TRUCK LOADS DUMPED IN PLACE (PREFERRED)

CONVEYOR TUNNEL

CONED STOCKPILE WITH DUAL FEED CONVEYOR TUNNEL (BETTER)

CONED STOCKPILE WITH ROCK LADDER (PREFERRED)

Figure 2.15 (*Continued*)

its ability to perform as needed and not on cost alone. It is extremely shortsighted to accept inferior materials that will not do the job just to try to save money initially. The initial savings and much more usually will be consumed in rework, repairs, and maintenance.

The construction industry is greatly hampered today by having a multitude of specifications promulgated by a variety of organizations, and these specifications are not consistent with one another and are often badly outdated. Table 2.14 shows four sample specification requirements for base course aggregates in flexible (asphalt) pavements from different organizations. The Federal Aviation Administration and the Corps of Engineers are oriented toward airfield requirements, which with their wheel loads in the 30,000- to 50,000-lb category and tire pressures of 200 to 300 psi would be expected to be more stringent than the highway loads (4500 lb and 100 psi) such as anticipated by the North Carolina specification. The

TABLE 2.14   Selected Base Course Requirements from Different Organizations

Requirement	FAA*	COE†	ASTM D 2940‡	N.C. type A§
Gradation (% passing)				
2 in	100	100	100	—
1½ in	95–100	70–100	95–100	100
1 in	70–95	45–80	—	75–97
¾ in	55–85	—	70–92	—
½ in	—	30–60	—	55–79
⅜ in	—	—	50–70	—
No. 4	30–60	20–50	35–55	35–55
No. 10	—	15–40	—	25–45
No. 30	12–30	—	12–25	—
No. 40	—	5–25	—	14–30
No. 200	0–8	0–10	0–8	4–12
Atterberg limits (%)				
Liquid limit	≤25	≤25	≤25	≤35
Plasticity index	≤4	≤5	≤4	≤6
Aggregate quality				
Crushed particles (%)	≥90	≥50	≥75	—
LA abrasion (%)	≤45	≤50	—	≤55
Flat and elongated particles (%)	≤15	≤30	—	—
Sulfate soundness, 5 cycles (%)	≤12	—	—	≤15

*Federal Aviation Specification Item P-209, *Crushed Aggregate Base Course,* AC 150/5370-10A.
†Corps of Engineers Guide Specification, Military Construction CEGS-02241, *Stabilized-Aggregate Base Course*; stabilized refers to compaction, not chemical stabilization.
‡*Standard Specification for Graded Aggregate Material for Bases or Subbases for Highways or Airfields.*
§*North Carolina Standard Specification for Roads and Structures,* Section 1010, Aggregate for Non-Bituminous Flexible Type Bases.

ASTM specification nominally addresses both highway and airfield applications.

Each specification essentially addresses three areas: gradation, plasticity characteristics, and quality/durability. As discussed earlier, gradations are typically based on the specifying organization's experience and some consideration of the power grading law. Because these requirements are primarily experience-based, the four specifications have significantly different requirements for the sizes specified, the amounts passing each specified sieve, and the allowable/required amount of fines. Similarly, the Atterberg limit requirements to control plasticity follow the traditionally accepted division between plastic and nonplastic behavior, with the airfield-oriented applications drawing the line a little more conservatively than highway applications. The aggregate quality basically addresses aggregate shape (crushing and amount of flat and elongated particles) and durability (LA abrasion and sulfate soundness). These are valid areas of interest to the specifying agency. However, they are also very difficult to define and measure, and once measured, the current methods of measurement are not necessarily reflected in actual field performance. There is often a lack of consistency in specification of aggregates, and our ability to measure all the desired properties of aggregates is still limited, so careful judgment in assessing aggregates is needed.

Good-quality aggregates are becoming increasingly scarce and expensive for a variety of reasons (particularly limitations on opening new sources, urban development, and exhaustion of readily available supplies). The extent of the problem in regional terms is illustrated in Fig. 2.16. This figure is based on the results of a survey of highway departments concerning the availability of aggregates for pavement construction (Witczak 1972). As availability of economical, good-quality aggregate decreases, there is a corresponding increase in pressure to use other marginal or substandard materials in construction. However, much of our past experience is with specific qualities of aggregates, and this past experience cannot be translated directly to new, substandard materials with which we have no experience. Use of such materials requires a significant increase in the engineering, testing, and evaluation of the materials before they are usable in construction (Rollings 1988). Another trend is to increase the amount of waste and recycled materials used in construction. This, unfortunately, is outside the scope of this book, but it is a topic area that will continue to grow in importance to the construction industry (American Society for Testing and Materials 1976, Goumans et al 1991, Waller 1993).

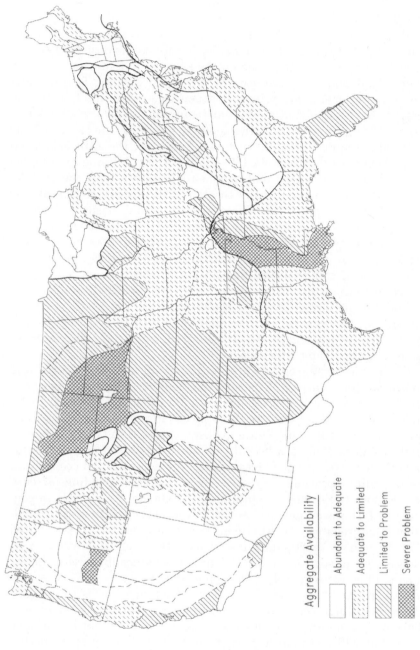

Aggregate Availability

Abundant to Adequate

Adequate to Limited

Limited to Problem

Severe Problem

Figure 2.16 Rating of aggregate availability (*Witczak 1972*).

2.6   References

Ahlrich, R. C. 1991a. "The Effects of Natural Sands on Asphalt Concrete Engineering Properties," technical report no. GL-91-3, USAE Waterways Experiment Station, Vicksburg, Miss.

Ahlrich, R. C. 1991b. "Investigation of Airfield Pavement Failure at Cairo East Air Base," miscellaneous paper no. GL-91-20, USAE Waterways Experiment Station, Vicksburg, Miss.

Ahlrich, R. C., and R. S. Rollings. 1994. "Marginal Aggregates in Flexible Pavements," DOT/FAA/CT - 94/58, Federal Aviation Administration, Washington, D.C.

Ahlvin, R. G. 1991. "Origin of Developments for Structural Design of Pavements," technical report no. GL-91-26, USAE Waterways Experiment Station, Vicksburg, Miss.

Al-Swailmi, S., and R. L. Terrel. 1992. "Evaluation of Water Damage of Asphalt Concrete Mixtures Using the Environmental Conditioning System (ECS)," *Journal of the Association of Asphalt Paving Technologists*, 61: 405–435.

American Concrete Institute. 1994. *ACI Manual of Concrete Practice*, 5 vols., ACI, Detroit, Mich.

American Concrete Institute. 1989. *Guide for Use of Normal Weight Aggregates in Concrete*, ACI 221R-89, ACI, Detroit, Mich.

American Concrete Pavement Association. 1988. *Ohio "D"-Cracking Test Road Study*, vol. 24, no. 4, ACPA, Arlington Heights, Ill.

American Concrete Paving Association. 1975. "Aggregate Handling," technical bulletin no. 21, Skokie, Ill.

American Society for Testing and Materials. 1978. *Significance of Tests and Properties of Concrete and Concrete-Making Materials*, STP-169B, ASTM, Philadelphia, Pa.

American Society for Testing and Materials. 1976. *Living with Marginal Aggregates*, STP-597, ASTM, Philadelphia, Pa.

Barksdale, R. D. 1991. *The Aggregate Handbook*, National Stone Association, Washington, D.C.

Barksdale, R. D. 1973. "Rutting of Pavement Materials," School of Civil Engineering, Georgia Institute of Technology, Atlanta, Ga.

Barksdale, R. D. 1972. "Laboratory Evaluation of Rutting in Base Course Materials," in *Proceedings of the Third International Conference on the Structural Design of Asphalt Pavements*, vol. I, University of Michigan, Ann Arbor, Mich., pp. 161–174.

Bureau of Mines. 1987. *Mineral Yearbook*, U.S. Department of the Interior, Washington, D.C.

Carter, M., and S. P. Bentley. 1991. *Correlations of Soil Properties*, Pentech Press, London.

Casagrande, A. 1948. "Classification and Identification of Soils," *Transactions of the American Society of Civil Engineers*, 113: 901–991.

Chisolm, E. E., and F. C. Townsend. 1976. "Behavioral Characteristics of Gravelly Sand and Crushed Limestone for Pavement Design," technical report no. S-76-17, USAE Waterways Experiment Station, Vicksburg, Miss.

Corps of Engineers. 1981. "Stabilized-Aggregate Base Course," in *Guide Specification for Military Construction*, CEGS-02241, Department of the Army, Washington, D.C.

Corps of Engineers. 1984. "Bituminous Intermediate and Wearing Course for Roads, Streets, and Open Storage Areas (Central-Plant Hot-Mix)," in *Guide Specification for Military Construction*, CEGS-022551, Department of the Army, Washington, D.C.

Crawford, C., 1989. "Overlay Water Damage" (discussion), *Journal of the Association of Asphalt Paving Technologists*, 58: 70–73.

Curtis, C. W., R. L. Lytton, and C. J. Brannan. 1992. "Influence of Aggregate Chemistry on the Adsorption and Desorption of Asphalt," transportation research record no. 1362, Transportation Research Board, Washington, D.C.

Curtis, C. W., D. J. Clapp, Y. W. Jeon, and B. M. Kiggundu. 1989. "Absorption of Model Asphalt Functionalities, AC-20, and Oxidized Asphalts on Aggregates," transportation research record no. 1228, Transportation Research Board, Washington, D.C.

Department of the Army. 1987. *Standard Practice for Concrete Pavements*, technical manual TM 5-822-7, Washington, D.C.

Department of the Army. 1985. *Standard Practice for Concrete,* EM 1110-2-20000, Washington, D.C.

Department of Commerce, 1962. *Aggregate Gradation for Highways, Simplification, Standardization, and Uniform Application and a New Graphical Evaluation Chart,* Washington, D.C.

Diamond, S. 1978. "Chemical Reactions Other Than Carbonate Reactions," in *Significance of Tests and Properties of Concrete and Concrete-Making Materials,* STP-169B, American Society for Testing and Materials, Philadelphia, Pa., pp. 708–721.

Dolar-Mantuani, L. 1978. "Soundness and Deleterious Substances," in *Significance of Tests and Properties of Concrete and Concrete-Making Materials,* STP-169B, American Society for Testing and Materials, Philadelphia, Pa., pp. 744–761.

Dunning, R. 1989. "Overlay Water Damage" (discussion), *Journal of the Association of Asphalt Paving Technologists,* 58: 74–75.

Federal Aviation Administration. 1989. *Standards for Specifying Airport Construction,* AC 150/5370-10A, Department of Transportation, Washington, D.C.

Ferguson, E. G. 1972. "Repetitive Triaxial Compression of Granular Base Course Material with Variable Fines Content," dissertation submitted in partial fulfillment of the requirements for the degree of Doctor of Philosophy, Iowa State University, Ames, Iowa.

Fookes, P. G. and I. E. Higginbottom. 1980. "Some Problems of Construction Aggregates in Desert Areas, with Particular Reference to the Arabian Peninsula: 2. Investigation, Production, and Quality Control," *Proceedings of the Institution of Civil Engineers,* London, 68 (part 1): 69–90.

Grau, R. W. 1979. "Utilization of Marginal Materials for LOC," technical report no. GL-79-21, USAE Waterways Experiment Station, Vicksburg, Miss.

Goumans, J. J., H. van der Sloot, and T. G. Aalbers. 1991. *Waste Materials in Construction,* Studies in Environmental Science 48, Elsevier, Amsterdam.

Haynes, J. H., and E. J. Yoder. 1963. "Effects of Repeated Loading on Gravel and Crushed Stone Base Course Material," highway research record no. 39, Highway Research Board, Washington, D.C.

Huhta, R. S., 1991. "Introduction to the Aggregate Industry," in *The Aggregate Handbook,* R. D. Barksdale, ed., National Stone Association, Washington, D.C., pp. 1.1–1.14.

Klieger, P., G. Monfroe, D. Stark, and W. Teske. 1974. "D-Cracking of Concrete Pavements in Ohio," report no. Ohio-DOT-11-74, Portland Cement Association, Skokie, Ill.

Lambe, T. W., and R. V. Whitman. 1954. *Soil Mechanics,* Wiley, New York.

Lambe, T. W. 1951. *Soil Testing for Engineers,* Wiley, New York.

Legget, R. F., and P. F. Karrow. 1983. *Handbook of Geology in Civil Engineering,* McGraw-Hill, New York.

Lottman, R. P. 1978. "Predicting Moisture-Induced Damage to Asphalt Concrete," NCHRP report no. 192, Transportation Research Board, Washington, D.C.

Marek, C. R. 1991. "Basic Properties of Aggregate," in *The Aggregate Handbook,* R. D. Barksdale, ed., National Stone Association, Washington, D.C., pp 3.1–3.81.

Mathewson, C. C. 1981. *Engineering Geology,* Bell & Howell Company, Columbus, Ohio.

Meininger, R. C. 1978. "Abrasion Resistance, Strength, Toughness, and Related Properties," in *Significance of Tests and Properties of Concrete and Concrete-Making Materials,* STP-169B, American Society for Testing and Materials, Philadelphia, Pa.

Mertens, E. W., and J. J. Borgfeldt. 1965. "Cationic Asphalt Emulsions," in *Bituminous Materials: Asphalts, Tars, and Pitches,* vol. 2, part 1, A. J. Hoiberg, ed., Wiley, New York.

Mitchell, J. K. 1993. *Fundamentals of Soil Behavior,* 2d ed., Wiley, New York.

Neville, A. M. 1981. *Properties of Concrete,* 3d ed., Pitman Books, London.

Nijboer, L. W. 1948. *Plasticity as a Factor in the Design of Dense Bituminous Road Carpets,* Elsevier, Amsterdam.

Peltier, L. 1950. "The Geographic Cycle in Periglacial Regions as It Is Related to

Climatic Geomorphology," *Annals of the Association of American Geographers,* 40: 214–236.

Petersen, J. C. 1984. "Chemical Composition of Asphalt as Related to Asphalt Durability: State of the Art," transportation research record no. 999, Transportation Research Board, Washington, D.C.

Reidenour, D. R., E. G. Geiger, and R. H. Howe. 1976. "Suitability of Shale as a Construction Material," in *Living with Marginal Aggregates,* STP-597, American Society for Testing and Materials, Philadelphia, Pa.

Roberts, F. L., P. S. Kandhal, E. R. Brown, D. Y. Lee, and T. W. Kennedy. 1991. *Hot Mix Asphalt Materials, Mixture Design, and Construction,* National Asphalt Pavement Association Education Foundation, Lanham, Md.

Rollings, R. S. 1988. "Substandard Materials for Pavement Construction," miscellaneous paper no. GL-88-30, USAE Waterways Experiment Station, Vicksburg, Miss.

Rollings, R. S., and M. P. Rollings. 1992. "Pavement Failures: Oversights, Omissions, and Wishful Thinking," *Journal of Performance of Constructed Facilities,* 5(4): 271–286.

Rollings, R. S., and Wong, G. S. 1992. "Investigation of a Concrete Blistering Failure," *Proceedings, Materials: Performance and Prevention of Deficiencies and Failures,* American Society of Civil Engineers, New York, pp. 16–30.

Scherocman, J. A. 1991. *Hot-Mix Asphalt Paving Handbook,* Transportation Research Board, Washington, D.C

Schwartz, D. R. 1987. "D-Cracking of Concrete Pavements," NCHRP synthesis of highway practice no. 134, Transportation Research Board, Washington, D.C.

Skempton, A. W. 1953. "The Colloidal Activity of Clays," in *Proceedings of the Third International Conference on Soil Mechanics,* vol, I, International Society for Soil Mechanics and Foundation Engineering, Zurich, Switzerland, pp. 57–61.

Sowers, G. F. 1981. "There Were Giants on the Earth in Those Days," *Journal of the Geotechnical Division,* ASCE, 107 (GT4): 385–419.

Stark, D. 1991. "The Moisture Condition of Field Concrete Exhibiting Alkali-Silica Reactivity," in *Durability of Concrete,* Second International Conference, SP-126, vol. II, V. M. Malhotra, ed., American Concrete Institute, Detroit, Mich., pp. 973–985.

Stark, D. 1976. "Characteristics and Utilization of Coarse Aggregate Associated with D-Cracking," research and development bulletin RD047.01p, Portland Cement Association, Skokie, Ill.

Stark, D. 1970. "Field and Laboratory Studies of the Effects of the Subbase Type on the Development of D-Cracking," highway research record no. 342, Highway Research Board, Washington, D.C.

Stark, D., and P. Klieger. 1973. "Effect of Maximum Size of Coarse Aggregate on D-Cracking in Concrete Pavement," highway research record no. 441, Highway Research Board, Washington, D.C.

Thompson, S. R., M. P. Olsen, and B. J. Dempsey. 1980. "Synthesis Report: D-Cracking in Portland Cement Concrete Pavements," report no. FHWA/IL/UI-187, University of Illinois, Urbana, Ill.

Transportation Research Board. 1967. "Effects of Different Methods of Stockpiling and Handling Aggregates," transportation research record no. 46, Washington, D.C.

USAE Waterways Experiment Station. 1963. "Forecasting Trafficability of Soils," technical memorandum no. 3-331, vol. II, Vicksburg, Miss.

U.S. Department of Agriculture. 1975. *Soil Taxonomy: A Basic System of Soil Classification for Making and Interpreting Soil Surveys,* USDA, Washington, D.C.

Waddell, J. J., and J. A. Dobrowolski. 1993. *Concrete Construction Handbook,* 3d ed., McGraw-Hill, New York.

Waller, H. F. 1993. *Use of Waste Materials in Hot-Mix Asphalt,* STP-1193, American Society for Testing and Materials, Philadelphia, Pa.

West, T. R., R. B. Johnson, and N. M. Smith. 1970. "Tests for Evaluating Degradation of Base Course Aggregates," NCHRP report no. 98, Highway Research Board, Washington, D.C.

Witczak, M. W. 1972. "Relationships Between Physiographic Units and Highway Design Factors," NCHRP report no. 132, Highway Research Board, Washington, D.C.

Yoder, E. J. 1974. "Review of Soil Classification Systems Applicable to Airport Pavement Design," report no. FAA-RD-73-169, Federal Aviation Administration, Washington, D.C.

Yoder, E. J., and K. B. Woods, 1946. "Compaction and Strength Characteristics of Soil Aggregate Mixtures," *Proceedings,* Highway Research Board, Washington, D.C.

Yoder, E. J., and M. W. Witczak. 1975. *Principles of Pavement Design*, Wiley, New York.

3

Natural Deposits

We talk of our mastery of nature, which sounds
very grand; but the fact is we respectfully adapt
ourselves, first, to her ways.   CLARENCE DAY

3.1 Introduction

Every engineered structure ultimately must support all its load on
the available natural deposits of soils and rock; moreover, we rely on
our natural deposits of soils and rock to provide the construction fill
and aggregates for many of our engineered structures. This chapter
will examine how surficial weathering and deposition processes affect
the properties of the soils and aggregates that form the backbone of
much of our engineering effort.

Soil characteristics will be largely determined by the environment
during the soil's formation. Soils can be broadly classified as either
transported or residual soils, as shown in Table 3.1. *Transported* soils
have been carried to their final resting place by some mechanism
such as wind, water, or ice, and the mechanism of transportation will
have a major influence on the soil properties. If the soil develops by
weathering in place, it is a *residual* soil, and its properties will be
largely a function of its parent material and the weathering factors
discussed in Chap. 2. The situation is further muddied when one considers that natural deposits are often intermingled and layered products of several processes occurring at different geologic times. All surficial soils undergo further weathering changes and develop a
distinctive profile with depth once they are in place.

TABLE 3.1   Soils from Different Transport Agents and Rock Sources

Transported soils			Residual soils		
Wind	Water	Ice	Sedimentary	Igneous	Metamorphic
Eolian	Alluvial	Till	Limestone	Granite	Quartzite
Dune	River	Moraine	Sandstone	Basalt	Slate
Loess	Outwash	Drumlin	Mudstones	Gabbro	Marble
			(includes shales)		Schists
					Gneiss

3.2   Soil Profile

A soil develops a distinctive profile as it weathers. This soil profile may vary from a few inches to 5 ft or more in thickness depending on weathering conditions and often obscures the nature of the parent material. Organic material develops from vegetation and accumulates on the surface, providing a source of organic acids to aid in the weathering process. Finely divided organic material is mixed with the underlying soil and develops a distinctive soil layer known as the *A horizon*. This horizon is a zone of *eluviation* where soluble constituents are removed by percolating water containing carbonic and humic acids and where colloidal particles of clay are mechanically removed by the water.

The material leached from the overlying A horizon begins to accumulate in the underlying soil along with weathering products from the A horizon such as iron oxide. This layer of *illuviation* is the *B horizon* and contains the maximum accumulation of clay and other secondary insoluble weathering products.

Beneath the B horizon is the original unaltered parent material known as the *C horizon,* and below the C horizon is the original unaltered underlying soil deposit or *bedrock.* The C horizon and bedrock do not have to be related materials. They would be for a residual soil developed by weathering in place. The material under a transported soil's C horizon could be unrelated bedrock or other soil deposit.

Figure 3.1 shows the relationships of the different horizons in a vertical soil profile. The upper region is composed of the A and B horizons and is called the *solum.* This is the zone that supports plant growth and is of primary interest to pedologists. The *regolith* is the uncemented mineral matter above the bedrock. Below this is the underlying bedrock, which is of primary interest to classic geologists. Geotechnical construction normally takes place throughout the regolith and solum and occasionally may encompass the upper reaches of the bedrock.

Time, in conjunction with the other weathering factors discussed in Chap. 2, determines the amount of vertical profile that develops in a soil. For example, soil profile development might take 100 to 200

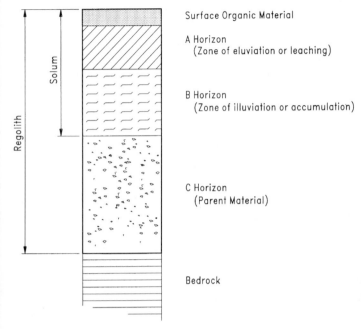

Surface Organic Material

A Horizon
(Zone of eluviation or leaching)

B Horizon
(Zone of illuviation or accumulation)

C Horizon
(Parent Material)

Bedrock

Figure 3.1   Development of a soil profile.

years in a loose, sandy soil under a forest canopy with a humid climate (Longwell et al. 1969), whereas development of such a profile on a glacial till in Michigan would be on the order of 3000 to 8000 years (Gerrard 1981).

Leaching and soil profile development are most prominent when annual rainfall exceeds 25 in per year. Below this, water for leaching is less readily available and contains less acid to aid in weathering. Under these conditions, soluble carbonates and salts that have leached out of the upper layers accumulate in the soil, making it alkaline. This contrasts with the soils in areas with abundant rainfall, where these soluble components are carried away with the water, and the soil is acidic.

In dry regions, evaporation may cause these soluble constituents to be precipitated in the soil capillaries. Here they cement the soil grains together, and under favorable conditions, the resulting cemented material may be quite thick. Such cemented material in arid areas is often referred to as a *duricrust* and may be further identified by its major constituent: Calcium carbonate–rich materials are referred to as *calcrete* (or more commonly *caliche* in the United States); iron-rich deposits are *ferricrete*; magnesium-rich deposits are *dolocrete* (from dolomite); sulfate-rich deposits are *gypcrete* (from gypsum); and silica-rich deposits are *silicrete*.

3.3   Soil Associations

3.3.1   Soil series

A soil deposit will develop a similar vertical profile of A and B horizons when environment, vegetation, drainage, and topography are similar. An area of soil that derives from the same parent material and develops a consistent identifiable vertical profile is a *soil series*. This is the basic soil mapping unit used by the U.S. Department of Agriculture in preparing county soil survey reports. Table 3.2 compares four different soil series from different parts of the United States. The different modes of origin, parent material, age, and local drainage and vegetation developed distinctly different soils at each location with distinctly different textural patterns and plasticity characteristics. In each case, the leaching of the A horizon has resulted in an increase of the clay-sized fraction and an increase in the plasticity index of the B horizon.

3.3.2   Catenas

The physics of soil development create consistent soil profiles under similar conditions, which is the key to identifying and mapping surficial soils by soil series. Under differing drainage conditions, separate soil series will develop from the same parent material. The transformation of soil series as a function of slope results in distinct textural and constituent changes. However, these soil series will always be associated with one another, and they can be expected to occur together on a regional level as a function of topography. Such associations of

TABLE 3.2   Sample Pedologic Soil Profiles

Soil description	Horizon	Depth (in)	Percent finer No. 200 sieve	Percent finer 0.002 mm	LL	PI
Helena fine sandy	A	3–10	40	11	11	3
loam, Virginia,	B	15–27	75	55	75	43
residual soil: gneiss	C	36–48	34	11	31	7
Hastings silt loam,	A	10–15	99	32	38	18
Nebraska, transported	B	20–37	100	39	52	28
soil: loess (eolian)	C	48–60	100	27	37	14
Decauter silty clay loam,	A	1–6	78	23	32	11
Tennessee, residual soil:	B	26–46	89	57	62	35
limestone	C	46–60	96	78	83	40
Eustis loamy sand,	A	0–6	21	6	NP	NP
Florida, transported	B	12–23	21	8	NP	NP
soil: coastal plain	C	30–46	21	8	NP	NP

soil series that are from the same parent material and differ only in soil profile due to topographic conditions are called *catenas*.

The concept of a catena is illustrated in Fig. 3.2. Two parent materials are present over the bedrock: One is either a transported or residual soil identified as soil A, and the other is alluvium transported from upper reaches of the river. Four different soil series are developed on soil A, and these are unrelated to any soil series in the alluvium because of the difference in their parent material. In the highlands of soil A, vertical percolation allows formation of a vertical soil profile with good development of the A and B horizons, and this soil is identified as soil series 1. Adjacent to the highlands, a less distinct profile develops because of limited water infiltration, and this becomes soil series 2. In the level plains, there is more time for water infiltration, and more water is available from runoff of the surrounding areas. Here a distinct profile develops, and it will have more extensive horizon formation than soil series 1 because of the increased weathering and leaching that can occur. This becomes soil series 3. Finally, in the low areas where water accumulates and the water table is high, a fourth soil series develops reflecting the changed weathering and leaching regime. These four soil series form one catena, and where one series is found, the others can be inferred to exist in adjacent areas of the same parent material as a function of the local topography.

Soil color changes are often apparent in the soil transitions in a catena. Gerrard (1981) describes the following common sequence in West African catenas. Upland, well-drained soils are typically reddish from unhydrated iron oxides. In the lower slopes, where the soil remains moist longer than in the upland areas, hydration of the iron compounds increases, leading to a gradual transition of color from red to orange to yellow. In still lower areas, where there is a waterlogged soil profile, reduction occurs rather than oxidation, and the soils are bluish to greenish gray. In the zone where the water table fluctuates, characteristic mottling occurs.

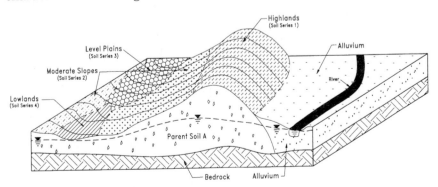

Figure 3.2 Example of a soil catena.

3.4 Residual Soils

Residual soils are formed in situ by weathering, so the parent material and modes of weathering are dominant factors in determining the
engineering characteristics of residual soils. In warm humid areas,
chemical weathering develops deep residual soil deposits, while, conversely, a cool, dry terrain only supports a thin mantle of residual
soil. Figures 3.3 through 3.6 show the general distribution of residual
soils for different parent materials. The residual soil is a product of
the parent material's composition and also retains the relic structure
of the parent material's jointing and bedding structure. Soils developed from in situ weathering of rocks are called *saprolites*. The
intense weathering action of the tropics can cause extensive changes
in the residual soil and develop a special residual soil known as a *laterite*. Saprolites and laterites are both residual soils, but they have
some significantly different characteristics.

3.4.1 Saprolites

Problems with residual saprolitic materials were once largely ignored
in the conventional geotechnical literature. This has been redressed
to a considerable extent by the efforts of Professor Sowers (e.g.,
Sowers 1963, Sowers and Richardson 1983) and recent specialty conferences on the topic (American Society of Civil Engineering 1982,
1987). Figure 3.7 shows idealized residual soil weathering profiles
that may develop in different rock types. In soluble rocks such as
limestone, the transition from intact rock to residual clay is sharp
and distinct, although the rock surface may be irregular from preferential weathering along joints and other fractures. In more resistant
rocks, the transition from intact material to soil is gradual, moving
erratically from weathered and weakened rock to soil with fragments
of unweathered rock to soil consisting only of weathered products and
then finally to the surficial A and B horizons. A highly foliated metamorphic rock such as gneiss or schist poses a particularly nasty problem because each narrow band of the rock will weather into a different material, and adjacent bands will weather at different rates.

The erratic and gradual transition from soil to weathered rock to
sound rock causes major geotechnical construction problems. The
thickness of each zone of weathering often will be highly variable, and
the materials may show considerable variation. Borings for site investigations are difficult to interpret because of the existence of boulders
that may be mistaken for bedrock. Relic joint structures of the parent
rock exist in the weathered soil and may control the soil's strength,
permeability, or slope stability. Fragments of weathered rock that

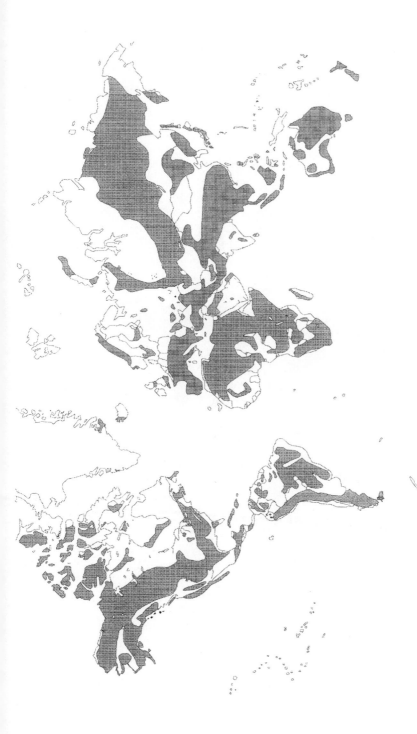

Figure 3.3 Residual soils developed from sedimentary rocks (*USAE 1963*).

Figure 3.4 Residual soils developed from intrusive igneous rocks, primarily granitic shields and mountains (*USAE 1963*).

Figure 3.5 Residual soils developed from extrusive igneous rocks, primarily basaltic plateaus and mountains (*USAE 1963*).

Figure 3.6 Residual soils developed from metamorphic rocks (*USAE 1963*).

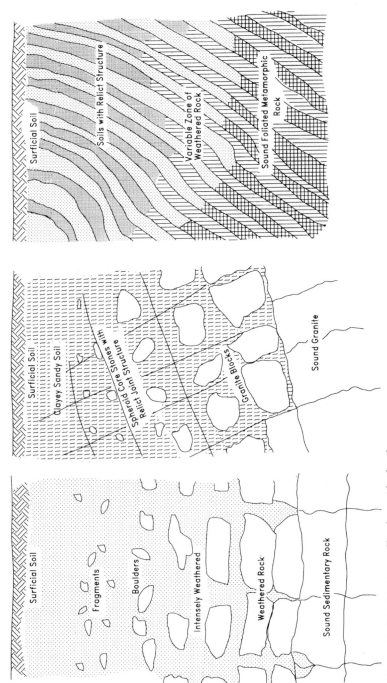

Figure 3.7 Weathering profiles for residual soils.

appear sound may prove to be weak and easily broken when they are handled. Construction claims are common for excavations and foundations in residual soils because it is difficult to predict ahead of time the condition of the weathered rock and the actual thicknesses of soil, weathered rock, and sound rock that will be encountered.

White and Richardson (1987) surveyed consultants who were experienced with work in the Blue Ridge and Appalachian Piedmont region of the eastern United States. The survey respondents' experience indicated that typically standard penetration test N values (ASTM D 1586) of 80 to 100 or compression-wave velocities of 3500 ft/s from seismic refraction marked the boundary between soils and weathered rock where ripping was needed for excavation. N values of 100 and compression-wave velocities of 6000 ft/s were typical of the interface between weathered and sound rock where blasting would be needed for excavation. However, recommendations for identifying these boundaries between soil, weathered rock, and intact rock overlapped considerably, and it was apparent that there was no hard and fast rule.

Sedimentary rocks comprise 75 percent of the rocks exposed at the earth's surface. Of this group, sandstone, limestone, and shale are the most common. Sandstone typically weathers into sand or silty sand. Limestones weather into fine plastic soils, but the final product depends on the type and amount of impurities in the parent limestone. For example, the limestone in Dekalb County, Tennessee, contains large amounts of chert, a highly resistant rock composed of silica. Under the warm, moist Tennessee climate, this limestone weathers into a cherty silt loam containing 35 to 75 percent chert fragments. Limestone residual soils are common in large areas south of the Ohio and Missouri Rivers in the United States.

Shales, which for this discussion will include all indurated, fine-grained, sedimentary rocks such as nonfissile mudstones, siltstone, and claystone, as well as conventional fissile shales, are a highly variable and troublesome rock. Some are hard and very stable, while others are soft and degrade into clay soon after exposure to the atmosphere. Resistant shales tend to develop a relatively thin soil mantle with a silty texture, whereas the softer shales develop thicker, more clayey soils. Clays developed from shales are often highly plastic and troublesome.

Shales are very common in a number of areas, so there have been a number of studies to try to determine how to predict the durability of shales in embankments and similar earthworks (Chapman 1975, Reidenour et al. 1976, Noble 1977, Strohm et al. 1978). These studies have developed a variety of testing and classification schemes; however, they have not been directly correlated with field performance

(Rollings 1988). Shales often will weather into a clayey material within the life of an engineered structure, so the residual soil developed on the shale probably provides the best estimate of the future strength and other properties of the shale if it is used in construction (Smith 1986).

Igneous rocks formed from solidification of magma can weather into sandy, clayey soils of varying properties depending on rock composition and climate. Granite weathers by progressive decomposition of minerals in accordance with the inverse of Bowen's reaction series in Fig. 2.2, producing a spheroidal solid fragment in a matrix of weathering products. The weathered granite profile in Fig. 3.7 shows a progressive decrease in number and size of decomposing rock fragments toward the surface and corresponding increase in finer weathered product. The final weathered soil will be a variably clayey, silty, sandy material. It often contains mica, which makes the soil compressible and difficult to compact. Basalt, on the other hand, tends to weather into relatively pure and often highly plastic clay under favorable conditions.

The silty gravel from Oman in Fig. 2.3 is an example of a residual soil developed from the igneous rock gabbro under arid conditions. From Fig. 2.1, one would surmise that under low rainfall and high temperatures weathering would be limited and would be predominately mechanical rather than chemical. The thickness of the silty gravel was only a few feet, which was consistent with limited weathering. During construction of a military airfield, this silty gravel was to be used extensively as a subbase for road and airfield pavements, and more limited quantities of good-quality aggregate were reserved for use as base course, asphalt concrete, and portland cement aggregates. The silty gravel was excavated from borrow areas with scrappers, run through a portable screening plant to remove oversized material (greater than $1\frac{1}{2}$ in), hauled to the construction sites, and then graded and compacted. Original testing of the borrow area found that the material generally met the specification requirements for a subbase material if some areas that had slightly high fines content and plasticity index values were avoided. However, as shown in Table 3.3, construction quality-control samples of the material found that the aggregate deteriorated progressively as it was handled. The material as placed was unacceptable and had to be rejected. The contractor was finally able to blend the material with a local sound beach sand to make a material that would meet the Atterberg limit requirements after placement, although a quarter of the samples were a few percent high on the fines content. This was accepted because of the general lack of alternative materials and the need to get the base operational. The pavements have been in operation without problems for

TABLE 3.3   Degradation of an Aggregate during Construction

When tested	Percent of samples failing specification requirements for	
	Atterberg limits	Fines
In borrow area	19	4
At screening plant	62	47
After placement	Not tested	58
After blending and placement	0	25

10 years, including intense use during the 1991 Gulf War. Degradation of seemingly sound aggregate and rock materials, as seen in this example, is not an uncommon problem with residual weathering products.

Soils derived from metamorphic rocks are often very complex. Metamorphic rocks are subjected to intense heat and pressure during formation. These factors often vary dramatically within a local area, so the metamorphic rock and resulting soil will vary also. As shown in Fig. 3.7, metamorphic rocks can have very unfavorable weathering profiles. These rocks often also contain mica, which is highly resistant to weathering. The residual soil containing this mica will be compressible and difficult to compact. The Piedmont province of the United States contains some major population centers, including Atlanta, Greenville, Raleigh, Richmond, and Washington, D.C., and is largely composed of residual soils derived from metamorphic rocks with granite intrusives.

Santi (1994) gives an extensive list of tests used with weathered rock, references dealing with classification of weathered rock, and evaluations of diagnostic tests for weathered rock. There is no generally accepted standard for dealing with weathered materials at this time.

3.4.2   Laterites

The term *laterite* was derived from the latin *later* meaning "brick" and was originally used to describe a soil in India that was "soft enough to be readily cut into blocks by an iron instrument, but which upon exposure to air quickly became hard as brick and is reasonably resistant to the action of air and water" (Buchanan 1807). Since then, the term has been used so indiscriminately to describe a variety of tropical and subtropical red soils that its usefulness has almost been destroyed.

Laterites form from intense leaching in well-drained soils under tropical and subtropical climates. This leaching removes almost all silicates and bases and leaves secondary oxides and hydroxides of aluminum and iron with variable amounts of kaolinite and quartz. This process is termed *laterization,* and Bawa (1957) proposed using the ratio of SiO_2 to Al_2O_3 as a measure of laterization: laterites <1.33, lateritic 1.33 to 2.00, and nonlaterite >2.00. Krinetsky et al. (1976) proposed descriptions of tropical deposits as follows:

Tropical red soil: Reddish soils that are not self-hardening and do not contain sufficient hardened laterite to be classified as laterite soil.

Laterite soil: A reddish soil developed by laterization that is self-hardening or contains already hardened laterite rock or laterite gravel.

Laterite gravel: Unconsolidated deposits of hardened, concretionary gravel formed by the laterization process.

Laterite rock: Lateritic material that has hardened either as large masses, boulders, or crusts or has cemented other materials into rocklike masses.

Townsend (1985), noting that the terminology for laterites remained confusing even after Bawa (1957) had made the same observation 27 years before, recommends adopting the terminology of Krinetsky et al. (1976). Figure 3.8 shows a generalized distribution of laterites and related deposits.

Often test methods and specifications developed for soils in temperate climates are not adequate to describe the behavior of lateritic materials. Lateritic materials often are sensitive to mechanical reworking and to drying, which may cause irreversible changes in the properties. Table 3.4 shows the effect of remolding on the Atterberg limits of several laterites, and Fig. 3.9 shows the effect of air drying on the plasticity index of a wide selection of laterites. Clearly, remolding field samples or allowing samples to dry either in handling or, as is commonly done, to process soils prior to running many standard laboratory tests may result in completely misleading results.

Laterite gravels (also commonly referred to *lateritic gravel* or locally sometimes as *ridge gravels*) are an important source of pavement construction materials in many parts of the world. However, specifications and tests developed in temperate areas do not adequately assess the properties of laterite gravels and often would incorrectly exclude their use in pavements as a base or subbase. Local experience and a comprehensive series of strength tests must be used to evaluate these materials. Experience has found that often the limitations on fines, durability, and Atterberg limits in conventional specifications from temperate regions can be relaxed. Krinetsky et al. (1976) proposed a

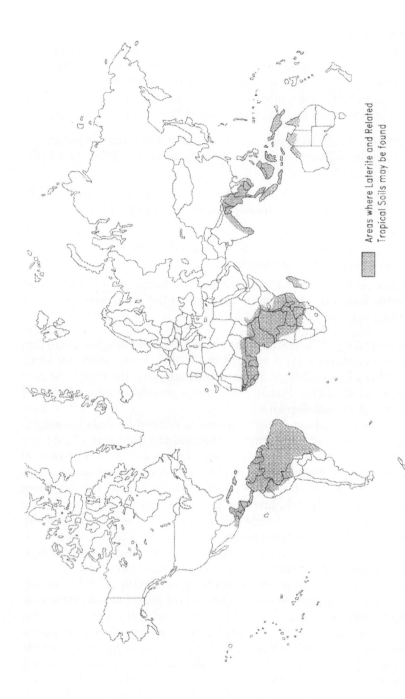

Figure 3.8 Distribution of laterites and related deposits (*adapted from Morin and Todor 1975 and Gidigasu 1975*).

Areas where Laterite and Related Tropical Soils may be found

TABLE 3.4   Effect of Remolding on Atterberg Limits of Laterite Soils

Soil	Liquid limit		Plasticity index		Source
	Natural	Remolded	Natural	Remolded	
Red clay, Kenya	74	84	36	45	Newill, 1961
Red clay, Kenya	77	91	16	32	Newill, 1961
Lateritic, Cuba	46	53	15	22	Winterkorn and Chandrasekhaven, 1951
Lateritic, Panama	60	70	21	30	Townsend et al., 1969

SOURCE: Townsend, 1985.

revised specification that divided laterite into three classes (I to III) with design California Bearing Ratio (CBR) values of 100, 70, and 50. This proposed laterite gravel base course specification widened the allowable gradation and allowed more plastic and finer materials than conventional specifications such as those in Table 2.13.

A trial pavement was constructed in Queensland, Australia, using laterite gravels that corresponded to the class I and III laterite gravels of Krinetsky et al. Table 3.5 compares a typical temperate climate base course requirement with the classification of Krinetsky et al. and with the two laterites used in the pavements. In the dry season after construction, the class I laterite test sections were trafficked with 2.1 million equivalent 18,000-lb single-axle loads, and the class III section was trafficked with 935,000 equivalent 18,000-lb single-axle loads. Both sections had a ⅛-in rut at the end of traffic. The section was allowed to sit through the monsoon season, which allowed

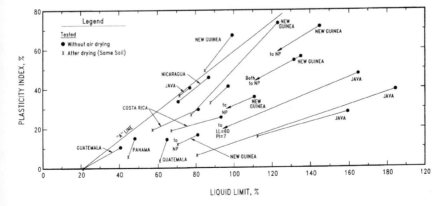

Figure 3.9   Effect of air drying on the plasticity index of laterites (*adapted from Morin and Todor 1975*).

TABLE 3.5   Specification Requirements for Conventional and Laterite Base Course Aggregates

		Laterite class requirements			Cook's Pit laterite	
Property	Temperate	I	II	III	I	III
CBR	100	100	70	50	100+	80
Passing No. 200 sieve	≤10	≤15	≤15	≤25	14	22
LL	≤25	≤40	≤40	—	—	20
LL × (% − No. 200)	—	≤600	≤900	≤1200	—	440
PI	≤5	≤10	≤12	—	—	8
PI × (% − No. 200)	—	≤200	≤400	≤600	—	160
LA abrasion	≤50	<65	<65	—	—	—
Linear shrinkage (%)	—	—	—	—	2.0	5.1
Crushed particles	Yes	—	—	—	No	No

NOTES: Temperate specification is from Table 2.13. The class I through III requirements are from Krinetsky et al. (1976). The Cook's Pit laterite is from Queensland, Australia. The two qualities of Cooke's Pit laterite reflect different locations on a ridge and correspondingly different weathering histories.

water infiltration and weakening of the lateritic gravel bases. When the sections were retrafficked, the class I section developed a 1-in rut after 2.3 million equivalent 18,000-lb single-axle loads, and the class III section developed a similar failure condition after 1.1 million equivalent 18,000-lb single-axle loads. The Cook's Pit laterites were both gap graded, contained excess fines, and were more plastic than would be allowed by temperate climate specifications. They both were able to carry significant traffic after becoming soaked in the monsoon season. This behavior would not have been predicted with conventional temperate climate specifications.

Laterite gravel also has been used for aggregate in asphalt and portland cement concrete. In an example from Gambia described by Swartzman et al. (1986), a local laterite aggregate and beach sand were used as asphalt concrete aggregate. The laterite was soft and absorptive, and it had an LA abrasion (see Chap. 4) ranging from 45 to 50 percent and a plasticity index (PI) of 9 percent. The laterite was processed, crushed, and blended with 20 percent beach sand, which reduced the plasticity index to 4. Some pavements constructed with asphalt concrete using this aggregate were up to 8 years old at the time of the report, and they had given usable service during this time with some significant aggregate wear under the local traffic. The pavements were not built to conventional standards, but the local materials could be adapted economically to provide usable service.

Laterites have been used successfully in construction of roads, earthwork dams, and foundations in the tropics. They can give either good or bad service, so proper engineering for their use is necessary.

3.5 Transported Soils

The process of transportation results in natural sorting of soil particles in many cases. This sorting is a function of the transporting medium's velocity. A flash flood in the mountains can move boulders, but as the water slows, the size of the particles it can carry decreases. Consequently, one may find boulders in flash flood debris, gravel in fast-flowing streams, sands in slowly meandering streams, and fine silts and clays in quiet lake waters. A similar phenomenon occurs with wind, but it is effective primarily in transporting and depositing fine sand- and silt-sized particles. Glacial ice is capable of transporting any size material, so it leaves a heterogeneous mixture of particle sizes. Consequently, the present characteristics of a transported soil deposit reflect its origin, mode of transportation, and postdepositional weathering.

3.5.1 Lacustrine and other still-water deposits

As water from a stream enters a lake or other body of water, its velocity decreases and its ability to carry material by bed shear or in suspension decreases. Fine sands, silt, and clay now settle to the bottom, forming a fine-textured deposit. Such deposits are often very uniform and are commonly weak and compressible. These materials may be found in lake bottoms (*lacustrine*), areas where floodwaters lie still, *playas* (desert dry lake beds), and river cutoffs that no longer have flowing water. The backswamp deposit in Fig. 2.3 is an example of the fine texture of soils deposited in still water. Figure 3.10 shows a general distribution of these still-water deposits that includes lacustrine lake deposits, flood plains, playas, and coastal deposits.

Some glacial lakes left lacustrine deposits that consisted of alternating thin laminae of fine sand or silt and clay. Such a deposit is called *varved* silt and clay. The fine sand and silt settled out in the summer, and the clay settled out during the winter when no meltwater entered the lake (Terzaghi 1955). These laminae are horizontal and typically vary from a fraction of an inch to several inches thick. Laminae can vary in thickness within the same deposit. The sand and silt laminae have different engineering properties from the clay (Terzaghi 1955, Parsons 1976). Typically, the horizontal permeability is much higher than the vertical because of the sand and silt layers. Also, the deposit is often sensitive and will lose strength on remolding. Excavations in these varved materials with their variable layered

Figure 3.10 Distribution of still-water deposits to include lacustrine, flood plains, playa, and coastal deposits (*USAE 1963*).

properties, high ratio of horizontal to vertical permeability, normally high water content, and loss of strength on remolding are particularly tricky.

Clays deposited in a marine environment that are later uplifted can turn into extrasensitive clays that drastically lose strength when disturbed or remolded (see Chap. 8). Such high sensitivity has been attributed to leaching of their salt content after uplift, and famous examples exist in Quebec and Norway (Terzaghi 1955, Bjerrum 1967).

3.5.2 Alluvial deposits

Soils deposited from moving water are called *alluvium,* or *alluvial soils.* Their grain-size distribution is a direct function of the transporting water's velocity. As the velocity of water decreases, the maximum-sized particle of sediment that can be held in suspension also decreases. Alluvial soils develop in any situation where soil is deposited by moving water, and their distribution is shown in Fig. 3.11.

In arid regions such as the U.S. Southwest, infrequent rains cause massive torrential flows in the mountains capable of moving large boulders. As the canyon or arroyo widens or the slope decreases, the water velocity drops, and deposition begins, forming alluvial fans as the water moves away from the mountains. Near the mountain are boulders, cobbles, and gravel left by the water. Progressively finer materials are left behind as the distance from the mountains increases. Finally, in the center of the basin adjoining the mountain, the flood waters pond, forming shallow lakes. Here fine particles settle out, and as the water evaporates, dry lake beds or playas of clays and silts are left behind.

Rivers and streams develop characteristic sinuous paths as they move in a meandering pattern. Velocity of the water is not constant across the river (Straub 1935), so the ability of the water to transport material is similarly not constant across the width of the channel. This variation in velocity, coupled with variable flow, results in erratic grain-size distribution from point to point in river deposits. Typically, grain size will tend to increase with depth. If there is ample sediment available but limited water to transport it, a braided stream forms with water flowing in multiple interconnected channels around islands of alluvium. Typical examples of this can be seen in rivers at low-water stage, in glacial outwash streams, and in arid regions where water is lost rapidly by evaporation.

Rivers characteristically constantly shift location within a meander belt that may be many miles wide and leave deposits hundreds of feet deep in the case of major rivers such as the Mississippi. The water slows on the inside of a bend in the river's meander, and coarse mate-

Figure 3.11 Distribution of alluvial deposits to include delta deposits (*USAE 1963*).

rial is deposited to form a point bar. These coarse deposits are often potential sources of construction aggregate. As a river meanders, the point-bar deposits are progressively abandoned, leaving behind a series of sandy ridges. The swales between these ridges eventually fill with finer material, making it difficult to separate material during construction.

When the river abandons a channel, the entrance and exit of the old channel become plugged, forming an ox-bow lake. This lake slowly fills with fine-grained material, forming abandoned channel deposits of normally consolidated clays and silts at high moisture contents.

As flood waters break over a river's banks, the water velocity slows, and coarse material is dropped at the bank. This progressively builds a natural levee along the river's banks. These natural levees may provide an elevated well-drained location with a good subgrade for a pavement or other structure in an area otherwise plagued with high water tables and soft soils. The flood waters continue past the natural levees to fill surrounding low-lying areas. Fine material in the now still flood water settles out to form very uniform backswamp deposits with a large amount of clay. Figure 3.12 shows the complex interrelationships that can develop between these alluvial deposits in a meander belt of a river.

When a river flows into standing water as in a lake or ocean, the water velocity slows, and sediment is deposited. Progressive outbuilding of these sediments forms *deltas*. The ocean or lake can then rework these deposits, and associated deposits such as sand or shell beaches, marshes, swamps, and bay-sound deposits develop. These may then be buried by further delta growth, leaving a complex soil stratification behind.

Alluvial deposits from rivers are complex and can pose difficult problems for the engineer. The complex interrelationships of deposits such as shown in Fig. 3.12 make it very difficult to plan and interpret site investigation borings. Even with relatively closely spaced borings, crucial features such as swales with fine-grained material or abandoned channel deposits could be easily missed, and they would have a significant impact on construction activities. Some typical soil properties from alluvial deposits in the lower Mississippi Valley are shown in Table 3.6. Very soft deposits such as swamps and marshes are difficult and expensive to cross. Other deposits such as backswamps or abandoned channels exist at high moisture contents, are soft, and may consolidate under the weight of any structure. As can be seen in Table 3.6, there is a shortage of gravel in the deposits in the lower Mississippi River. Often a river's gradient decreases as it approaches the ocean so that the gravel deposits such as the alluvial gravelly

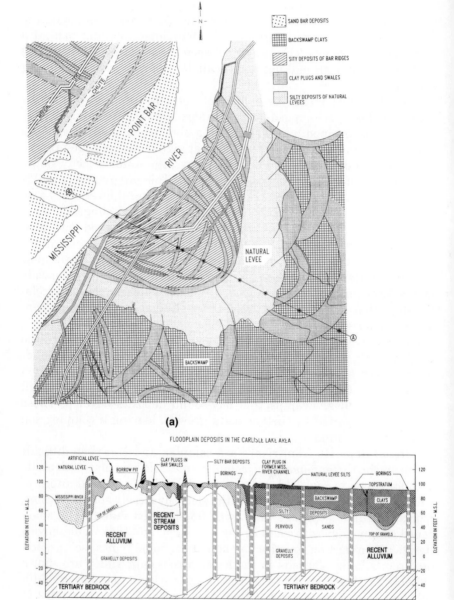

(a)

FLOODPLAIN DEPOSITS IN THE CARLISLE LAKE AREA

CROSS SECTION ALONG LINE A–A

(b)

Figure 3.12   Complex alluvial deposits in the Carlisle Lake area of the Lower Mississippi Valley (*modified from Fisk 1944*).

TABLE 3.6 Representative Alluvial Deposits for the Lower Mississippi River Valley

Deposit	Soil components					Texture	Nat. w, %	Atterberg limits		Typical strength parameters	
	Gravel	Sand	Silt	Clay	Organic			LL	PI	c	ϕ
Braided stream	0	25	30	45	0	Clayey sand, SC, silty clay, CL, and sands, SP	25–40	30–75	10–55	0.2–1.0	30
Natural levee	0	15	40	45	0	Clays, CL, and silts, ML	— 25–35 15–35	NP 35–45 NP–35	NP 15–25 NP–5	0 0.3–1.2 0.2–0.7	30–40 0 10–35
Point bar (ridges)	0	42.5	45	10	2.5	Silt and silty sand, ML to SM	25–45	30–55	10–25	0.0–0.7	25–35
Abandoned channel	0	5	20	72.5	2.5	Clays, CL & CH	30–95	30–100	10–65	0.3–1.2	0
Backswamp	0	5	5	80	10	Clays, CH	25–70	40–115	25–100	0.4–2.5	0
Swamp	0	5	10	50	35	Organic clay, OH	?–265	135–300	100–165	Very low	Very low
Marsh	0	0	0	25	75	Peat, Pt	100–465	250–500	150–400	Very low	Very low
Delta	0	12.5	12.5	62.5	12.5	Clay	20–120	25–155	10–80	0.2–0.3	0
Lacustrine	0	0	5	87.5	7.5	Clay, CH	45–115	85–165	65–95	0.1–0.2	0
Beach	0	87.5	10	5	0	Sand, SP	Saturated	NP	NP	0	30
Bay-sound	0	25	35	40	5	Clays, CL & CH	20–70	40–80	25–65	0.2–0.7	15–25
Substratum	0	70	5	0	0	Sand, SP	Saturated	NP	NP	0	30–38

SOURCE: Adapted from Kolb and Shockley, 1959.
NOTES: c is cohesion, kips/sf; ϕ is friction angle in degrees; NP is nonplastic; Nat. w is natural water content.

sand in Fig. 2.3 are left upstream. Even though river deposits are considered good potential sources of gravelly construction materials, the lower reaches of a river may be short of good aggregates with which to build. Many major cities in the world are located along the reaches of rivers, so there is a high demand for construction in alluvial areas.

3.5.3 Glacial deposits

As seen in Fig. 3.13, much of the world north of the 40th parallel has been covered by large glaciers in the past. Many of the populous cities and regions of North America and Europe are within the glaciated zones, so glacial deposits have received considerable study in the past. These glaciers moved in pulsating advances and retreats, excavating, transporting, and redepositing soils and rock in a highly complex pattern over the last $2\frac{1}{2}$ million years.

Materials deposited by a glacier are called *drift*, and some nomenclature for these deposits is shown in Table 3.7. Those materials left by the ice are called *till* and are a heterogeneous, unsorted mix including everything from boulders to clay. They are analogous to materials mixed, heaped, and pushed aside by a bulldozer. The meltwaters flowing from the glaciers formed swift streams and rivers that deposited generally granular materials in broad sheets known as *outwash*. Meltwater trapped by ice or high ground formed lakes where lacustrine deposits developed. Consequently, glacial deposits consist of deposits from different mechanisms that are intermingled by the various advances and retreats of the glaciers. Because of these complex interactions, glacial deposits are highly unpredictable.

The glacial streams that formed glacial outwash deposits would form an alluvial fan with the coarsest material near the glacier and progressively finer gravel and sands deposited further away as the water velocity decreased. However interactions of the local topography, length of deposition time, and relative movements of the ice front are so complex that it is impossible to predict a general horizontal or vertical pattern of texture in these deposits (Jenkins et al. 1946). Outwash and other sorted deposits such as eskers formed by glacial meltwater are often sources of granular construction materials in glaciated areas.

The characteristics of a glacial till depend on the sediments and bedrock eroded by the glacier, but often tills of the same age show similar textures within an area (Peck and Reed 1954, Terzaghi 1955). Tills from New England are typically coarser and less plastic than those from the Midwest, for instance. Most tills show a complete range of particle sizes such as seen in the Illinois glacial till in Fig.

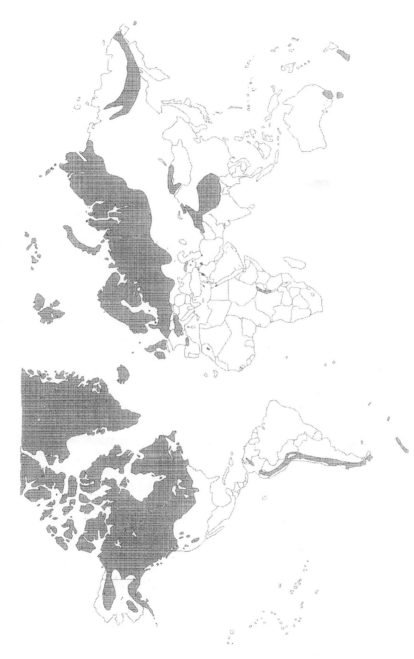

Figure 3.13 Glaciated regions (*USAE 1963*).

TABLE 3.7   Glacial Drift Deposits

Nomenclature	Characteristics
Unsorted deposits	Heterogeneous mixture of materials deposited by the glacial ice
Till	General term for ice-deposited material
Moraine	A ridgelike accumulation of till at the edges of a glacier, often identified more specifically by location relative to the glacier as an end moraine, lateral moraine, or terminal moraine; ground moraine is sometimes used for approximately horizontal till sheets
Drumlin	Elliptical hill of till or till over bedrock that was shaped by the moving ice; can be very large
Ice contact deposits	Deposits formed at edge of melting stagnant ice, often granular with erratic grain size and features
Eskers	Long, narrow, sinuous ridges of sorted material formed by water flowing in, adjacent to, or under the glacier
Kame	Short, steep-sided knolls and hummocks of water-sorted material
Kettle	Closed depression in glacial deposits created by melting of buried ice
Sorted deposits	Deposits from meltwater flowing from the glacier
Outwash	Alluvial deposits of granular material
Lacustrine	Fine-grained material deposited in glacial lakes

2.3, but the till also contains random pockets and seams of gravel, sand, silt, or clay. The thickness of the till varies considerably. In New England it might be 15 to 20 ft thick, while thicknesses of 50 to 115 ft are typical in the central Midwest, and thicknesses of 150 to 200 ft are found in Iowa (Krynine and Judd 1957). These are only averages, and local variations are common, as evidenced by a 763-ft-deep till deposit in Ohio (Krynine and Judd 1957). Older tills will show more weathering and profile development than younger tills. Also, if an old till was overridden by a later glacial advance, the thick glacial ice densely compacted the bottom till. This heavily compacted till may require blasting to excavate.

Legget (1979) provides an instructive description of problems brought on by the failure to appreciate the impact of glacial geology on construction. The St. Lawrence Seaway was built between 1954 and 1959. Major excavations were required as part of this construction, and competition for the excavation contracts was keen because of their magnitude and the prestige that would accompany them. The soils of the St. Lawrence Lowland in which excavations were to take place consisted of an upper stratum of marine clay underlain by two successive glacial till sheets. The older bottom till sheet had been overridden by the glacier depositing the upper till, so one could reasonably infer that this bottom till would be highly compacted. Ample geologic information was available at the time of bidding to recognize this stratigraphy, and borings for the project identified compact to very compact till as the material beneath the upper clay. One contractor on the job went bankrupt, another defaulted, and two others submitted claims totaling 31 and 85 percent of the original contract prices. The problems centered on the dense, "cemented" till that sometimes required ripping or blasting before it could be moved by scrappers and the sticky marine clay that could not be handled by normal earthmoving equipment. Similar problems had been encountered at the turn of the century digging a canal less than a mile away from part of the St. Lawrence Seaway project and at two other canals built in the 1920s in these deposits. These earlier projects had resulted in bankruptcies and large claims that foretold the outcome of the St. Lawrence excavations.

This work resulted in large claims, litigation, conflicting views of who was to blame, and very intense feelings. Relatively small settlements ultimately were made on these claims, probably reflecting the contract requirement that the contractors were responsible for satisfying themselves as to the nature of the work (Legget 1979). No complete examination of the problems and responsibilities of this project has been published yet, but ample geologic information was available to foretell the problems that would occur. Everyone would have been well served had these conditions been clearly and unambiguously identified in the bid documents so that fair bids could have been prepared.

3.5.4 Eolian deposits

Wind has only a limited ability to transport soil particles. Large particles cannot be moved, and the interparticle forces of colloidal particles cannot be overcome by wind. Consequently, eolian, or wind-blown, deposits consist predominately of silt or sand and are remarkable for their uniformity and small range of grain sizes. Also, in arid regions, selective removal of sand particles results in the steady accumulation of coarser particles on the surface. This continues until the surface consists entirely of stones, gravel, and coarse sand that cannot be moved, and further sand and silt removal stops. This stony desert surface is sometimes referred to as *ablation* or *deflation armor* or *desert pavement*.

Eolian sands such as that shown in Fig. 2.3 are fine and uniform. They typically form dunes that can be hundreds of feet high. Major dune areas are shown in Fig. 3.14. Although sand dunes are normally associated with deserts, they are also found in both coastal and non-desert inland areas (e.g., in the Great Plains). The dune migrates as sand is redeposited in the lee of the dune. When roads, railroads, or other structures must be built through active dune regions, stabilization of the dunes with vegetation, fences, or bitumen and selection of favorable alignments become major design considerations to prevent the structures from being alternately buried and exposed by dune migration. Botanists are of major assistance in selecting appropriate shrubs, grasses, and trees that will survive in the sandy dunes and whose roots will help to stabilize the sand. This is environmentally far more preferable than some of the older techniques of stabilizing dunes such as spraying them with bitumens. Where possible, it is probably best to avoid areas of active dune movement.

These fine sands are usually poor construction materials and are not stable under traffic unless stabilized in some manner. For example, along the U.S. Gulf Coast, dredged oyster shell often has been mixed in a 50:50 blend with fine coastal sands to make an approximately 50 CBR low-quality base course for flexible pavements or to serve as an unsurfaced low-volume road. Trafficability of construction equipment in fine sands is poor; therefore, some sort of fill, base course, or stabilization may be needed to allow construction activities such as foundations for pavement forms, access by concrete trucks, stable platforms for cranes, etc.

Wind-blown silt, known as *loess*, composes about 11 percent of the world's surficial deposits and about 17 percent of the U.S. deposits (Turnbull 1968). Distribution of loess deposits is shown in Fig. 3.15. Loess is a uniform silt usually classifying as a CL or ML under the USCS. An example of a loess gradation is shown in Fig. 2.3. Loess consists of predominately fine sand- and silt-sized particles with some clay and calcium carbonate. The loss of calcium carbonate through leaching has been used effectively to measure the degree and depth of weathering in loess (Lamb 1990). The leached loess had higher void ratios, lower strength, and higher compressibility than the unleached soil (Lamb 1990). Loess's physical characteristics, such as gradation and Atterberg limits, are reasonably uniform worldwide, and some example characteristics are shown in Table 3.8.

The most unique characteristic of loess is its open structure in the undisturbed state. The wind-deposited silt and sand particles are coated with clay or a combination of clay and calcite, which cements the

Figure 3.14 Major sand dune areas (*USAE 1963*).

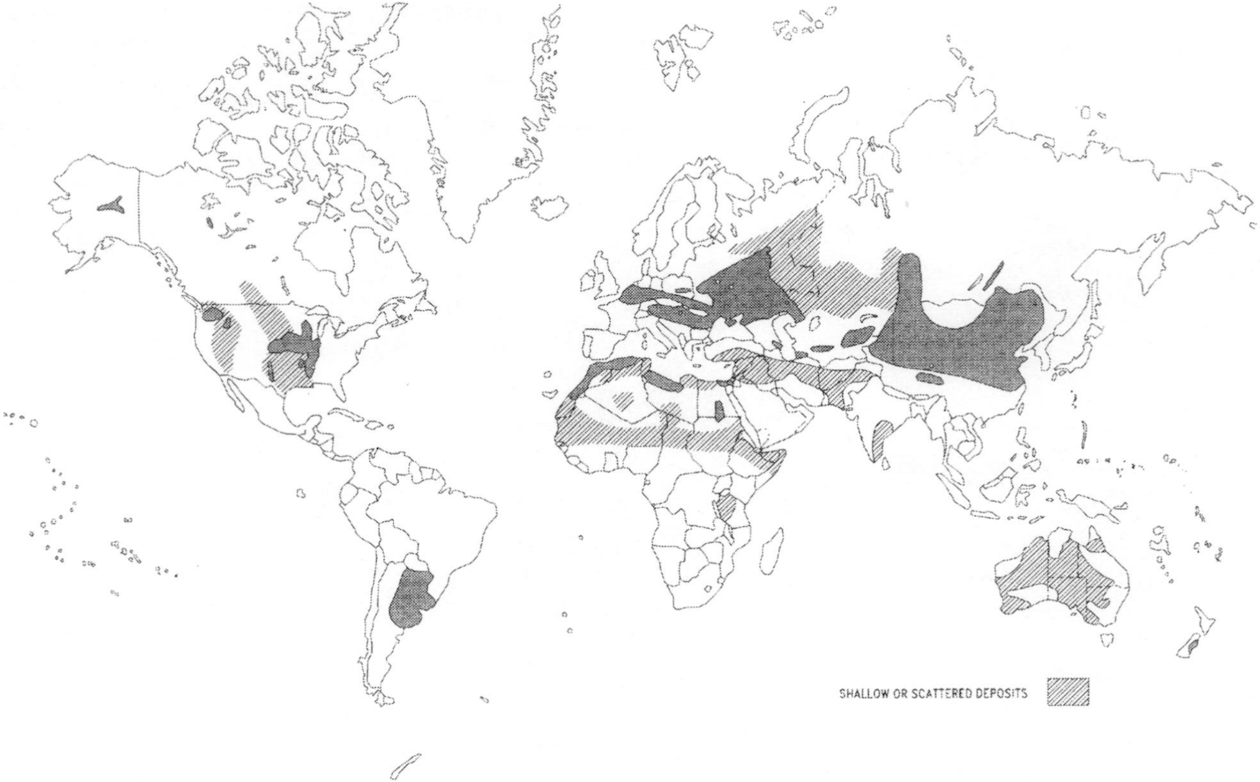

SHALLOW OR SCATTERED DEPOSITS

Figure 3.15 Distribution of loess (*USAE 1963*).

TABLE 3.8 Typical Properties of Loess

Origin	Mechanical analysis, %			G_s	Atterberg limits, %			e	γ_d, lb/ft³	Field moist, %	Sheer Strength§	
	Sand	Silt	Clay		LL	PL	PI				c, psi	ϕ, degrees
Alaska*	2–21	65–93	3–20	2.67–2.79	22–32	19–26	0–8			11–49		
Arkansas*			11					0.828	92.4	1.4		
Colorado*	30	50	20		37	20	17		76–95	8–10		
Illinois*†	1–4	48–64	35–49		39–58	18–22	17–37	0.885	89.3	1		
Indiana†			3					0.879	89.9	1		
Iowa*	0–27	56–85	12–42	2.68–2.72	24–53	17–29	3–34	0.742	66–99	4–31	0–8	28–31
Kansas†			17					0.742	88.6	2		
Mississippi*†	0–8	75–85	0–25	2.66–2.73	23–43	17–29	2–20	0.922 0.782	80–104	19–38	3–8	26
Missouri†			15–17					0.742 1.044	82–97	2–3		
Nebraska*†	0–41	30–71	11–49	2.57–2.69	24–52	17–28	1–24	0.944 1.137	79–87	3		
Tennessee*†	1–12	68–94	4–30	2.65–2.70	27–39	23–26	1–15	1.030	82	2–25		
Washington*	2–10	60–90	8–20		16–30		<8					
China‡												
Lanchow‡					24–32	16–19	8–13	0.96–1.30		16–29	2–4	16–18
Xian‡					30–32	17–19	13	0.96–1.20		19–23	3	22
Lowchuan‡					31	20	11	1.00–1.20		4–12		32–34

*As summarized by Sheeler (1968) from various sources.
†Lutenegger and Saber, 1987.
‡Zhi-hui et al., 1987.
§Consolidated, undrained triaxial test.

loess into its characteristic open structure. This open structure in combination with frequent root holes throughout the deposit leads to high vertical permeability and rapid consolidation under load when the soil is wetted. This collapsible structure when wetted led Karl Terzaghi (1951) to refer to loess as the most treacherous of foundation materials. Figure 3.16 shows the dramatic effect of applying load and soaking a low-density loess. Sheeler (1968) observed that large settlements were uncommon for loess with a dry density over 90 lb/ft^3, which is also consistent with the behavior of the two loess soils in Fig. 3.16.

When reworked and compacted, loess makes a reasonably good subgrade with design CBR values on the order of 10 to 12. The reworking and recompacting destroy the original open structure, and its collapsible characteristics are eliminated. However, loess is highly erodible, so considerable attention to cut faces, water control, and possible piping from utility leaks or other sources is needed. The loess structure allows it to stand vertically, as seen in Fig. 3.17. These vertical faces are commonly cut to heights of 12 to 30 ft. Cutting loess to any standard slope angle rapidly leads to massive erosion. When benched cuts are used, the water runoff on each bench should be collected in a ditch several feet from the face and then drained off at a tangent to the face. Placing the drainage ditch directly at the toe of the cut saturates the loess at the base, leading to large arched caving failures (Turnbull 1968). Obviously, culvert outlets and drainage ditches also require careful attention to detail to prevent erosion. Recently, the Bureau of Reclamation (Cast et al. 1987) tried several different designs for canal slopes to be cut in Nebraska loess and determined, based on performance of slopes and construction and maintenance costs, that the most promising approach was as follows:

1. Slope $\frac{1}{4}$ vertical to 1 horizontal ($\frac{1}{4}$:1) for cuts less than 25 ft deep.

2. Slope $\frac{1}{2}$:1 for deeper cuts.

3. Use 1:1 slope 5 ft high at the toe of the cut to promote vegetation growth, catch debris falling down slope, and avoid saturation at toe of the slope.

4. Directing water away from the cut face prior to excavation and away from the base afterwards is crucial.

5. Design is an ongoing process, and further adjustments will be made based on slope performance.

3.6   Investigating Soil Deposits

Each soil deposit is a unique product of its parent materials, alteration by geologic processes, topography, and ongoing weathering

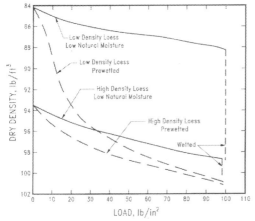

Figure 3.16   Collapse behavior of loess (*Clevenger 1958*). [*"Experience with Loess as a Foundation Material,"* W. A. *Clevenger*, Transactions, *1958, reproduced by permission of American Society of Civil Engineers.*]

Figure 3.17   Vertical standing cuts in loess.

processes. We can broadly characterize them by process of origin, as was done in this chapter, and can broadly identify where we expect to find them. However, within these broad categories, many variations of natural materials will be encountered, and the deposits will be mixed and interbedded reflecting past geologic history, of which we are often largely ignorant.

Tables 3.9 and 3.10 show the variety of different soil classifications from samples taken at 0- to 12-in depths in different geologic and

TABLE 3.9 Distribution of Residual Soil Types from Different Geologic Sources

Geologic origin of parent material	n^*	OH	OL	CH	CL	MH	ML	CL-ML	SM	SC	SM-SC	SP	SP-SM	GM
Igneous rocks														
Granite	2								100					
Basalt	3		33		67									
Metamorphic rocks														
Slate	2				50		50							
Gneiss	2			50	50									
Schist	2				50				50					
Sedimentary rocks														
Calcareous†	10			10	80		10							
Sandstone	5				60		20		20					
Shale	5				40		20		20		20			
Interbedded‡	26	4		4	39		25	7	7		4	4		

*Number of samples, e.g., two samples of residual soil from granite bedrock, both of which (100 percent) classified as SM by USCS; three samples from basalt bedrock, one of which classified as OL and two of which classified as CL. Each sample has similar topography, climate, and geology and may consist of 1 to 10 individual measurements.

†Calcareous rocks are limestone and dolomite.

‡Interbedded rocks are alternating layers of different sedimentary rocks, e.g., limestone and shale, sandstone and dolomite, etc.

§Use of dual symbols differs somewhat from ASTM convention.

SOURCE: Compiled from data reported by USAE Waterways Experiment Station, 1963.

TABLE 3.10 Distribution of Transported Soil Types from Different Environments

Geologic origin	n																	
							Percent of samples classifying in different USCS categories											
	n	Pt	OH	OL	CH	CL	MH	ML	CL-ML	SM	SC	SM-SC	SW	SW-SM	SP	SP-SM	GP	GM
Eolian																		
Sand dunes	16	6						6		25					44	25		
Loess	23	12	9	9	4	30		39	4									4
Glacial																		
Till	52		13	10	10	21	4	12		8	4							
Outwash	8		12			12		12		12	12	6			4	25		
Alluvial																		
Fans	12				8	33		17		17	8	17						
Flood plains	12		17	8	17	17	17	17	8									
Deltas	4		25			25				25					25			
Terraces	20		10	10	10	20	10	5		25	5							
Lacustrine	35	3	14	9	11	17	6	11		11	9					9	3	
Coastal																		
Swamps	3	33	33									33						
Terraces	15					13	7	13	13	13	13			7	13	7		
Beach ridge	11	9	9		9					45	18				13	9		

NOTES: See notes in Table 3.9. Glacial outwash includes eskers. Coastal terraces include undifferentiated coastal deposits. The GM sample under the loess category is from erosion surface penetrating to lower nonloess deposit.

SOURCE: Compiled from data reported by USAE Waterways Experiment Station, 1963.

environmental locations throughout the continental United States. The number of samples in these tables is too small to allow any generalizations about the soils and their general occurrence in specific geologic environments. Even though two soils may classify with the same USCS classification in Tables 3.9 and 3.10, it does not imply that they will behave the same during construction. There are distinct limits on the soil classification without additional information.

Church (1981) provides an excellent example of geologic appreciation in construction bidding. Bids had been requested for an 850,000-yd^3 excavation to depths up to 140 ft as part of a freeway project on the east slope of the Sierra Nevada in California. Bare granite surrounded the proposed site of the excavation. The site itself was covered with a surficial gravelly soil, and limited exploratory borings encountered rock at shallow depths. A number of contractors prepared their bids on the belief that this would be predominately a rock excavation requiring blasting. However, a consultant to one contractor found that geologic maps showed the excavation site to be a geologically young terminal moraine from the Wisconsin glacial stage (approximately 18,000 years ago). Further seismic studies by the contractor revealed uniformly low wave velocities to finished grade, consistent with a deposit of till rather than rock. Boulders in the till were probably the rock identified in the borings. Consequently, the excavation was successfully bid and carried out with normal equipment suitable for dealing with bouldery till material rather than by much more expensive blasting in rock.

A basic technique for determining the soil conditions at a site is to take borings at some prescribed interval and to test selected soil samples for specific characteristics of interest. If this boring and testing program is conducted in a vacuum without appreciation of the geologic conditions of the site, completely erroneous conclusions can result—as in the preceding example. Consider a site 150 ft long by 100 ft wide by 20 ft deep to be examined for a building foundation, borrow pit, excavation, or other purpose requiring knowledge of the material within this zone. The area to be investigated encompasses 300,000 ft^3 of material. Each typical 3-in-diameter boring samples only 0.0003 percent of the deposit. Over 3000 borings would be needed to examine 1 percent of the deposit, which may still not be enough to identify the variations that might exist in natural deposits. Consequently, it is impossible to know with certainty what the subsurface conditions will be prior to excavating the site. Therefore, some intellectual sleuthing is needed to gain an understanding of the site. Simply making a pin cushion of the site with borings may still leave us ignorant of the important parameters of the site. Geology will be the key factor to allow us to organize and interpret our findings.

Before any effort is made at site investigation, the local regional geology should be understood. If the economics of the project allow it, a competent engineering geologist familiar with the area should be added to the project team. State and national geologic surveys and studies provide readily available geologic information, and further information is often available from local universities and organizations such as the Tennessee Valley Authority and highway departments.

The division and study of areas in terms of their physiographic units provide an effective method of dealing with regional geographic, geomorphic, and environmental information. Mallott (1922) describes a physiographic unit as

> ...an area or division of land in which topographic elements of altitude, relief, and types of landforms are characteristic throughout and as such is set apart or contrasted with other areas or units with different sets of characteristic topographic elements.

An example of physiographic division for the continental United States is shown in Fig. 3.18, where there are 5 physiographic divisions that are further subdivided into 20 provinces. These sections are then further divided into 96 physiographic sections (not shown). Within each physiographic unit there are regionally related landforms, environment, and geology that control the types and variants of the soils found in the unit. There are a number of good references describing physiographic regions, and these are extremely helpful in getting oriented to conditions in an area with which one is not familiar (e.g., Fenneman 1931, 1938, Woods et al. 1962, Thornbury 1965, Witczak 1972).

Once the larger-scale regional geologic and physiographic information is collected, localized information is needed. This may be found in archival material of cities and government agencies, local universities, or local USDA soil maps, or it may be gathered from the experiences of local groups that must deal with the local soils (well drillers, utility companies, construction firms, engineers, etc.)

Once this material is digested, the site-investigation program can be planned based on the geologic conditions at the site. For instance, if the site is part of an ancient lake bed, a relatively uniform lacustrine clay might be expected, and a limited number of scattered borings might be appropriate. If, on the other hand, the site is a complex alluvial deposit such as shown in Fig. 3.12 and the existence of the clay swales is important to the planned construction, a series of closely spaced borings to find and identify the extent of the swales will be needed. The pattern of borings also will have to be adjusted in the field as swales are located. If the site is a residual soil over weathered rock, diamond-core drilling to penetrate weathered rock boulders will

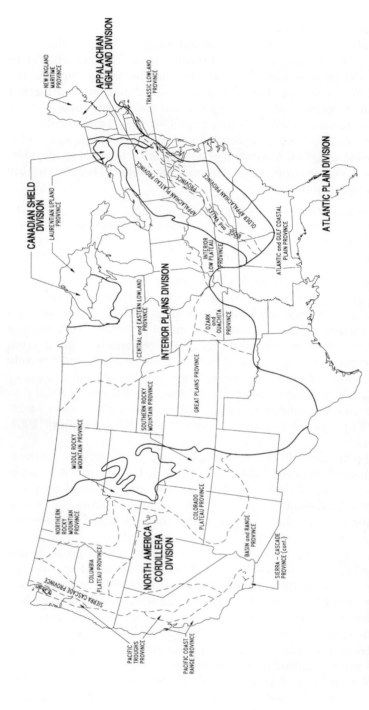

Figure 3.18 Physiographic divisions and provinces of the continental United States (*modified from Witczak 1972*).

be needed, and geophysical testing may prove useful to supplement the borings. Historical maps and records are also helpful. For instance, the city boundary of Charleston, South Carolina, has been steadily expanded by filling adjoining marshes since the city's settlement in 1672. Knowledge of this fact would be useful in interpreting borings in the filled areas. A number of references describe specific site-investigation methods in detail [e.g., Hvorslev (1949) is a classic, and Clayton et al. (1982) provides a comprehensive overview].

Any program of borings, test pits, trenches, etc., will result in specific items of information on the site soils and rocks at scattered locations. Commonly, boundaries between soil layers are extrapolated between borings, and material properties are assigned to these layers. This may be convenient for organizing and visualizing the site information, but it has little likelihood of conforming to the real site stratigraphy. Whenever such extrapolations are done, they should be looked on as a first-generation model of the site that will be updated and adjusted as site information is revealed during construction.

An understanding of the geologic processes of soil development and its impact on soil and aggregate properties is crucial to construction with geotechnical materials. It provides a key to understanding the range of problems that geotechnical sites may pose and how to use the site and materials effectively. In the Thirteenth Terzaghi Lecture, Legget (1979) quoted Karl Terzaghi on the importance of geology in dealing with geotechnical materials, and the quote is well worth repeating:

> The geological origin of a deposit determines both its pattern of stratification and the physical properties of its constituents....Therefore, the knowledge of the relation between physical properties and geological history is of outstanding practical importance (Terzaghi 1955).

3.7 References

American Society of Civil Engineers. 1982. *Proceedings of the Specialty Conference on Engineering and Construction in Tropical and Residual Soils, ASCE, Honolulu, Hawaii.*

American Society of Civil Engineers. 1987. *Foundations and Excavations in Decomposed Rock of the Piedmont Province,* geotechnical special publication no. 9, American Society of Civil Engineers, New York.

Bawa, K. S. 1957. "Laterite Soils and Their Engineering Characteristics," *Journal of Soil Mechanics and Foundation Engineering,* 83 (SM4): 1428–1443.

Bjerrum, L. 1967. "Engineering Geology of Norwegian Normally Consolidated Marine Clays as Related to Settlement of Buildings," *Geotechnique,* 17 (2): 81–118.

Buchanan, F. 1807. *A Journey from Madras Through the Countries of Mysore, Canava and Malabor,* vol. 3, East India Company, London.

Cast, L. D., T. J. Casias, and R. A. Baumgarten. 1987. "USBR Case History of Loess Cut Slopes in Nebraska," in *Engineering Aspects of Soil Erosion, Dispersive Clays, and Loess,* C. W. Lovell and R. L. Wiltshire, eds., geotechnical special publication no. 10, American Society of Civil Engineers, New York, pp. 98–114.

Chapman, D. R. 1975. "Shale Classification Tests and Systems: A Comparative Study," Joint Highway Research Project JHRP-75-11, Purdue University, West Lafayette, Ind.

Church, H. K. 1981. *Excavation Handbook,* McGraw-Hill, New York.

Clayton, C. R., N. E. Simmons, and M. C. Matthews. 1982. *Site Investigation: A Handbook for Engineers,* Granada Publishing, Ltd., London.

Clevenger, W. A. 1958. "Experience with Loess as a Foundation Material," *Transactions ASCE,* 123: 151–180.

Fenneman, N. M. 1938. *Physiography of Eastern United States,* McGraw-Hill, New York.

Fenneman, N. M. 1931. *Physiography of Western United States,* McGraw-Hill, New York.

Fisk, H. N. 1944. "Geological Investigation of the Alluvial Valley of the Lower Mississippi River," Mississippi River Commission, Vicksburg, Miss.

Gerrard, A. J. 1981. *Soils and Landforms: An Integration of Geomorphology and Pedology,* Allen & Unwin, London.

Gidigasu, M. D. 1975. *Laterite Soil Engineering,* Elsevier Scientific Publishing, Amsterdam.

Hvorslev, M. J. 1949. *Subsurface Exploration and Sampling of Soils for Civil Engineering Purposes,* reprinted by American Society of Civil Engineers, New York.

Jenkins, D. S., D. J. Belcher, L. E. Gregg, and K. B. Woods. 1946. "The Origin, Distribution, and Airphoto Identification of United States Soils," technical development report no. 52, U.S. Department of Commerce, Civil Aeronautics Administration, Washington.

Kolb, C. R., and W. G. Shockley. 1959. "Engineering Geology of the Mississippi Valley," *Transactions ASCE,* 124: 633–656.

Krinetsky, E. L., D. M. Patrick, and F. C. Townsend. 1976. "Geology and Geotechnical Properties of Laterite Gravel," technical report no. S-76-5, USAE Waterways Experiment Station, Vicksburg, Miss.

Krynine, D. P., and W. R. Judd. 1957. *Principles of Engineering Geology and Geotechnics,* McGraw-Hill, New York.

Lamb, R. O. 1990. "Geotechnical Aspects of Leaching of Carbonates from Loessial Soils," in *Physico-Chemical Aspects of Soil and Related Materials,* K. B. Hoddinott and R. O. Lamb, eds., STP 1095, American Society for Testing and Materials, Philadelphia, Pa.

Legget, R. F. 1979. "Geology and Geotechnical Engineering," *Journal of the Geotechnical Engineering Division, ASCE,* 105 (GT3):342–391.

Longwell, C. R., R. F. Flint, and J. E. Sanders. 1969. *Physical Geology,* Wiley, New York.

Lutenegger, A. J., and R. T. Saber. 1987. "Pore Structure of Loess Using Mercury Intrusion Porosimetry," in *Engineering Aspects of Soil Erosion, Dispersive Clays, and Loess,* C. W. Lovell and R. L. Wiltshire, eds., geotechnical special publication no. 10, American Society of Civil Engineers, New York, pp. 115–128.

Mallott, C. A. 1922. *Handbook of Indiana Geology,* part II, publication no. 21, Indiana Department of Conservation, Indianapolis, Ind.

Morin, W. J., and P. C. Todor. 1975. "Laterite and Lateritic Soils and Other Problem Soils of the Tropics," U.S. Agency for International Development, report AID/CSD 3682, Washington.

Newill, D. 1961. "A Laboratory Investigation of Two Red Clays from Kenya," *Geotechnique,* 11: 802–816.

Noble, D. F. 1977. "Accelerated Weathering of Tough Shales, Final Report," VHTRC 78-R20, Virginia Highway and Transportation Research Council, Charlottesville, Va.

Parsons, J. D. 1976. "New York's Glacial Lake Formation of Varved Silt and Clay," *Journal of the Geotechnical Engineering Division, ASCE,* 102 (GT6): 605–638.

Peck, R. B., and W. C. Reed. 1954. "Engineering Properties of Chicago Subsoils," Engineering Experiment Station bulletin no. 423, University of Illinois, Urbana, Ill.

Reidenour, D. R., E. G. Geiger, and R. H. Howe. 1976. "Suitability of Shale as a Construction Material," in *Living with Marginal Aggregates,* STP 597, American Society for Testing and Materials, Philadelphia, Pa.

Rollings, R. S. 1988. "Substandard Materials for Pavement Construction," in *Proceedings of the 14th Australian Road Research Board Conference*, vol. 14, part 7, Canberra, Australia, pp. 148–161.

Santi, P. M. 1994. "Classification of Weathered Rock Materials," *AEG News*, 97 (1): 35–40.

Sheeler, J. B. 1968. "Summarization and Comparison of Engineering Properties of Loess in the United States, highway research record 212, Washington, pp. 1–9.

Smith, R. B. 1986. "Evaluation of Sydney Shales for Use in Road Construction," in *Proceedings of the 13th ARRB–5th REAAA Combined Conference*, vol. 13, part 5, Australian Road Research Board, Adelaide, Australia, pp. 133–145.

Sowers, G. F. 1963. "Engineering Properties of Residual Soils Derived from Igneous and Metamorphic Rocks," in *Proceedings of the 2nd Pan American Conference on Soil Mechanics and Foundation Engineering, Brazil*, Am. Geophysical Union, pp. 39–62.

Sowers, G. F., and T. L. Richardson. 1983. "Residual Soils of the Piedmont and Blue Ridge," transportation research record 919, Washington, pp. 10–16.

Straub, G. 1935. "Some Observations on Sorting of River Sediments," in *Transactions of the American Geophysical Union 16th Meeting*, Am. Geophysical Union, Washington.

Strohm, E. E., G. H. Bragg, and T. W. Ziegler. 1978. *Design and Construction of Compacted Shale Embankments*, vol. 5: *Technical Guidelines*, FHWA-RD-78-141, Federal Highway Administration, Washington.

Swartzman, F. L., L. M. Sanneh, and W. J. Mason. 1986. "Local Laterite Aggregates in Pavement Construction," in *Proceedings of the 6th International Road Federation African Highway Conference*, International Road Federation, Cairo, Egypt.

Terzaghi, K. 1955. "Influence of Geological Factors in the Engineering Properties of Sediments," *Economic Geology*, Fiftieth Anniversary Volume: 557–618; also reprinted as Harvard Soil Mechanics Series no. 50, Harvard University, Cambridge, Mass.

Terzaghi, K. 1951. Discussion of "Consolidation and Related Properties of Loessial Soil," ASTM special publication no. 126, American Society for Testing and Materials, Philadelphia, Pa.

Thornbury, W. D. 1965. *Regional Geomorphology of the United States*, Wiley, New York.

Townsend, F. C. 1985. "Geotechnical Characteristics of Residual Soils," *Journal of Geotechnical Engineering*, 111 (1): 77–94.

Townsend, F. C., P. G. Manke, and J. V. Parcher. 1969. "Effects of Remolding on the Properties of Laterite Soil," bulletin 284, Highway Research Board, Washington.

Turnbull, W. J. 1968. "Construction Problems Experienced with Loess Soils," highway research record 212, Washington.

USAE Waterways Experiment Station. 1963. "Forecasting Trafficability of Soils, Airphoto Approach," technical memorandum no. 3-331, report 6, vol. II, Vicksburg, Miss.

White, R. N., and T. L. Richardson. 1987. "Investigation of Excavatability in the Piedmont," in *Foundations and Excavations in Decomposed Rock of the Piedmont Province*, geotechnical special publication no. 9, American Society of Civil Engineers, New York, pp. 15–27.

Winterkorn, H. F., and R. E. C. Chandrasekhaven. 1951. "Lateritic Soils and Their Stabilization," bulletin 44, Highway Research Board, Washington.

Witczak, M. W. 1972. "Relationships Between Physiographic Units and Highway Design Factors," NCHRP report 132, Highway Research Board, Washington.

Woods, K. B., R. D. Miles, and C. W. Lovell, Jr. 1962. "Origin, Formation, and Distribution of Soils in North America," in *Foundation Engineering*, G. A. Leonards, ed., McGraw-Hill, New York.

Zhi-hui, W., X. Ding-yi, L. Qu-yao, and C. Dong-yai. 1987. "Dynamic Characteristics of Loess," in *Engineering Aspects of Soil Erosion, Dispersive Clays, and Loess*, C. W. Lovell and R. L. Wiltshire, eds., geotechnical special publication no. 10, American Society of Civil Engineers, New York, pp. 148–166.

4

Materials Testing

I came to the United States and hoped to discover the philosopher's stone by accumulating and coordinating geological information....It took me two years of strenuous work to discover that geological information must be supplemented by numerical data which can only be obtained by physical tests carried out in a laboratory.   KARL TERZAGHI, 1936

4.1 Introduction

Because of the infinite variety of geotechnical materials available on the earth, testing of these construction materials is essential if projects are to be completed successfully and if the final product is to perform to the owner's expectations. In order to properly design and construct projects of, on, or in geotechnical materials, it is essential to first determine what materials are present or will be used at the project site and second to measure the engineering characteristics of these materials.

Laboratory testing is necessary prior to design by the engineer, since it provides characterization of the raw materials and allows prediction of their performance as they will be used in construction. Such properties as strength, durability, volume-change potential, and permeability often must be determined in addition to material classification tests; different tests and testing procedures are necessary for soils and for aggregates. Laboratory testing often is needed also for proper control of ongoing construction projects. For example, grain-size distribution tests may be needed periodically as sand is taken from different sections of a borrow pit to maintain consistency of the designed concrete mixture. Both quality control and quality assurance may require various types of laboratory and field testing during construction.

Field testing is critical to ensure that the placed material meets specifications and will be capable of performing as required under the expected loading conditions. Such testing typically involves determination of moisture, density, and/or permeability conditions upon material placement and also may include in-place strength testing. Field testing also may be required to determine native (in situ) soil or rock characteristics prior to construction on the deposit.

This chapter discusses the most common laboratory and field tests used to characterize soils and aggregates. The uses and importance of each test are emphasized; advantages and limitations of the tests are presented. Although the procedures of the individual tests are not provided in detail herein, standardized testing procedures [such as those of the American Society for Testing and Materials (ASTM)] are readily available and should be consulted prior to initiation of testing. The engineer or other responsible party should determine the purpose of the testing, review the pertinent test procedures, and request any modifications to testing needed for the particular project. The importance of this aspect of project planning and design cannot be overemphasized.

For convenient reference, Tables 4.1 and 4.2 provide the ASTM designation of each test discussed herein. They also cross-reference the ASTM test procedures with corresponding American Association of State Highway and Transportation Officials (AASHTO), U.S. Army Corps of Engineers (COE), and Department of Defense (DOD) test procedures. (ASTM is an organization whose mission is to establish and publish standard test procedures for use by any organization involved in testing. The other organizations are ones with large construction missions and published test procedures that are accepted in much of the construction industry.) In some cases, there is not an accepted ASTM procedure for a particular test, although ASTM has the most comprehensive list of procedures. The other organizations (AASHTO, COE, and DOD) have only those procedures for which they have a specific need. A more comprehensive list of commonly used ASTM soil and aggregate testing procedures is provided in Appendix B.

4.2 Laboratory Tests for Soils

For laboratory testing to be meaningful, testing should be conducted on representative samples of the materials of concern. Sufficient samples from appropriate locations should be taken to adequately characterize the site. Materials for laboratory testing may be obtained from

TABLE 4.1   Standard Laboratory Test Procedures

Test	ASTM	AASHTO	COE*	DOD†‡	Comments
SOILS:					
Water content	D 2216	T265	I	Method 105, 2-VII	—
Grain size	D 422	T88	V	2-III, 2-V, 2-VI	—
Atterberg limits	D 4318	T89 T90	III	Method 103, 2-VIII	—
Classification	D 2487	—	—	—	Engineering
	D 3282	—	—	—	Highway
Specific gravity	D 854	T100	IV	2-IV	
Unit weight	—	T233	II	2-X	—
Relative density	D 4253 D 4254	—	XII	—	—
Compaction	D 698	T99	VI	Method 100, 2-IX	Standard
	D 1557	T180		Method 100, 2-IX	Modified
CBR	D 1883	T193	—	Method 101, 2-XI	—
Resilient modulus	—	T274	—	—	—
Permeability	D 2434	T215	VII	—	—
Unconfined compression	D 2166	T208	XI	2-XIII	—
Direct shear	D 3080	T236	IX	—	—
Triaxial	D 2850	T234	X	—	—
Consolidation	D 2435	T216	VIII	—	—
AGGREGATES:					
Abrasion resistance	C 131 C 535	T96	—	3-V	LA abr.
Freeze/thaw	—	T103	—	4-II	—
Wetting/drying	—	—	—	—	—
Specific gravity and absorption	C 127 C 128	T84 T85	—	3-V, 4-II 3-V, 4-II	Fine Coarse
Unit weight and voids	—	T19	—	—	—
Grain size	C 136	T27	—	4-II	F & C
	C 117	T11	—	4-II	<No. 200
Sampling stone	—	T2	—	—	—
Moisture	—	T255	—	—	—
Durability	D 3744	—	—	—	—
Sulfate soundness	C 88	T104	—	4-II	—
Aggregate expansion from hydration	D 4792	—	—	—	—

*Dept. of the Army *Laboratory Soils Manual,* EM 1110-2-1906.
†Dept. of Defense *Military Standard,* MIL-STD-621A (Method 100, etc.).
‡Dept. of the Army *Materials Testing Field Manual,* FM 5-530 (2-III, etc.).

TABLE 4.2   Standard Field Test Procedures

Test	ASTM	AASHTO	DOD*	Comments
		Designation		
Water content				
Oven	D 2216	T265	—	—
Nuclear	D 3017	T239	—	—
Microwave	D 4643	—	—	—
Speedy moisture	—	T217	—	—
Density				
Nuclear	D 2922	T238	—	—
Drive cylinder	D 2937	T204	Method 102	—
Sand cone	D 1556	T191	Method 106	—
Water balloon	D 2167	T205	—	—
Plate load test	D 1195	T221	Method 104,	Repetitive
		T222	2-XII	Nonrepetitive
Dynamic cone				
penetrometer	—	—	—	—
Permeability	—	—	—	—

*Dept. of Defense *Military Standard,* MIL-STD-621A (Method 100, etc.), Dept. of the Army *Materials Testing Field Manual,* FM 5-530 (2-III, etc.).

subsurface investigations (undisturbed borings, excavations, etc.), test pits, or grab samples from borrow pits or stockpiles. Regardless of the sampling method used, the material used in the laboratory testing program *must* be representative of the material to be used in the construction project. Not only is the location of sampling important, but the number of samples tested is also important and should be "sized" to the project. For instance, on a $15 million dollar project covering tens of square miles, numerous samples (not just one sample, as was done on a recent large construction project) should be tested to determine the (multiple) soils' strengths and load-carrying capabilities. Guidance is available to determine the appropriate methods of sampling and the number and locations of samples needed for testing at various types of projects (Hvorslev 1948, Sanglerat 1972, Peck et al. 1974, Winterkorn and Fang 1975).

4.2.1   Moisture content

The *moisture content* ω is defined in geotechnical engineering as the ratio of the weight of water W_w to the weight of solids W_s in a given mass of soil; it is usually expressed as a percentage. It should be noted that many other scientific/engineering disciplines use a volumetric moisture content (volume of water V_w divided by the total volume of the sample V_t) instead of the gravimetric (weight-based) one used in geotechnical engineering; therefore, care should be used in comparing moisture contents determined by different scientific/engineering organizations. If the dry unit weight (dry density) of the soil is known, a simple conversion can be made between the gravimetric and volumetric moisture contents:

$$V_w = \omega \frac{\gamma_d}{\gamma_w} \times 100$$

where V_w = volume of water in the soil; also referred to as the *volumetric moisture content*

ω = gravimetric moisture content

γ_d = dry density of soil

γ_w = unit weight of water, 62.43 pcf

The moisture content of a soil is important because it gives a first indication of the condition or consistency of the material. For instance, a moisture content of 15 percent in a lean clay (CL) is somewhere near optimum moisture content for compaction, whereas a value of 350 percent likely indicates a sediment that has recently been deposited in a water body or dredged from one. As this example indicates, the water content can and does vary tremendously in different materials.

Moisture condition of a soil is extremely important for geotechnical design and construction purposes. In design, the moisture conditions of in situ soil masses are critical for design of stable slopes, settlement determinations, etc. In construction, compaction control and material performance require careful control not only of moisture content but also of moisture variation. Thus it is necessary to determine initial moisture content on almost any soil to be used on a project.

The moisture content is determined by taking a representative sample of soil, weighing it, drying it in an oven at a specific, controlled temperature (usually 105 to 110°C), and then weighing it again after thorough drying. To obtain an accurate moisture content, it is necessary to carefully and accurately weigh the sample while preventing any loss or addition of water to the sample. ASTM D 2216 gives standard procedures for determining the moisture content of soils.

Possible sources of error include

- Nonaccurate or poorly calibrated balance

- Loss of soil between first and second weighings

- Loss of moisture from sample before first weighing

- Addition of moisture to sample after drying and before second weighing
- Incorrect oven temperature
- Specimen too small
- Incorrect tare weight
- Specimen removed from oven before reaching a constant oven-dry weight
- Weighing dry specimen while still hot

4.2.2 Grain-size distribution

The *grain-size distribution* provides insight into the engineering behavior of soils and is crucial to engineering classification. It is normally one of the first tests conducted on a laboratory soil sample. For coarse-grained soils, the distribution of grain sizes is determined by a mechanical, or sieve, analysis. This may be performed as either a dry or a washed (wet) sieve analysis, with the wet analysis required for samples containing cohesive material (see ASTM D 422). The grain-size distribution for fine-grained soils is determined by hydrometer analysis.

Proper sample preparation is essential to good testing. Samples of cohesionless material may be prepared by oven drying and breaking down all lumps into individual grains either before or during sieving. Cohesive soil must be dispersed in water, possibly using a dispersing agent, before passing it through the sieves.

A sieve analysis is performed by pouring the prepared sample into the top sieve of a nest of sieves. The nest, or set, of sieves is usually selected so that the opening in any sieve screen is double that of the next-finer (lower) sieve screen. The finest sieve used is normally the No. 200 U.S. Standard sieve with an opening size of 0.075 mm. This represents the boundary between coarse- and fine-grained soils. The largest sieve used is determined by the material to be tested; normally, one sieve should be used that has 100 percent of the material passing through it.

For a dry sieve analysis, the nest of sieves is generally placed in a mechanical shaking device that shakes and taps the sieves causing the particles to fall through the sieves until they are retained on a particular sieve, thus separating the size fractions. In a wet analysis, the dispersed suspension must be washed through the nest of sieves with clean water until the various size fractions are retained on the proper sieves. The material on each sieve must then be oven dried and weighed to determine the percent passing each sieve. Care should be taken to use the minimal amount of water needed to completely separate the size fractions; this is helpful in minimizing the amount of fluidized sample that must be handled and dried.

The grain-size distribution of the fine-grained portion of a sample is determined by a hydrometer analysis. For this test, a 50-g sample is dispersed in a liter of distilled water and is poured into a standard cylinder. A deflocculating agent such as sodium hexametaphosphate is normally used to disperse the soil particles. The cylinder containing the soil-water mixture is then agitated for 1 minute. A calibrated hydrometer is then inserted in the suspension, and density readings are taken at specified time intervals (typically at 2, 4, 8, 15 minutes, etc.). From these readings, the grain-size distribution can be calculated based on the assumption that the particles are spherical. Since (1) soil particles are not spherical, and in fact, the smallest particles are flat and platelike, and (2) the behavior of fine-grained soils is determined largely by characteristics other than particle size, the grain-size distribution of fine-grained soils generally has limited engineering usefulness. However, the percentage of a soil that is fine grained (smaller than the No. 200 sieve) is very important for soil classification, in evaluation of material for drainage layers, and in determining aggregate acceptability for use in base courses, portland cement concrete, and asphalt concrete (see Chap. 2).

The results of the sieve and/or hydrometer analyses are normally presented in the form of a grain-size distribution curve. This curve is presented on a semilog plot with the percentage of the material finer than a given sieve size (percent passing) plotted on the ordinate (y axis) and the particle diameter plotted to a log scale on the abscissa (x axis). The shape of the curve may vary significantly and will give a visual indication of whether the material is well-graded, gap-graded, etc. Some typical grain-size distribution curves are shown in Fig. 2.3. Several useful soil characteristics can be calculated from the grain-size distribution curve, as described in Chap. 2.

Possible sources of error in grain-size distribution tests include

Sieve analysis
- Failure to adequately disperse fine-grained material in the wet mechanical analysis (may need to use a dispersing agent such as sodium hexametaphosphate or less commonly trisodium phosphate)
- Failure to break down agglomerations of material into individual grains
- Loss of material during testing

- Overloading of sieves
- Broken or distorted sieve screens
- Inadequate shaking of sieves

Hydrometer analysis
- Oven drying of soil before test
- Inadequate type or quantity of dispersing agent
- Incomplete dispersion of soil
- Insufficient shaking in cylinder at start of test
- Too much soil in suspension
- Disturbance of suspension during insertion or removal of hydrometer
- Hydrometer not kept clean
- Temperature variation during testing
- Loss of material during testing

4.2.3 Atterberg limits

The *Atterberg limits,* as introduced in Chap. 2, are useful in defining the limits of various stages of consistency in fine-grained soils. They provide a convenient index of the plastic behavior of soils. The Atterberg limits are very useful in soil classification and as a crude indicator of engineering behavior. The liquid and plastic limits are used routinely in soils work. These Atterberg limits are particularly useful in construction, where they are used as an inexpensive initial screening tool to select soils for project use and as a construction quality-control and quality-assurance test to identify changes in material in a borrow area during construction.

The liquid limit LL is determined in a special liquid-limit device consisting of a brass cup of specified dimensions attached to a carriage that controls its drop onto a rubber base. This device must be checked and calibrated prior to use to obtain repeatable results. Critical dimensions for the liquid-limit device and procedures for calibrating the equipment, as well as test procedures, are given in ASTM D 4318.

A 150- to 200-g soil sample is prepared for testing by mixing soil at its natural water content with distilled or demineralized water to reach a consistency requiring 15 to 20 blows of the liquid-limit device to close a $\frac{1}{2}$-in-long groove. A portion of the prepared soil is spread and smoothed into the brass liquid-limit cup to a depth of 10 mm, and a standard groove is cut through the soil using a specific grooving tool (described in ASTM D 4318). The cup is then caused to drop at a rate of 1.9 to 2.1 drops per second until the soil flows together for a distance of $\frac{1}{2}$ in along the groove. The number of drops required to close the groove is recorded; a sample of soil is removed along the closed portion of the groove, and its water content is determined. The remaining soil is remixed to decrease the water content; then the procedure is repeated at least three additional times to produce two results requiring 25 or more blows and two at 25 or fewer blows.

The plastic limit *PL* is determined by selecting a 20-g sample of material prepared for the liquid-limit test. The water content should be reduced by spreading and mixing on a dry glass plate until the soil has a consistency that will allow rolling without sticking to the hands. When this consistency is achieved, the plastic-limit test is performed by selecting a 1.5- to 2-g sample of the prepared soil and rolling it into an elongated ball (ellipsoidal mass). The soil mass is placed on the glass plate and rolled by hand to form a thread of a uniform diameter of $\frac{1}{8}$ in (3.2 ± 0.5 mm). The thread is then re-formed into an elongated ball and rerolled to $\frac{1}{8}$ in diameter. This process is repeated until the thread crumbles and can no longer be rolled into a thread. At this point, the plastic limit has been reached, and the crumbled thread should be placed in a preweighed airtight container until sufficient additional threads can be rolled to the plastic limit and placed in the same container to total approximately 9 g of soil. The entire process should be repeated so that 9 g of soil is placed in a second container. The water content of the soil in the containers should then be determined, and the average of the two calculated. If the difference in water contents of the two samples is greater than 2 percent, the test should be repeated.

It is preferable that the soil used in the Atterberg limits determination be at its natural water content prior to the test, since drying (either natural or oven drying) can affect the results. Organic soils, laterites, and soils derived from volcanic rocks are the most sensitive to drying. The effect of drying on a particular soil can be determined by testing undried, air-dried, and oven-dried soil samples and comparing the Atterberg limits obtained for each.

The soil used in the Atterberg limit tests should pass the No. 40 (425-μm) sieve. If there are particles retained on the No. 40 sieve, a sufficiently large sample should be selected and washed over the No. 40 sieve to produce about 150 to 200 g of minus No. 40 material for testing.

After the liquid and plastic limits have been determined, the plasticity index *PI* should be calculated as follows:

$$PI = LL - PL$$

Both the liquid limit and the plastic limit are whole numbers. If either limit could not be determined, or if the plastic limit is equal to or greater than the liquid limit, then the soil should be reported as nonplastic (NP). Results from the Atterberg limit tests should be plotted on a plasticity chart (see Fig. 2.6). Use of this chart allows determination of the classification category and can help detect errors in determination of the limits.

Possible sources of error in determining the Atterberg limits include

- Nonrepresentative sample (must be the same for both liquid and plastic limits)
- Improperly prepared and cured sample
- Incorrect water content determination

Liquid limit
- Improper or unadjusted/uncalibrated test device
- Worn parts on test device
- Soil on contact point between the cup and rubber base of test device
- Loss of moisture during test

Plastic limit
- Incorrect final thread diameter
- Stopping the rolling process too soon

4.2.4 Specific gravity

The *specific gravity* of solids is the ratio of the weight in air of a given volume of soil particles to the weight in air of an equal volume of distilled water, all at a given temperature. The specific gravity of solids applies to the minus No. 4 sieve material, i.e., to soil not rock. (Other types of specific gravity—apparent and bulk specific gravities—are applicable to aggregates, i.e., materials larger than the No. 4 sieve, and are discussed later in this chapter.) Typical values for specific gravity of solids are 2.65 for sands and 2.7 for clays, with values for various clays ranging from 2.5 to 2.9.

The specific gravity of solids test is performed in a calibrated 500-ml volumetric flask that is attached to a vacuum pump or system. The equivalent of 50 to 80 g of oven-dried material is placed into the calibrated flask. The flask is then filled about half full of distilled water, and the sample is allowed to sit overnight. If soil is used at its natural water content, it is placed in a dish, mixed with distilled water to form a slurry, and poured into the flask; this material does not have to stand overnight. When the test is to proceed, the vacuum line is connected to the flask, and a vacuum of about 29 in Hg is applied for 2 to 4 hours for oven-dried soils or 6 to 8 hours for nondried soils. During the evacuation period, the flask should be agitated periodically to help remove trapped air bubbles. After vacuuming, the flask is filled with deaired distilled water to about $\frac{3}{4}$ in below the 500-cc graduation line, and a vacuum is applied. The suspension is usually considered to be deaired if the water level drops $\frac{1}{8}$ in or less upon release of the vacuum. The flask should then be filled to the calibration line, thoroughly dried outside and in the neck, and weighed. The temperature of the suspension should then be determined. The flask contents should then be placed in an evaporating dish and the oven-dried weight determined, if it was not done prior to testing. The specific gravity can then be calculated as follows:

$$ G_s = \frac{W_s K}{W_s + W_{bw} - W_{bws}} $$

where W_s = weight of oven-dried soil, g
  K = correction factor based on density of water at 20°C
  W_{bw} = weight of flask plus water, g
  W_{bws} = weight of flask plus water plus soil, g

The soil sample used for the specific gravity determination should be composed of material passing the No. 4 sieve. The soil may be at its natural water content or oven dried, although some soils, particularly organic soils, may be difficult to rewet after oven drying. If the material to be tested contains material both larger and smaller than the No. 4 sieve, the material should be separated on the No. 4 sieve; the specific gravity of solids should be determined on the fine fraction and the apparent specific gravity on the coarser fraction (see Sec. 4.3.4). The specific gravity of the sample will then be calculated based on the relative percentages of the solid volume of the components.

Possible sources of error in determining the specific gravity include

- Imprecise weights of flask and its contents
- Temperature of flask and contents not uniform at time of weighing and temperature measurement
- Flask not clean

- Moisture on outside of flask or on inside of neck
- Meniscus not level with mark on neck of flask
- Use of tap water or other water that is not distilled or demineralized
- Incomplete removal of entrapped air from soil suspension—the most serious error, which will tend to lower the computed value
- Gain in moisture of oven-dried soil before it is weighed
- Loss of material from oven-dried specimen

4.2.5 Unit weight

The *unit weight* of soil is defined as the density or weight per unit volume of the soil mass; in equation form, it is expressed as

$$\gamma = \frac{W}{V}$$

where W = weight of soil, including the weight of the moisture
V = total volume occupied by the soil mass

Unit weight is one of the most important physical properties of a soil. It is used to calculate stresses in the soil for determination of overburden pressures and earth pressures behind retaining walls. It is also needed to assess the adequacy of field construction. Unit weight is a critical value in calculating numerous parameters associated with soils engineering. Depending on the moisture conditions in a soil, the unit weight calculated may be either the wet unit weight or the dry unit weight of the soil. Sometimes one may encounter the term *submerged or buoyant unit weight*; this is a reference to an effective weight and is actually the wet unit weight minus the unit weight of water. In dealing with compaction, the term *unit weight at zero air voids* is encountered; this is the density of a soil in which the voids are completely filled with water; i.e., there is no (zero) air in the void spaces.

In addition to unit weight, other relationships based on weight and volume relationships that are important in laboratory and field calculations are *void ratio, porosity,* and *degree of saturation.* The quantities that must be known to calculate these relations are weight and volume of the wet specimen, weight of the dry specimen, and the specific gravity of solids. The weights can be determined easily; the volume of the specimen can be determined by either the displacement or volumetric methods. The specific gravity is determined as discussed previously.

The volumetric method involves accurately measuring the specimen dimensions and calculating the volume. This method is applicable to regularly shaped specimens, i.e., normally, those extruded from a sampling tube or those cut or trimmed from a soil mass into a cylindrical or rectangular shape. (ASTM D 2937 recommends using a drive cylinder to obtain soil densities.) After the specimen is formed into a regular shape, multiple measurements of each dimension are made and averaged (for a cylindrical sample, typically three diameter measurements are taken at each of three locations—-top, middle, and bottom—and three to six height measurements are taken at various points around the sample). These measurements are used to calculate volume of the specimen. The specimen is then weighed and dried in the oven. After drying, it is weighed again to get the oven-dried weight.

The displacement method of determining soil volume requires determination of the volume or weight of water displaced by the soil sample. This procedure is applicable to irregularly shaped samples and those which cannot be trimmed to have smooth sides because of inclusions of rocks, shells, etc. This procedure involves trimming the soil into a fairly regular shape, weighing it, and coating it with two coats of melted wax. (The wax must be a specified type that will resist shrinkage cracking upon cooling, and its specific gravity must be known.) The wax-coated specimen is then weighed in air and in water. The wax is then removed, and the water content of the soil is determined.

After the wet and dry weights and the volume have been determined, the following quantities can be calculated:

Dry unit weight γ_d: The weight of oven-dried soil solids per unit of total volume of soil mass. This value is also called *dry density* and is usually expressed in pounds per cubic foot.

Wet unit weight γ_w: The weight of solids plus water per unit of total volume of soil mass, expressed in pounds per cubic foot. This value is also called *wet density,* regardless of the degree of saturation of the soil.

Dry unit weight at zero air voids γ_{zav}: The weight of solids divided by the volume of water plus the volume of solids. This value is used in assessing compaction.

Void ratio e: Ratio of the volume of voids to the volume of solid particles in the soil mass, expressed in decimal form.

Porosity n: Ratio of the volume of voids to the total volume of the soil mass. This is usually expressed as a percentage.

Degree of saturation S: Ratio of the volume of water in a soil mass to the total volume of voids, expressed as a percentage.

The physical significance and method of calculating each of these quantities are shown in Table 2.6.

Possible sources of error in determining the volume of a soil sample include

Volumetric method
- Inaccurate measurement of sample

- Voids on sides of specimen from improper trimming or from sampling of soil with inclusions of large particles

- Loss of material while extruding specimen from sampling tube

Displacement method
- Voids on surface of specimen that are not filled with wax

- Air bubbles formed under wax coating

4.2.6   Relative density

Neither the void ratio nor the dry density of cohesionless soils gives a complete picture of how such soils will behave in the field and/or under load. They do not indicate whether the material is loose or dense because the values are greatly influenced by particle shape and gradation. Therefore, a different relationship has been developed that will provide a better indication of cohesionless material performance. This relationship is called the *relative density,* and it applies only to cohesionless materials.

The relative density indicates the compactness of a cohesionless soil relative to its loosest and densest possible conditions. Relative density is calculated by

$$D_d = \frac{e_{max} - e}{e_{max} - e_{min}} \times 100$$

where e_{max} = void ratio of the soil in the loosest state attainable in the laboratory

e_{min} = void ratio of the soil in the densest state attainable in the laboratory

e = void ratio of the soil in place

or in terms of dry density,

$$D_d = \frac{\gamma_d - \gamma_{dmin}}{\gamma_{dmax} - \gamma_{dmin}} \times \frac{\gamma_{dmax}}{\gamma_d} \times 100$$

where γ_d = dry density of the soil in place, or *in-place density*

γ_{dmin} = dry density of the soil in the loosest state attainable in the laboratory, or *minimum density*

γ_{dmax} = dry density of the soil in the densest state attainable in the laboratory, or *maximum density*

The relationship expressed in terms of density is usually used because it is easier to obtain density values than values for void ratio.

The in situ density of a soil can be determined using one of several methods in the field (see Sec. 4.4.2) or from undisturbed samples tested in the laboratory (discussed above). The minimum density is determined by careful placement or "raining down" of the material to fill a mold at the loosest possible density. The maximum density is obtained by carefully placing loose material to fill the mold and subsequently subjecting the specimen to simultaneous static and vibratory loads.

The equipment (mold, vibratory table, and surcharge assembly) used in this testing procedure must meet certain specifications and should be carefully calibrated before use. ASTM D 4253 and ASTM D 4254 give these requirements as well as detailed test procedures. The molds used are either 0.1-ft^3 capacity (6 in ID) for smaller-particled soils or 0.5-ft^3 capacity (11 in ID) for soils with larger particles.

The soil to be tested should be oven dried and cooled in an airtight container. Any lumps or aggregations of smaller particles should be thoroughly broken down. A representative sample should then be obtained for testing using a sample splitter or riffle; the size of the sample needed (either 25 or 100 lb) depends on the particle size of the material being tested.

After equipment calibration and sample preparation, the minimum density is determined by filling the mold in layers using a special pouring device (or hand scoop for larger materials) and allowing the material to free fall 1 in. The spout or scoop should be moved in a spiral motion from the outside edge of the mold toward the center while maintaining a continuous, steady flow of material. The mold should be filled to just above its top, and the soil surface is then trimmed level with the top of the mold. The weight of the material required to fill the mold is then determined. This procedure should be repeated until consistent results (within 1 percent) are obtained.

To obtain a maximum density, the same filling procedure is followed as for minimum density, but after the mold is filled and the surface has been trimmed, the mold is attached to the vibratory table (if not done previously) and the surcharge assembly is attached to the mold. The sample should be vibrated for 8 minutes at the rheostat setting previously determined to give the maximum density. The surcharge weight is removed, and the displacement Δh of the soil surface is measured. The weight of material is then determined. The procedure is repeated until consistent results are obtained.

After the maximum and minimum dry densities (or void ratios) are calculated, the relative density is determined using the in situ density (or void ratio) and the equations given previously. Since several procedures are available to determine e_{min} and e_{max}, there are no definitive numerical values for these functions, and therefore, it is necessary to specify which procedure was used in their determination.

In the field, relative density is often assessed by indirect means such as static and dynamic penetrometers and load tests.

Possible sources of error in the relative density test include

- Use of inappropriate (cohesive) soil
- Segregation of material while processing soil
- Increase in soil moisture before or during test
- Inaccurate calibration of molds

Minimum density
- Disturbing the mold during filling
- Segregation of material while filling the mold
- Loss of material from the mold before weighing

Maximum density
- Improper amplitude of vibratory table
- Loss of fines (through dust) during vibration
- Misalignment and binding of surcharge weight

4.2.7 Compaction

Compaction is the process of mechanically densifying a soil or aggregate. Densification is accomplished by pressing the particles into closer contact while expelling air from the soil mass. (This term is not to be confused with the agricultural use of the term *compaction* which is synonymous with the construction/engineering term *consolidation*.) Compaction of a soil or aggregate will increase its density and strength and will reduce its permeability. These changes are all desirable, and compaction is the simplest and most effective way to improve a soil's or an aggregate's engineering properties.

In the laboratory, soils are typically compacted in several layers in a cylinder using a set number of blows with an impact hammer. The test is repeated with the soil at different water contents until enough points have been defined to establish a relationship between water content and density. Typically, five well-selected points will adequately define the curve. There are several other laboratory approaches for compaction, including kneading compactors, gyratory compactors, and vibratory compaction methods, but the impact compactor is the most common laboratory test method used for geotechnical construction.

Because the laboratory compaction test is intended to simulate field conditions, and since heavier construction equipment has been produced over time, different laboratory tests have been needed to replicate various field compactive efforts. Table 4.3 compares several common laboratory compaction test methods. There are minor differences in mold size and methods of handling large aggregate particles, but there are two basic tests. One is commonly referred to as *standard density, standard Proctor, standard AASHTO,* or *CE 12 compaction* and imparts approximately 12,000 ft · lb/ft^3 of energy in compaction of the sample. The other, commonly referred to as *modified density, modified Proctor, modified AASHTO,* or *CE 55 compaction,* uses approximately 55,000 or 56,000 ft · lb/ft^3 of energy to compact the sample. The Corps of Engineers also includes an intermediate compaction level, CE 26, with 26,000 ft · lb/ft^3 of energy. The major difference between these tests is the amount of energy applied to the sample by using different weight hammers and different numbers of blows per layer.

Figure 4.1 illustrates typical compaction test results for a clayey gravel (GC). At a given compaction energy, the soil's dry density increases up to a maximum point as the moisture content at which the soil is compacted increases. This maximum point is the optimum dry density and optimum moisture content for the soil at this specific energy of compaction. As the compaction moisture content increases past optimum, the dry density decreases. As the compaction energy increases (as from CE 12 to CE 26 to CE 55), the optimum dry density increases and the optimum moisture content decreases. A line known as the *line of optimums* can be drawn to connect the optimum dry densities for all compaction energies. Figure 4.1 also shows a *zero air voids line* plotted alongside the compaction curves. This is the maximum density that can be achieved at a given moisture content. It is a function of the material's specific gravity and represents the point where all voids in the material are filled with water (hence the name *zero air voids line*). Compaction curves physically cannot plot to the right of the zero air voids line.

TABLE 4.3 Summary of Compaction Procedures

Test method	Mold diameters, in	Hammer weight, lb	Drop height, in	Number of layers	Blows per layer	Energy, ft · lb/ft^3	Over-size correction
1. ASTM D 698 Standard Proctor							
Method A	4	5.5	12	3	25	12,375	a
Method B	6	5.5	12	3	56	12,320	a
Method C	6	5.5	12	3	56	12,320	b
Method D	6	5.5	12	3	56	12,320	c
2. ASTM D 1557 Modified Proctor							
Method A	4	10.0	18	5	25	56,250	a
Method B	6	10.0	18	5	56	56,000	a
Method C	6	10.0	18	5	56	56,000	b
Method D	6	10.0	18	5	56	56,000	c
3. Corps of Engineers (EM 1110-2-1906, "Laboratory Soils Testing")							
Standard	4	5.5	12	3	25	12,375	d
	6	5.5	12	3	56	12,320	e
Modified	4	10.0	18	5	25	56,250	d
	6	10.0	18	5	56	56,000	e
4. Corps of Engineers CBR Samples (TM 5-530, "Materials Testing")							
CE-12	6	10.0	18	5	12	12,000	f
CE-26	6	10.0	18	5	26	26,000	f
CE-55	6	10.0	18	5	55	55,000	f

[a]Discard material retained on No. 4 sieve. If >7 percent retained on No. 4, use method C.
[b]Discard material retained on 3/4-in sieve. If >10 percent retained on 3/4-in sieve, use method D.
[c]Replace material retained on 3/4-in sieve but passing 3-in sieve with an equal percentage by weight of material passing the 3/4-in sieve and retained on the No. 4 sieve. Discard the material retained on 3-in sieve. If 30 percent or more is greater than 3/4-in sieve, this test method is not appropriate.
[d]Discard material retained on No. 4 sieve. If >10 percent retained on No. 4 sieve, use 6-in mold (see note e).
[e]Discard material retained on 3/4-in sieve. Amount retained on the 3/4-in sieve should not exceed 10 percent.
[f]Replace material retained on 3/4-in sieve with an equal percentage by weight of material passing the 3/4-in sieve and retained on the No. 4 sieve.

Illustrative Example 4.1 Determine the point on the zero air voids line for a material at 5 percent moisture and having a specific gravity of 2.62.

solution This can be solved using the basic unit weight and volume relationships described in Sec. 4.2.5. On the zero air voids line the material is completely saturated; i.e., the volume of air is zero. The corresponding dry unit weight of the soil for a moisture content of 5 percent can be determined as follows:

$$w = W_w/W_s = 0.05$$
$$W_w = 0.05 W_s$$

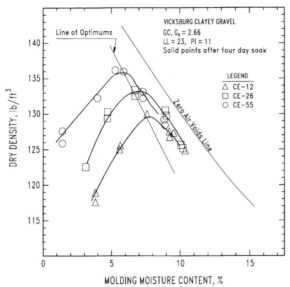

Figure 4.1 Typical compaction curves for a clayey gravel at different compactive efforts.

Let the total wet unit weight be 100 lb.

$$W = 100 = W_w + W_s = 1.05W_s$$
$$W_s = 95.24 \text{ lb}$$
$$W_w = 4.76 \text{ lb}$$
$$\gamma_w = W_w/V_w = 62.4 \text{ lb/ft}^3$$
$$V_w = 0.0763 \text{ ft}^3$$
$$G_s = \gamma_s/\gamma_w = 2.62$$
$$(W_s/V_s)/62.4 \text{ lb/ft}^3 = 2.62$$
$$V_s = 0.5825 \text{ ft}^3$$

Remembering that the volume of air is zero,

$$V = 0.0763 + 0.5825 = 0.6588 \text{ ft}^3$$
$$\gamma_d = W_s/V = 95.24 \text{ lb}/0.6588 \text{ ft}^3$$
$$= 144.6 \text{ lb/ft}^3$$

The zero air voids line corresponds to complete saturation; it is sometimes desirable to plot lines for other degrees of saturation (e.g., 90 percent, 80 percent, etc.). The dry density corresponding to any moisture content and degree of saturation can be calculated from the following equation:

$$\gamma_d = \frac{G_s \gamma_w}{1 + \dfrac{w G_s}{S}}$$

where γ_d = dry density
      G_s = specific gravity
      γ_w = density of water $\approx$ 62.4 lb/ft^3
      w = moisture content expressed in decimal form (not percent)
      S = saturation in decimal form (e.g., 100 percent is 1.00)

Possible sources of error that could cause inaccuracies in compaction curves (for any compactive effort) include

- Lumps or aggregations of dry particles

- Water not thoroughly mixed into soil

- Soil is reused for multiple points on the compaction curve. Some soils are significantly affected by recompaction; therefore, it is recommended that fresh material be used for each specimen. Recompaction may increase the maximum dry weight and thus decrease the optimum water content.

- Insufficient points to accurately define the compaction curve. The water content should vary by about 1½ percent between specimens and must cover the range in which the optimum water content and maximum density occur; generally, five points will be sufficient.

- Improper foundation for compaction mold

- Incorrect volume used for compaction mold

- Mechanical compactor not accurately calibrated

- Human variation in operation of hand compactor

- Excessive variation in total depth of compacted sample prior to trimming

- Water content not representative of entire specimen

Chapter 5 discusses compaction in more detail.

4.2.8 California bearing ratio

This test was developed by the California Highway Department and was first used as the basis for an empirical flexible pavement design procedure in 1929. The *California bearing ratio* (CBR) design procedure was studied extensively by the U.S. Army Corps of Engineers during World War II and soon became (and has remained) the basis for one of the leading flexible pavement design procedures in the world. It is still used for pavement design and as a standard basis of comparison for validating less well-accepted design procedures in many countries around the world.

The CBR test is both a laboratory test and a field test. It can be used to test laboratory-compacted samples or samples taken from the field; it also can be run on in situ field samples. Laboratory-compacted samples are tested to determine the design CBR for a particular material; thus a suite of tests must be performed to properly identify the correct CBR value to use. The purpose of running the CBR test on field samples is to determine as-placed or in situ conditions in compacted fill. The test is most appropriate and gives most reliable results for fine-grained soils, although it is also used to characterize the strength of soil-aggregate mixtures (e.g., subbases) and unbound aggregate base courses. In cohesionless soils, especially ones that include large particles, the reproducibility of the test is poor.

The CBR test provides an index of strength, but it is not a strength measurement, as are the unconfined compression, triaxial, and direct shear tests (which are discussed later). The CBR test measures a soil's resistance to penetration and compares those results with a standard set of values to determine the CBR value for that particular soil. The test involves pushing a piston of 3-in^2 end area into a soil specimen at the specified rate of 0.05 in/min. The unit load is recorded at each 0.10 in of penetration up to a total deformation of 0.5 in. The loads at 0.1 and 0.2 in of deformation are then compared with the loads required to cause equal penetrations into a standard well-graded crushed-stone specimen containing $\frac{3}{4}$-in maximum-sized particles. The values for the standard material are as follows:

Penetration, in	Load, psi
0.1	1000
0.2	1500
0.3	1900
0.4	2300
0.5	2600

To calculate the CBR as a percentage, the load at each penetration value for the test specimen is divided by the corresponding standard load and multiplied by 100. For example, results from a CBR test showed the load needed to cause 0.1 in of penetration in the sample to be 660 psi, and at 0.2 in of penetration the load was 940 psi. Calculation of CBR values gave $CRB_{@0.1} = (660 \div 1000)(100) = 66$ and $CBR_{@0.2} = (940 \div 1500)(100) = 63$.

When evaluating the CBR test, the CBR at 0.1 in of penetration into the soil is usually taken for use in design. This is typically the largest value obtained for a particular test, since the CBR generally decreases as penetration increases. If the CBR at 0.2 in of penetration is larger, the test should be rerun. If the CBR at 0.2 in is again larger, that value should be taken as the CBR for the sample.

CBR test specimens should be prepared to duplicate existing or anticipated field (moisture and density) conditions. The most critical condition for most materials exists when the moisture content is highest, i.e., when the material is saturated, causing the strength to be lowest. For this reason, the CBR test is often run on specimens that have been soaked for 4 days while confined in a mold under a surcharge load equal to the weight of the planned overlying layers. The exact method used to prepare the test specimen depends on the type of material to be tested; there are different procedures for cohesionless sands and gravels, cohesive soils, and swelling soils. Procedures for specimen preparation and CBR testing are given in ASTM D 1883.

To conduct a proper laboratory CBR determination, three laboratory compaction curves should be developed, one each using 12, 26, and 55 blows per layer (see Sec. 4.2.7 and Chap. 5 for a general discussion of compaction). For the CBR test, a spacer is placed in the mold, resulting in a shorter final specimen. After the sample is prepared but while still in the compaction mold, it is submerged in water for 4 days under surcharge weights equal to the anticipated overlying pavement weight. The penetration test is then run strictly according to established procedures, and the penetration results are plotted. If the curve of stress (unit load) versus penetration does not pass through the origin of the plot, the curve must be shifted graphically (see Fig. 4.2). If this shift is required, then the CBR is referred to as the *corrected CBR*. The final results of the CBR test should be plotted as shown in Fig. 4.3 to provide needed moisture-density-CBR values for design. More details on selecting CBR values for design are given in Sec. 5.3.5. Typical CBR values for some common geotechnical construction materials are given in Table 4.4.

Field CBR test procedures are essentially the same as in the laboratory except that the sample is tested at in situ moisture conditions. For the field test, it is critical that the testing device be plumb and that the deflection dial is fastened well outside the loaded area.

Potential errors in the CBR test include

- Inclusion of large (oversized) particles
- Use of the wrong rate of loading
- Poor selection of surcharge weight. This should be chosen to represent anticipated field loads.

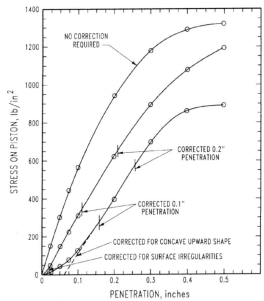

Figure 4.2 Stress-penetration curve for CBR test (*from ASTM D 1883*). [Courtesy ASTM. Reprinted with permission.]

- Wrong size of loading piston
- Deviation from standard size of compacted specimen used for testing
- Improper curve correction
- Disturbance of the specimen surface (after soaking) during removal of free water.
- Improper calibration of proving ring used for collecting load data
- Improper calibration of extensiometer used for determining rate of loading

4.2.9 Resilient modulus

The *resilient modulus* of a material is a specific type of modulus of elasticity that is based on recoverable strain instead of total strain. It is calculated as

$$M_r = \frac{\sigma_d}{\epsilon_r}$$

where M_r = resilient modulus, psi
σ_d = maximum repeated deviator stress, psi
ϵ_r = elastic (recoverable or resilient) strain, in/in

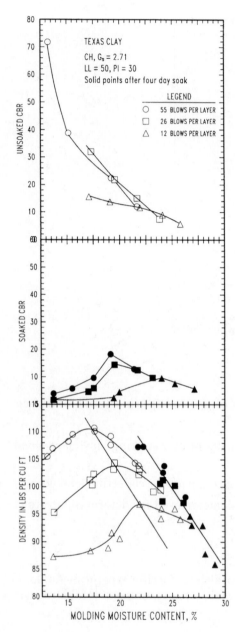

Figure 4.3 Typical CBR test results as needed for design.

TABLE 4.4   Typical CBR Values for Different Materials

Description of material	CBR, %
Wet clay soil where thumb penetration into the soil is	
Easy	<1
Possible	1
Difficult	2
Impossible	3+
A trace of a footprint is left by a walking man	1
Typical base and subbase materials	
Well-graded crushed aggregates	100
Dry-bound and water-bound macadam	100
Well-graded natural gravels	80
Limerock	80
Shell and sand mixtures	50–80
Coral	80
Shell rock (marine deposit of hard, cemented shells)	80
Gravelly sands (predominately sand)	20–50
Silty or clayey sands	10–40
Fine clean sands	10–20
Classification by Unified Soil Classification	
GW: gravel or sandy gravel	60–80
GP: gravel or sandy gravel	35–60
GM: silty gravel or silty, sandy gravel	40–80
GC: clayey gravel or sandy, clayey gravel	20–40
SW: sand or gravelly sand	20–50
SP: sand or gravelly sand	10–25
SM: silty sand	20–40
SC: clayey sand	10–20
CL: lean clays, sandy clays, gravelly clays	5–15
ML: silts, sandy silts, diatomaceous soils	5–15
OL: organic silts, lean organic clays	4–8
CH: fat clays	3–5
MH: plastic silts, micaceous clays, or diatomaceous soils	4–8
OH: fat organic clays	3–5
Pt: peat and highly organic soils	<1

The resilient modulus is determined from a repeated-load triaxial test that is intended to measure the elastic response of a flexible pavement structure to a moving wheel load. The test is applicable to surface (asphalt concrete), base, subbase, and subgrade materials; stabilized materials also may be tested. It is important that the loading waveform used simulate as nearly as possible the vehicular loading (in both shape and duration) that will occur in the field. The haversine wave is most often used with a 0.1-s duration and is repeated 15 to 60 times a minute. The duration of loading affects the resilient modulus less in granular materials, more in fine-grained soils, and most significantly in asphalt concrete.

The equipment used to determine the resilient modulus is basically a triaxial cell capable of applying a repeated load and maintaining a constant temperature during testing. The specimen tested is 4 in in diameter and 8 in high. Linear variable differential transformers (LVDTs) are typically clamped to the specimen at its upper and lower quarter points to measure deformations.

The test is performed by compacting a specimen to required conditions, placing it in the triaxial cell, "conditioning" the specimen, and then applying the repeated loads. Conditioning of the specimen is necessary "to eliminate initial loading effects" (Departments of the Army and Air Force 1989), i.e., to cause the specimen to undergo most of the plastic deformation that will occur under the maximum anticipated loads. It is accomplished by applying a specified combination of confining pressures and deviator stresses to result in roughly 1000 load applications. The conditioning process is somewhat different for asphalt concrete, granular materials, and fine-grained materials, with less conditioning required for the fine-grained materials. After conditioning is complete, resilient modulus values can be calculated after 150 to 200 load repetitions at each stress state.

Specific test procedures for the resilient modulus are under revision at present. AASHTO had a standard test procedure (method T-274) that was withdrawn in 1990, and a new procedure is under consideration. Another procedure is under development for the Strategic Highway Research Program (SHRP). ASTM has had a resilient modulus test procedure under study for several years, but since interest in this test has waned, it appears doubtful that a standard will be adopted. There are two main reasons that the resilient modulus is losing favor: (1) the test has some inherent problems and often gives inconsistent results, and (2) the test has limited usefulness, since it cannot be used for construction control and is useful for only a limited number of accepted design procedures.

Potential errors in determining the resilient modulus of materials include

- Measurement of deformations outside the chamber, which may include deformation of the load piston and/or end platens
- Improper sample conditioning
- Method of laboratory compaction used in specimen preparation not simulating field soil structure

4.2.10 Permeability

Permeability is a measure of the rate at which water flows through a soil mass. The coefficient of permeability k (often called *permeability*) is defined as the "rate of discharge of water at a temperature of 20°C under conditions of laminar flow through a unit cross-sectional area of a soil medium under a unit hydraulic gradient" (Department of the Army 1986). It is this value of k that is determined in laboratory permeability tests. Throughout this book we will use the term *permeability* in this, the common soil mechanics sense. The term *hydraulic conductivity* is coming into more common engineering usage to avoid confusion with the coefficient of permeability k with units of length squared. This latter coefficient is used in some fields of science and depends on the pore structure of the material but is independent of the fluid passing through the material.

Two basic types of permeability tests may be run in the laboratory: the constant-head test and the falling-head test. The constant-head test is used for coarse-grained materials (clean sands and gravels). In this test, the head of water maintained on the specimen remains constant throughout the test (see Fig. 4.4). This permeability test is the simplest to run, and the measurement of required quantities (quantity of water flowing through the soil Q, length of specimen L, head of water h, and the elapsed time t) is most straightforward. Because relatively little water will flow through a fine-grained soil and Q would be very small and difficult to measure accurately, this test is not used for materials with low coefficients of permeability (less than about 10×10^{-4} cm/s).

The falling-head test is normally run on finer-grained soils (fine sands to fat clays). It should be noted that fine sands are *not* free draining, as is often assumed; they should be treated as less permeable materials, similar to silts. In this test, the head of water is not constant but falls as water flows through the soil mass; the upper surface of the water must remain above the top of the specimen (see Fig. 4.5). Readings of the remaining head are taken throughout the test and are used to calculate the quantity of water flowing through the specimen. The head of water is usually maintained in a small-diameter burrette or standpipe so that changes in head can be determined easily.

The coefficient of permeability can be determined in the laboratory on either laboratory-prepared specimens or on undisturbed samples from the field. In either case, the specimens must be prepared and/or handled very carefully. Loss of material resulting in a rough side to the specimen will allow a path for water flow along the side of the specimen instead of through it. This will cause the calculated coefficient of permeability to be too high. Any densification of the specimen before permeability testing will cause the permeability value to be too low. The density may accidentally be increased by any of a number of actions:

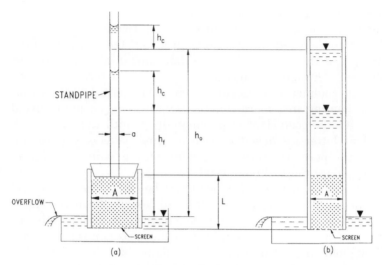

using setup shown in (a), the coefficient of permeability is determined as follows:

$$k = \frac{La}{At} \ln \frac{h_o}{h_f} = 2.303 \frac{La}{At} \log_{10} \frac{h_o}{h_f}$$

using setup shown in (b), the coefficient of permeability is determined as follows:

$$k = \frac{L}{t} \ln \frac{h_o}{h_f} = 2.303 \frac{L}{t} \log_{10} \frac{h_o}{h_f}$$

WHERE: h_c = height of capillary rise

a = inside area of standpipe

A = cross-sectional area of specimen

L = length of specimen

h_o = height of water in standpipe above discharge level minus h_c at time, t_o

h_f = height of water in standpipe above discharge level minus h_c at time, t_f

Figure 4.4 Constant-head permeability test device.

- Improper field coring—use of sampling tubes with a dull cutting edge, driving the sampling tube instead of smoothly pushing it into the soil

- Improper sample extraction from coring tube

- Use of too great a head of water during testing

Permeability testing can be accomplished in any of several types of equipment. It also can be accomplished in conjunction with other test-

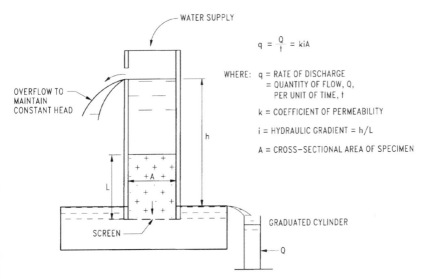

WATER SUPPLY

$$q = \frac{Q}{t} = kiA$$

WHERE: q = RATE OF DISCHARGE
= QUANTITY OF FLOW, Q,
PER UNIT OF TIME, t

OVERFLOW TO
MAINTAIN
CONSTANT HEAD

k = COEFFICIENT OF PERMEABILITY

i = HYDRAULIC GRADIENT = h/L

A = CROSS−SECTIONAL AREA OF SPECIMEN

h

L

GRADUATED CYLINDER

SCREEN

Q

Figure 4.5 Falling-head permeability test device.

ing. The general types of permeameters, or containers, used for this testing are permeameter cylinders, sampling tubes, (triaxial) pressure chambers, and consolidometers. Selection of the type of permeameter will depend on material type (fine- or coarse-grained), state of saturation (saturated or unsaturated), and sample type (undisturbed, remolded, or compacted). Remolded cohesionless materials are commonly tested in permeameter cylinders. Undisturbed cohesionless soils can be tested in the sampling tube, but this will yield only a vertical permeability; horizontal permeability of these soils can be approximated by the permeability of remolded cohesionless soils. Fine-grained soils, either undisturbed, remolded, or compacted, can be tested in either the pressure chamber or the consolidometer; these samples can be oriented to obtain either vertical or horizontal permeabilities. Typical values of coefficient of permeability for various void ratios, materials, and soil deposits are given in Fig. 4.6 and Tables 4.5 and 4.6.

Possible errors causing inaccurate determinations of coefficient of permeability are

- Stratification or nonuniform compaction of specimen

- Incomplete saturation of specimen at beginning of test

- Excess hydraulic gradient. The gradient must remain small enough to maintain laminar (not turbulent) flow within the specimen. If the gradient is too great, it also may cause seepage consoli-

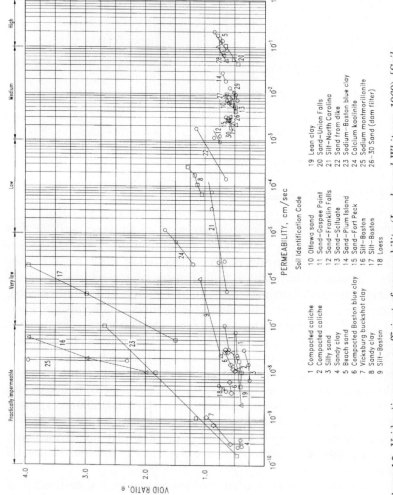

Figure 4.6 Void ratio versus coefficient of permeability (*Lambe and Whitman 1969*). [*Soil Mechanics*, W. T. Lambe and R. V. Whitman, Copyright © 1969. Reprinted by permission of John Wiley and Sons.]

Soil Identification Code

1 Compacted caliche	10 Ottawa sand	19 Lean clay
2 Compacted caliche	11 Sand–Gaspee Point	20 Sand–Union Falls
3 Silty sand	12 Sand–Franklin Falls	21 Silt–North Carolina
4 Sandy clay	13 Sand–Scituate	22 Sand from dike
5 Beach sand	14 Sand–Plum Island	23 Sodium–Boston blue clay
6 Compacted Boston blue clay	15 Sand–Fort Peck	24 Calcium kaolinite
7 Vicksburg buckshot clay	16 Silt–Boston	25 Sodium montmorillonite
8 Sandy clay	17 Silt–Boston	26–30 Sand (dam filter)
9 Silt–Boston	18 Loess	

TABLE 4.5   Typical Permeability Values for Construction Materials

Material	Permeability, cm/s
Portland cement concrete	$<10^{-8}$
Asphalt concrete	4×10^{-3} to 4×10^{-8}
Compacted clays	$<10^{-7}$
Compacted silts	7×10^{-6} to 7×10^{-8}
Silty and clayey sands	10^{-5} to 10^{-7}
Concrete sand with fines	7×10^{-4} to 7×10^{-6}
Clean concrete sands	7×10^{-2} to 7×10^{-4}
Well-graded aggregate without fines	10^{-1} to 10^{-3}
Uniformly graded coarse aggregate	10^{2} to 10^{-1}

SOURCE: Adapted from Carter and Bentley, 1991.

dation of the specimen or sand boils/piping; in either case the specimen is destroyed and permeability test results are invalid.

- Dissolved air in water flowing into the specimen

- Leakage of water between sides of specimen and container. This potential problem is minimized by use of the triaxial compression chamber for permeability tests.

- Densification of the specimen before permeability is determined

TABLE 4.6   Coefficient of Permeability of Common Natural Soil Formations

Formation	Value of k, cm/s
River deposits	
Rhone at Genissiat	Up to 0.40
Small streams, eastern Alps	0.02–0.16
Missouri	0.02–0.20
Mississippi	0.02–0.12
Glacial deposits	
Outwash plains	0.05–2.00
Esker, Westfield, Mass.	0.01–0.13
Delta, Chicopee, Mass.	0.0001–0.015
Till	<0.0001
Wind deposits	
Dune sand	0.1–0.3
Loess	0.001 ±
Loess loam	0.0001 ±
Lacustrine and marine offshore deposits	
Very fine uniform sand, C_u = 5 to 2	0.0001–0.0064
Bull's liver, 6th Ave., N.Y., C_u = 5 to 2	0.0001–0.0050
Bull's liver, Brooklyn, C_u = 5	0.00001–0.0001
Clay	<0.0000001

SOURCE: From *Soil Mechanics in Engineering Practice*, K. Terzaghi and R. Peck, Copyright © 1965. Reprinted by permission of John Wiley and Sons.

4.2.11   Shear strength

While the general concept of strength is simple, definition and determination of material strength, particularly soil strength, are not. Because of the particulate nature of soils, deformations in this material occur mainly through slippage (or shear) between particles; therefore, shearing resistance in a soil is synonymous with soil strength. *Shear strength* is a fundamental soil property that is affected by numerous variables and is difficult to determine rigorously. It depends on the type of material (mineralogic composition, grain size, and grain shape), moisture conditions (saturated, unsaturated, drained, and/or undrained), past stress history, state of stress at failure, and test method.

Shear strength is important because it largely controls bearing capacity of soil masses, slope stability, pile foundation design, resistance to rutting under moving wheel loads, and almost any other assessment of the "steadfastness," or stability, of soil under any type of loading.

The shear strength of a soil is composed of two basic components: cohesion (or adhesion between particles) and friction (a resistance to movement caused by particle-to-particle contact). These parameters are usually determined through laboratory testing. Soil samples are tested at various stress states, and test results are analyzed using Mohr's circles of stress to define the strength envelope, shown as a solid line in Fig. 4.7a. This line is known alternatively as the *strength envelope, failure envelope, strength line,* or *rupture line.* It represents a soil's strength at failure (under the specific conditions of test) as a function of normal stress. The strength envelope is generally a curved line and may have an intercept c on the vertical axis. It is often approximated in the region of interest by a straight line (Coulomb's equation) as

$$\tau = c + \sigma_n \tan \phi$$

where τ = shear strength
    c = cohesion
    σ_n = normal (perpendicular) stress on the failure plane
    $\tan \phi$ = coefficient of internal friction

The graphic representation of this equation is shown as the dashed line in Fig. 4.7a. The figure also illustrates several shapes that the strength envelope may take. Shear strength of soils is treated at length in all standard soil mechanics textbooks.

As shown in Fig. 4.7, three individual test results (for tests run at different normal stresses) are typically plotted together, a line is drawn tangent to all "Mohr's failure circles," and the values of c and ϕ are determined graphically. If pore pressure measurements are made

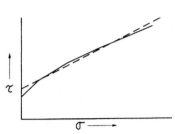

a. Straight Line Approximation of Shear Strength

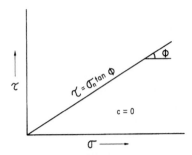

b. Ideal Frictional Soil (Sand)

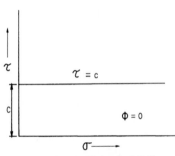

c. Purely "Cohesive" Soil (Saturated Clay, Undrained Loading)

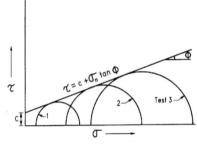

d. Typical Soil Strength Envelope with "cohesive" and "frictional" components (Partially Saturated Clay)

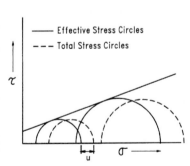

e. Effective Stress = Total stress minus pore pressure $(\sigma' = \sigma_T - u)$

Figure 4.7 Typical shear-strength relationships for various testing conditions and soils.

during the test, then either total stress or effective stress (total stress minus pore pressure) circles (see Fig. 4.7e) can be plotted and used in the subsequent engineering analyses. In many cases, it is critical to perform an effective stress analysis to truly predict what will happen in the field; in this case, use of the triaxial consolidated, undrained test with pore pressure measurement is strongly recommended.

The shear strength of soils is usually determined in the laboratory by one of the following tests: direct shear test, triaxial compression test, or unconfined compression test. These are all destructive tests in that the sample is caused to fail during testing. The particular variation of the test method used will make the test simulate different drainage and consolidation conditions in the field; Table 4.7 presents the combinations that can be used in the three shear tests. The test method used for a particular project should be selected to most closely simulate anticipated field conditions. The following discussion will briefly describe each test method and the situations in which it should or should not be used.

Direct shear. The direct shear test is the oldest test used to determine the shear strength of soils. It is used to measure the shear strength of soil under drained conditions. Because a thin sample is used, the drainage path for water to exit the specimen is short, allowing excess pore water pressures to be dissipated relatively quickly. This test is very good for simulating slow failures (without pore pressure buildup) that occur after field consolidation is finished; e.g., failure of the soil occurs slowly 20 years after construction of a building was completed on the site. It is also an excellent test for assessing shear strength along known weak planes within a soil mass; in this case, the weak

TABLE 4.7   Types of Laboratory Shear Tests

Test name by test conditions	Other names	Abbreviated name	Lab method used to perform test
Unconsolidated, undrained	Quick test	UU, Q	Triaxial, unconfined
Consolidated, undrained	Consolidated quick test	CU, R, Q_c	Triaxial
Consolidated, undrained with pore pressure measurement	"R bar" consolidated quick w/pore pressure measurement	CU′, R', Q_c'	Triaxial
Consolidated, drained	Slow	S, Q_s	Triaxial, direct shear

plane of the soil is placed in the test device with the proper orientation to force failure to occur along that plane.

The direct shear test is run by placing a thin, square specimen in the test device that is divided horizontally into two pieces and then applying a predetermined normal stress while allowing the specimen to drain and consolidate. The test device is then unlocked, and a shear force is applied to the specimen, causing it to fail horizontally (creating two thin wafers of soil). Three tests are normally performed at different normal stresses to define the relationship between shear strength and normal stress; the normal stresses used in testing should be representative of the anticipated field stresses. This test can be run on any soil type; the specimen may be undisturbed or remolded.

Limitations of the direct shear test that restrict its usefulness include the following: (1) prefailure stress conditions are not known accurately enough to allow analysis of the soil's stress-strain characteristics, (2) water flow into and out of the specimen cannot be controlled or measured so the water content varies during the test, (3) pore pressures cannot be measured during the test, and (4) the distribution of shear stress-strain across the failure plane is not uniform. All these items make it difficult or impossible to accurately analyze the test results in sufficient detail for many projects.

Triaxial. The triaxial test is the most versatile test used to determine shear strength of soils. The behavior of soil, both during laboratory testing and under field loading, is tremendously influenced by the drainage/pore water conditions that exist. For this reason, several versions of the triaxial test were developed to obtain the material strength under a variety of (controlled) drainage conditions, as shown in Table 4.7. When selecting the type of triaxial test to run, the field conditions to be simulated must be assessed. For example, if the strength of a simple earthen slope exposed to rapid reservoir drawdown conditions is to be evaluated, the Q test might be used. However, if the analysis is for the soil under a large visitor's center located just downstream but still in the drawdown area, the R' test would be preferable. Further, if the reservoir water level is not rapidly drawn down, but instead a multiyear drought causes a slow drop in water level of 15 ft, an S test might be the appropriate test choice.

In the triaxial test, a cylindrical specimen enclosed in a rubber membrane is subjected to an all-around confining pressure and possibly a backpressure to force complete saturation of the specimen and is then loaded axially to failure. The drainage conditions in the specimen are controlled and/or measured throughout the test. Three specimens are typically tested to establish the relationship between shear strength and normal stress, as shown in Fig. 4.7.

The triaxial test is an excellent tool for use in assessing the shear strength of soils for numerous field conditions. Since a specimen can be set up once in the triaxial test device, saturated, consolidated, permeability determined, and sheared (either with or without pore pressure measurements), this test can provide a multitude of data with minimal specimen handling. The test conditions can be controlled very precisely, and measurements can be made very accurately. The triaxial test is commonly performed by most competent soils testing laboratories and should be run to obtain accurate shear strength(s) of underlying soil(s) before a project is designed.

Unconfined compression. The unconfined compression test is used to quickly get "approximate quantitative values of the compressive strength of soil possessing sufficient coherence to permit testing in the unconfined state" (ASTM D 2166). Both undisturbed and remolded samples can be tested.

In this test, a cylindrical specimen is placed vertically in the test device, and a gradually increasing axial (vertical) load is applied while no lateral support is provided. (Since there is no confining pressure used in this test, there is no need to run tests on multiple specimens to relate shear strength to confining stress as is done in other forms of shear testing.) Unconfined compressive strength q_u is defined as the load per unit area at which the specimen will fail; this is taken as either the maximum load per unit area or the load at a specified percentage of axial strain (usually 20 percent). The undrained shear strength s_u is assumed to be equal to one-half the unconfined compressive strength.

The unconfined compression test is a form of triaxial test in which the confining pressure is zero. This is a rapidly run, inexpensive test that can give a feel for material strength. There is also a vast amount of existing data that have been correlated with field strength and can be used for comparison with current test results. The soil tested must able to stand freely without slumping, cracking, crumbling, etc. This test is not appropriate for use with dry or crumbly soils, fissured or varved soils, sands, or silts. Since unconfined compression testing does not simulate the stress conditions, or stress state, that the soil would feel in the ground, this test is not a substitute for the triaxial Q test.

Possible errors that can cause inaccurate shear strength determinations include the following:

Direct shear test

- Loss of moisture through evaporation during specimen preparation
- Top and bottom of specimen not flat and parallel
- Gap between top and bottom of shear box too large or too small

- Inaccurate measurement of small shear stresses
- Porous stone not permeable enough
- Galvanic currents in shear box during long-duration tests
- Stopping test too soon—data must be plotted during test to establish when test can be stopped

Triaxial test

Test device

- Leakage of chamber fluid into specimen
- Leakage of pore water out of specimen
- Permeability of porous stones too low
- Membrane too thick, i.e., too strong, for use on particular soil, thus making soil appear stronger
- Friction on loading piston making load at failure appear larger than it is

Soil specimen

- Specimen disturbed during trimming for test
- Specimen disturbed while putting on membrane
- Dimensions not accurately measured

Q test

- Changes in specimen dimensions during application of chamber pressure
- Rate of strain too fast
- Water content determination after test not representative

R test

- Application of backpressure done in increments that are too large
- Backpressure applied too rapidly
- Chamber pressure and backpressure not precisely maintained during consolidation
- Specimen not completely consolidated before shearing
- Rate of strain too fast
- Excessive temperature variations during shear testing
- Specimen absorbed water from porous stones after test

S test

- Rate of strain or rate of loading too fast

- Volume change measurements during shear not accurate

Unconfined compression test
- Test run on inappropriate soil
- Specimen disturbed during trimming
- Loss of moisture through evaporation during specimen preparation
- Rate of strain or rate of loading too fast

4.2.12 Consolidation

Consolidation is the gradual squeezing together of soil particles under an applied static load. During this process, applied loads are gradually transferred from the pore water to the soil solids as pore water is squeezed out of the voids. As a soil consolidates, any overlying structures will move downward, or "settle," with the surface of the soil. Therefore, one of the main objectives for determining the consolidation properties of soil is to predict the ultimate magnitude of settlement as well as the rate of settlement to be expected under a structure. This test is also used to determine the amount of any differential settlement (or variation in amount of settlement) that may be expected on a construction site; this normally occurs because of differences in material type or conditions. The magnitude of settlement, or consolidation, depends on the magnitude of the applied load and the compressibility of the soil mass; the higher the water content of a soil, the greater is the compressibility. The rate of consolidation is a function of the permeability; the higher the permeability (the faster the water can get out), the quicker consolidation will occur. The consolidation test is a small-scale model test, not a destructive test, as are the shear tests.

The one-dimensional consolidation test is typically used to measure the amount and rate of consolidation that will occur under given load(s). In this test a small soil sample is placed in a lateral confining ring, and a vertical load is placed on the specimen. Porous stones on top and beneath the specimen allow two-directional vertical drainage for relief of excess pore pressures caused by the load(s). The maximum loads used in this test should be large enough to simulate expected field loads and to establish the straight-line portion of the void ratio–effective stress curve. Typically each incremental load used in the test doubles the previous load; the following loading sequence is often used: 0.25, 0.5, 1.0, 2.0, 4.0, 8.0, and 16.0 ton/ft^2. Smaller increments are sometimes used to more accurately define the void ratio–effective stress curve at stresses near the preconsolidation stress. The times after loading at which deformation readings normally should be recorded are 0.1, 0.2, 0.5, 1, 2, 4, 8, 15, and 30 minutes and 1, 2, 4, 8, and 24 hours.

The laboratory technician conducting this test often will request instructions regarding the loading sequence, since this is frequently modified by the responsible engineer to better reflect field loading. For instance, when testing very soft soils such as dredged material or river sediments, a much lighter set of loads must be used; the Corps of Engineers typically recommends 0.005, 0.01, 0.025, 0.05, 0.10, 0.25, 0.5, and 1.0 ton/ft^2 for these soft soils (Department of the Army 1987b).

Consolidation test results are commonly presented in the form of a void ratio–pressure (or effective stress) plot, as shown in Fig. 4.8; these plots are also referred to as "$e-\log p'$" curves. Values are typically shown on most $e-\log p'$ curves include existing overburden pressure p and preconsolidation pressure p_p. These values are usually determined by the design engineers (not testing laboratories), and their values, relative to the load to be applied, can significantly affect the computed total settlement (Department of the Army 1986, ASTM 1994). Other plots that also may be developed from consolidation test data are time versus percent consolidation and void ratio versus permeability.

Many soft materials are too fluidlike to be tested easily in the standard consolidometer, and this test cannot be conducted on materials at high void ratios (at void ratios greater than about 5 or 6). Therefore, much recent work has been conducted by mining industries (e.g., phosphate mining), dredging organizations (e.g., U.S. Army Corps of Engineers and Dutch and Japanese entities), and many researchers to develop appropriate consolidation tests for slurries. Results from these slurry consolidation tests are often used in finite (large) strain consolidation analyses; this type of analysis is much more appropriate for use with extremely soft soils than is the normal (small strain) consolidation analysis (Schiffman et al. 1984, Martinez et al. 1988, Townsend and McVay 1990, Rollings 1994). At present, no test procedures have been widely adopted or published by ASTM, but this should happen in the future.

Possible errors in determining the consolidation characteristics of soils include

- Disturbance of the specimen during trimming
- Specimen not snugly fitting into and filling consolidometer ring
- Galvanic currents in consolidometer
- Permeability of porous stones too low
- Friction between specimen and ring
- Inappropriate load-increment factor
- Improper specimen height

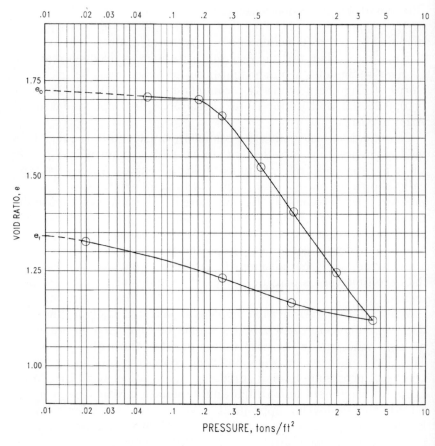

PRESSURE, tons/ft^2

PROJECT: CORPUS CHRISTI SHIP CHANNEL
DREDGED MATERIAL CONTAINMENT AREA FOUNDATION SOIL

SAMPLE ID: Boring #84 – 82, Sample #5   SAMPLE PROPERTIES:
              Depth: 8.0 – 10.0 ft   Classification: Plastic Clay (CH), Gray
              Type Specimen: Undisturbed   LL = 81, PL = 23, PI = 58
              2.65"ɸ; 1.170" high G_s = 2.65

TEST DATA:	BEFORE TEST	AFTER TEST
Water Content, %	63.9	50.6
Dry Density, pcf	60.9	70.7
Saturation, %	98.8	100+
Void Ratio	1.714	1.339
Back Pressure	–	–

Figure 4.8 Void ratio–effective stress curve from consolidometer test.

4.3   Laboratory Tests for Aggregates

Testing of aggregates may be performed on in situ materials in an attempt to identify acceptable quarries or aggregate sources prior to development of the aggregate source, or it may be done on bulk (stockpiled) materials that have already been removed from their original source. In either case, correct sampling techniques must be used. Appropriate sampling of aggregates is essential for obtaining laboratory testing results that accurately reflect the properties of the aggregates of interest. Without proper sampling to ensure testing of representative materials, even the most meticulous and thorough testing program cannot provide data to ensure successful design and construction.

4.3.1   Abrasion resistance

The most widely used test to specify abrasion resistance and mechanical soundness is the *Los Angeles abrasion test* (ASTM C 131 and C 535). In this test, a sample of aggregate is placed in a hollow revolving drum with a charge of steel balls. As the drum rotates, a shelf inside the drum lifts and then drops the aggregate and steel balls on each revolution. The Los Angeles resistance is expressed as the percentage loss finer than the No. 12 sieve after 500 revolutions of the drum.

This test gives an indication of the impact toughness of the aggregate. In a general sense, it identifies aggregates that may be susceptible to abrasion under traffic or physical degradation during excavation, handling, and placement. Generally, a Los Angeles abrasion test result of 40 percent or less is considered satisfactory for aggregates, although sometimes a more restrictive value of 30 percent is used for aggregates in surfacings for critical facilities such as airports. Aggregates within the pavement structure such as in base courses are sometimes allowed to have Los Angeles abrasion values as high as 50 percent. Even though aggregates within the pavement structure may not be exposed directly to the abrasion of traffic, they may degrade on handling, placement, and compaction with a resulting increase in fines and reduction in strength if the Los Angeles abrasion results are too high.

The Los Angeles abrasion test exposes the aggregate to considerable impact, and brittle aggregates may degrade more in this test than they would under actual construction and traffic conditions. For example, blast furnace slag and some granites often perform poorly in this test but perform well in practice. Also, lateritic gravels perform better than indicated by this test (Morin and Todor 1975).

Consequently, results of this test should be correlated with local experience. An aggregate meeting this test is probably sound and resistant to physical degradation and abrasion. An aggregate failing this test may still be suitable, and further evaluation of its past history of use, hardness, and petrographic characteristics may be justified. Further information may be found in Meininger (1978) and Barksdale (1991).

4.3.2   Freezing and thawing

If the pores of an aggregate are filled with water such that there is insufficient space to allow the expansion of the water upon freezing, the aggregate is considered *critically saturated*. If frozen in this condition, degradation of the aggregate may occur. The vulnerability of aggregates to freezing is a function of the saturation, aggregate size, pore structure, permeability, and tensile strength of the aggregate (American Concrete Institute 1989, Barksdale 1991, Powers 1975). Certain rock sources such as shale, porous cherts, laminated limestones, and some sandstones are associated with freezing and thawing durability problems and should be evaluated carefully before use.

Some agencies expose an aggregate to repeated freezing and thawing (AASHTO T103) to evaluate durability. Results are variable, and specific test procedures between agencies may vary. The *sulfate soundness test* (ASTM C 88) is also widely used as a durability test. This test subjects aggregate samples to repeated wetting and drying with either a sodium or magnesium sulfate solution. Conceptually, the growth of the sulfate crystal in the aggregate pores due to repeated wetting and drying is analogous to the growth of an ice crystal in the aggregate pores due to freezing and thawing. However, the actual performance of the aggregate in the field is complex, and the sulfate soundness test cannot clearly discriminate between aggregates that are susceptible to freezing and thawing and those which are not (Dolar-Mantuani 1978). Although the direct freezing and thawing test of aggregates and the sulfate soundness test cannot directly predict field durability, they may be useful for evaluating relative aggregate performance, at least on a regional basis. An aggregate should not be rejected for use on the basis of these tests alone if it has a proven local record of performance under conditions similar to its anticipated use.

4.3.3   Wetting and drying

Most nonargillaceous aggregates do not deteriorate when exposed to alternate wetting and drying. However, aggregates with appreciable clay content such as shales may be very vulnerable. Shales (as discussed in Chap. 3) are widely distributed and vary from soft clay shales to tough calcareous or silica-cemented shales. Many unsatisfactory experiences with shales in construction have led to their exclusion or very conservative use in many specifications. However, some shales are satisfactory for construction, and some are not. The Federal Highway Administration and a number state departments of transportation have conducted extensive investigations of the suitability of shale for use in embankments, but work evaluating shale as a subbase, base course, or aggregate source for pavement surfacing is less extensive.

Evaluation of a shale's durability should follow three steps: a review of its past performance and its field and weathering characteristics, initial slaking tests, and physical degradation tests. In the United States, many shale formations that perform poorly in construction have already been identified, so problems with these specific formations can be anticipated. The weathering characteristics of exposed shale outcrops provide an indication of the material's durability. Shales with laminae thickness greater than $\frac{1}{8}$ in also have been associated with poor durability (Reidenour et al. 1976). This type of initial survey will provide an estimate of the shale's durability characteristics.

Next, shale samples should be subjected to simple slaking tests. Typically, for this test, shale fragments are placed in a liquid for 2 to 40 hours, and their deterioration to soaking is assessed. Water is most commonly used as the liquid, although some investigators have used ethylene glycol (Reidenour et al. 1976) to react with montmorillonite clays in the shale and sulfuric acid (Noble 1977) to identify chlorite in the shale. Table 4.8 shows the rating system developed for the Federal Highway Administration for the jar-slake test, where oven-dried shale fragments are soaked in water for 2 to 24 hours. An I_J value of 1 or 2 is obviously associated with a soft, nondurable shale. The change in the shale liquidity index also can be used as a guide to the rate of slaking, as indicated in Table 4.9. The shale liquidity index is defined as

$$I_{SL} = \frac{w_f - w_i}{PI}$$

where I_{SL} = shale liquidity index
w_f = final shale moisture content after soaking
w_i = in situ shale moisture content
PI = Atterberg plasticity index

Several physical degradation tests have been developed that subject shale fragments to some mechanical action in the presence of

TABLE 4.8   Jar-Slake Index

I_j	Descriptive behavior
1	Degrades into a pile of flakes or mud
2	Breaks rapidly and/or forms many chips
3	Breaks rapidly and/or forms few chips
4	Breaks slowly and/or forms several fractures
5	Breaks slowly and/or develops few fractures
6	No change

SOURCE: From Strohm et al., 1978.

TABLE 4.9   Rate of Slaking Test

I_{SL}	Rate of slaking
<0.75	Slow
0.75–1.25	Fast
>1.25	Very fast

SOURCE: From Strohm et al., 1978.

water to further test their durability. In the slake-durability test, 10 oven-dried, unweathered, $^3/_4$- to 1-in shale samples are placed in a submerged wire screen drum that is rotated at 20 rev/min for 10 minutes. The material is oven dried, and the test is repeated. The slake-durability index I_d is then calculated as

$$I_d = \frac{\text{dry weight retained after two cycles}}{\text{dry weight before test}}$$

This index, together with the physical condition of the sample, can be used to classify the shale, as shown in Table 4.10.

Shales show considerable variation, so the guidelines presented here may have to be modified based on local experience. If the shale is soft and nondurable, or if it is hard and nondurable, it will weather into a residual soil. Consequently, it must be handled and compacted in thin lifts like a soil. This may require extra blasting or processing to ensure that the shale is broken into small enough particles to be placed and compacted in thin lifts. Figure 4.9 gives some tentative guidelines on the relation between lift thickness, slake-durability index, and construction problems. Smith (1986) evaluated several proposed methods for selecting a design CBR strength value for shale fills but found that they gave erratic results. He concluded that the residual clays derived from weathering of these shales in the field gave the best estimate for the design CBR value. Most work with

TABLE 4.10   Slake-Durability Test

I_d	Type of retained material*	Shale classification
<60%	T2, T3	Soil-like, nondurable
60–90%	T1S, T3	Soil-like, nondurable
	T1H, T2	Intermediate, nondurable
>90%	T1S, T3	Soil-like, nondurable
	T1H, T2	Rocklike, durable

*Type T1: no significant breakdown of original pieces; type T1S: soft, can be broken apart or remolded with fingers; type T1H: hard, cannot be broken apart; type T2: retained particles consist of large and small hard particles; type T3: retained particles are all small fragments.
SOURCE: From Strohm et al., 1978.

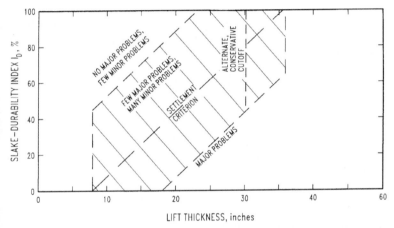

Figure 4.9 Guidance on use of compacted shale in embankments as a function of slake index and placement lift thickness (*Strohm et al. 1978*).

shales has concentrated on their use in embankments, and guidance on design, drainage, construction, causes of distress, and remedial repair measures have been published by the Federal Highway Administration (Strohm et al. 1978). There is less information available on the potential use of hard, durable shales in subbase course, base course, and surfacing construction. The major problem with all laboratory tests and ratings of shales is their very limited to nonexistent correlation with long-term field performance.

Another material that is susceptible to damage from wetting is slag from steel production. Blast furnace slags from iron production are widely and successfully used for aggregate in construction; however, steel slags contain magnesium oxides that hydrate slowly over time with a resulting increase in volume. There have been a number of instances where these materials have been used as fill or base material and later expanded with complete disruption of the surfacing (Gnaedinger 1987, Crawford 1969).

Slaking tests such as just described for shales also have been found to be useful for evaluating lateritic gravels (Morin and Todor 1975).

4.3.4 Specific gravity and absorption

Specific gravity is the ratio of the weight of a material of a given volume to the weight of a similar volume of water. It is a dimensionless number, and most natural aggregates will have a value of 2.6 to 2.7, although values of 2.4 to 3.0 may be encountered. Artificial aggregates will have a wide variety of specific gravities.

Figure 4.10 illustrates different moisture conditions for aggregates that are commonly used for determining specific gravity. Four different specific gravities can be defined depending on the weight and volume of the aggregate used in the calculation. These are

1. *Absolute specific gravity:* Volume of pores within the aggregate excluded from the volume used in the calculation.

2. *Apparent specific gravity:* Oven-dry aggregate weight used, permeable surface pores excluded from volume used in calculation.

3. *Saturated-surface dry bulk specific gravity:* Weight of aggregate plus weight of water in the permeable surface pores used, permeable surface pores included in volume.

4. *Bulk specific gravity:* Oven-dry aggregate weight used, permeable surface pores included in volume.

The specific gravity values above are in decreasing order of magnitude (absolute specific gravity is normally higher than apparent specific gravity, etc.). The absolute specific gravity is difficult to determine and is seldom used in construction.

Absorption is the amount of water that enters the permeable aggregate pores after 24 hours of soaking. It is expressed as a percentage of the aggregate dry weight. The 24-hour soaking may not be adequate to fill all permeable pores for some aggregates.

Calculations for specific gravity and absorption can be made as follows:

$$G_{sa} = \frac{A}{C - D + A}$$

or, alternatively,

Solid aggregate

Pores that are permeable to water

Pores that are not permeable to water

ABSOLUTE SPECIFIC GRAVITY
only soild portion used

APPARENT SPECIFIC GRAVITY
permeable surface pores not included

BULK SPECIFIC GRAVITY
volume of all pores included

Figure 4.10 Schematic diagram of various moisture conditions in aggregates.

$$G_{sa} = \frac{A}{A - E}$$

$$G_s = \frac{A}{C + B - D}$$

or, alternatively,

$$G_s = \frac{A}{B - E}$$

$$G_{s(ssd)} = \frac{B}{C + B - D}$$

or, alternatively,

$$G_{s(ssd)} = \frac{B}{B - E}$$

$$\text{Absorption, \%} = \frac{B - A}{A} \times 100$$

where G_{sa} = apparent specific gravity
$\quad\quad G_s$ = bulk specific gravity
$\quad G_{s(ssd)}$ = saturated surface dry specific gravity
$\quad\quad A$ = weight of oven-dry aggregate sample in air
$\quad\quad B$ = weight of saturated-surface dry aggregate sample in air
$\quad\quad C$ = weight of water of a given volume
$\quad\quad D$ = weight of water and aggregate of same volume as before
$\quad\quad E$ = weight of saturated aggregate in water

A pycnometer is used for the volume and weight measurements in the first form of the specific gravity formulas. This is the scheme used in ASTM C 127 for fine aggregates, and a detailed description of the procedures can be found there. The second form of the specific gravity equations uses the weight of the aggregate in air and in water and is the method of calculation used in ASTM C 128 for coarse aggregates. ASTM C 128 is not used for lightweight aggregates because of their generally high absorption potential. However, these aggregates are relatively seldom used in geotechnical construction work.

Often aggregates are divided into different size fractions, and the average specific gravity can be found as

$$G = \cfrac{1}{\cfrac{P_1}{100G_1} + \cfrac{P_2}{100G_2} + \cdots + \cfrac{P_n}{100G_n}}$$

where $\quad\quad\quad G$ = average specific gravity (Any consistent form of specific gravity can be used.)
$G_1, G_2, ..., G_n$ = specific gravity for each size fraction
$P_1, P_2, ..., P_n$ = each size fraction's percent by weight of the original sample

It also follows that if absorption is expressed in decimal form, then

$$G_{s(ssd)} = G_s(1 + \text{absorption})$$

Specific gravity is not an indication of quality in itself; however, it can provide an indication of specific potential problems. For example, many concrete specifications limit the chert content in the aggregate if the specific gravity is 2.4 or less, since this type of chert has had durability problems under freezing and thawing. Lightweight particles such as coal or lignite are often soft and will rise to the surface of asphalt or portland cement concrete if they are present. Consequently, some specifications will limit the content of particles with specific gravities less than 2.0.

Absorption also has been used as a crude indicator of a natural aggregate's durability to freezing and thawing. As absorption increases, the likelihood of encountering unsound aggregate increases (Wuerpel 1944). However, many durable aggregates have high absorption values, and low absorption does not guarantee that the aggregate is durable.

The perfect aggregate is seldom available, and engineers must build with the materials that are economically available. Therefore, specifications for aggregates should never be more restrictive than actually necessary. Many areas do not have abundant supplies of good-quality aggregates, and the use of substandard materials in construction under such conditions is an important topic (American Society for Testing and Materials 1976, Rollings 1988, Organization for Economic Co-operation and Development 1981, Metcalf 1988). Generally, use of substandard or marginal materials requires greater engineering effort in evaluation, design, mixture proportioning, and construction than use of standard materials, and project performance also usually suffers. Unless these extra engineering costs and potential loss in performance through shortened life or increased maintenance are fully recognized, any economic savings from using substandard materials may prove illusory.

Illustrative Example 4.2   An oven-dry sample of $\frac{1}{2}$-in limestone aggregate weighs 2096.7 g. In a saturated-surface dry condition it weighs 2112.5 g, and it weighs 1321.3 g in water. Determine the specific gravities and absorption.

solution

$$G_{sa} = \frac{2096.7}{2096.7 - 1321.3} = 2.70$$

$$G_{s} = \frac{2096.7}{2112.5 - 1321.3} = 2.65$$

$$G_{s(ssd)} = \frac{2112.5}{2112.5 - 1321.3} = 2.67$$

$$\text{Absorption} = \frac{2112.5 - 2096.7}{2096.7} \times 100 = 0.75\%$$

Illustrative Example 4.3   A sample of coarse aggregate is divided into three fractions, and saturated-surface dry bulk specific gravities and absorption are determined for each fraction as shown below. Determine the average specific gravity and absorption for the entire sample.

Size fraction	Percent of original sample	$G_{s(ssd)}$	Absorption, %
No. 4–¾ in	38	2.71	0.8
¼–½ in	32	2.64	1.1
½–¾ in	30	2.61	2.3

solution

$$\text{Average } G_{s(ssd)} = \cfrac{1}{\cfrac{0.38}{2.71} + \cfrac{0.32}{2.64} + \cfrac{0.30}{2.61}} = 2.66$$

$$\text{Average absorption} = 0.38 \times 0.8 + 0.32 \times 1.1 + 0.30 \times 2.3 = 1.3\%$$

4.4 Field Tests

Field testing is often required to determine whether construction with various geotechnical materials meets required specifications. Requirements for certain minimum (or maximum) moisture contents and/or dry densities are common on compacted soils, base courses, and subbase courses. Field testing also may be necessary to determine the condition of native materials prior to construction on these deposits. Strength testing may be needed to ensure that the placed materials will meet the future requirements put on them by either moving or stationary loads. Although there are numerous specialized tests that can and are conducted at field construction sites when warranted, the following tests are routinely conducted on many construction projects.

4.4.1 Moisture content

Field moisture contents may be needed in compaction control if a specified range of moisture contents is used, but more likely, moisture content will be needed to convert the wet densities measured in the field to the dry densities commonly used in field work. No matter what the reason for determining field moisture contents, several methods are available.

Oven. The most reliable method of determining an accurate moisture content is by oven drying a sample and then calculating the moisture content. This procedure is discussed in Sec. 4.2.1. For field work, the greatest disadvantage of this method is the time required to get final results. Because of this, several other methods of getting moisture content have been developed. Most other methods should be checked periodically by comparing their results with those obtained by oven drying a sample.

Nuclear. The nuclear moisture method provides a rapid, nondestructive method of getting in-place moisture and wet density data in soils, soil-aggregate mixes, and compacted aggregates. The nuclear gauge employed in this procedure consists of a single unit that houses an americium-beryllium source that provides fast neutrons for moisture measurement, slow-neutron detectors, and a readout device that operates the detectors and displays the results. This method is based on the principle of emitting fast neutrons that travel through the geotechnical material and slow in speed as they go, recording the rate at which "slow" neutrons arrive at the detector, and relating slow neutron activity to the concentration of moisture in the material.

The nuclear gauge must be "standardized" at the start of each day to offset the effect of aging of the nuclear gauge components. This is done by taking several readings on a standard material; specific instructions are given in ASTM D 3017 and by nuclear gauge manufacturers. When the gauge is used to take moisture readings, the test-site location must be prepared by removal of all loose, disturbed material, leveling an area large enough for the gauge to sit on, and filling any voids with fine sand or native fines. The gauge should be seated firmly on the ground, the source placed in the "use" position, and one or more readings taken. The moisture content must then be determined from the calibration curve.

A major advantage of the nuclear moisture method is the rapidity with which tests can be run. The technician is freed from collecting samples, and the results are obtained almost instantaneously. This is extremely valuable when field moisture tests are being used to control ongoing construction. This nuclear method of determining moisture content is widely accepted and used in construction.

Disadvantages of the nuclear moisture method include

- Equipment must be maintained and batteries charged.
- The volume of soil tested is "indeterminate" and varies with material type.
- About 50 percent of the measured count rate is determined by the upper 3 to 4 in of soils and soil-aggregate mixes.
- Because a nuclear source is used, only trained, certified operators can use the gauge, and transportation/shipment of the gauge is closely regulated and requires special handling.
- Meticulous records of radiation exposure must be kept.

When the moisture content is determined by the nuclear method in aggregates, the result often is not sufficiently accurate. It is thus advisable in these situations to use the nuclear gauge to get wet densities and then get an oven-dried moisture content from which to determine dry density.

Possible sources of error in the nuclear moisture test include

- Gauge not standardized at start of day

- Prepared surface not smooth
- Voids in prepared surface not filled with sand or native fines
- Placement too near (within 9 in) of adjacent vertical wall or side of test pit

Microwave. The microwave oven method of determining moisture content is another rapid technique that can be used when expedient results are needed or desired. This technique uses an incremental drying/weighing technique to get the moisture content of a soil sample. This method is not a replacement for but is a supplement to the normal oven-dried moisture content procedure, and microwave results should be checked periodically and confirmed in a conventional oven.

In this method, a moist soil sample is placed in a microwave-safe container and weighed. The sample and container are then placed in the microwave oven, dried for a specific period of time (3 minutes unless prior experience indicates differently), removed from the oven, and weighed. This process is repeated until there is no change in the weights before and after heating. When the weight is constant, the moisture content is determined from the change in weight of the specimen from the original moist weight to the final constant dry weight. The drying time required to reach a constant weight should be noted for future guidance; this time can then be used for the initial drying time for subsequent samples of the same material and sample size.

The incremental heating procedure described above has eliminated or significantly reduced the risk of overheating the soil and thereby obtaining a moisture content higher than the oven-dried one. Thus one of the major objections to this procedure has been overcome. Recent work at the USAE Waterways Experiment Station (Gilbert 1991) has led to development of a computer-controlled microwave oven-drying system that automatically uses incremental microwave heating and stops the procedure when a constant weight is obtained. A strong linear correlation was obtained between microwave moisture contents and oven-dried moisture contents for a variety of soils.

The microwave oven method can be used successfully on most soils, although it is best used on material passing the No. 4 sieve. (Larger particles may be shattered upon heating in the microwave, causing loss of material from the sample.) Some soils for which the moisture contents may not be determined accurately are those containing significant amounts of mica, montmorillonite, halloysite, gypsum, other hydrated materials, organics, or dissolved salts in the pore water.

Possible errors in moisture content determination using the microwave oven include

- Discontinuing the procedure before a constant weight is obtained
- Using the procedure on inappropriate soils
- Overheating the soil
- Inaccurate weight determinations
- Loss of material from exploding particles, oxidation, etc.

Speedy moisture tester. The speedy moisture tester is sometimes used when rapid determination of moisture content is needed during construction. This method uses calcium carbide reagent in a pressure vessel to determine the moisture content of soils.

In the speedy moisture method, approximately equal weights of calcium carbide and soil are put into the moisture tester. The tester is sealed and then inverted to bring the soil and reagent into contact. The apparatus should be shaken for 10 seconds and allowed to sit for 20 seconds. This process is repeated for a total of at least 1 minute for granular soils and up to 3 minutes for other soils to break up all lumps of soil and allow the calcium carbide and free moisture to react completely. The specimen weight and the dial reading are recorded. The moisture content is then determined from calibration curves and conversion charts. As with other rapid moisture content methods, this method should be checked for accuracy by comparing results with oven-dried values for the same soils.

Possible sources of error include

- Use of old calcium carbide reagent
- All soil lumps not broken down
- Soil and reagent brought into contact before tester is sealed
- Procedure stopped too soon

4.4.2 Density

In-place, or field, densities are often required for various reasons. The density of compacted soils in such projects as earthen dams, airfield and roadway construction, compacted clay landfill liners, and other compacted fills is required to determine conformance of construction to design specifications. Field densities are often obtained on natural soil deposits to determine the condition of these materials so that soil improvement techniques, e.g., densification or stabilization, can be used prior to construction if needed.

In-place densities can be determined in the field by any of several methods. Each of these methods, other than the nuclear method, involves determining the oven-dried water content of a soil sample of known volume. Since for each density method the water content of the sample is determined by the procedure described in Sec. 4.2.1, each test method simply varies in its approach to determining the volume of soil removed. The most common procedures for determining field densities are discussed below.

The density value obtained by the different field methods will vary somewhat depending on the combination of method used, material type tested, and operator care. Tables 4.11 and 4.12 give a comparison of the densities and their variability that may be expected.

Drive cylinder. The drive cylinder is a simple, easily used method of determining field density on an undisturbed specimen. In this test, a thin-walled, sharp-edged cylinder is driven into the soil mass. The cylinder with sample enclosed is then dug out, and the ends of the specimen are trimmed flush with the ends of the cylinder. The weight of the cylinder with soil is determined, the soil is removed, and its moisture content is determined from a sample taken from the center of the cylinder. Using the wet weight of the soil and the volume of the cylinder, the wet density can be determined. With the water content determined, the dry density can then be calculated [dry density = wet density $\times$ (1 + water content)]. Procedures for determining water content and density are presented in more detail in Secs. 4.2.1 and 4.2.5. Three separate samples are often tested, and the results are averaged.

The drive cylinder is a relatively quickly run test; it is easy and straightforward to conduct on natural or compacted inorganic, fine-grained soils. This method gives a direct measurement using a sample of constant volume. Disadvantages of this method are the small size of the sample, soils containing rocks cannot be sampled properly, and the procedure is not appropriate for use in noncohesive, friable, organic, saturated, or soft, plastic soils. Because of the limitations of this test, it is not used by some construction organizations, e.g., the U.S. Army Corps of Engineers, as a basis for accepting or rejecting a contractor's work.

Possible errors in determining soil density by the drive cylinder method may be caused by

- Inappropriate driving technique to advance the cylinder
- Dull or damaged cutting edge of cylinder
- Volume or weight of cylinder not accurately determined prior to testing
- Specimen disturbed during sampling or trimming
- Uneven surfaces (sides or ends) on specimen

TABLE 4.11 Comparison of Common Field Density Test Methods for Selected Soils

Material	Average dry unit weight, lb/ft³	Function	Dry density, lb/ft³ Sand cone MIL-STD	Sand cone ASTM	Water balloon	Drive cylinder	Nuclear gauge
Limestone	131.33	Unit weight	135.36	—	132.50	—	134.86
		Coefficient of variation	—	—	3.84	—	0.75
Limestone	118.35	Unit weight	—	125.98	114.85	—	123.24
		Coefficient of variation	—	—	—	—	0.81
Clay gravel	133.62	Unit weight	131.53	131.21	128.60	—	127.42
		Coefficient of variation	1.16	1.86	0.76	—	1.43
Silty clay	104.97	Unit weight	103.79	103.56	104.29	100.18	103.90
		Coefficient of variation	1.81	4.71	1.70	2.14	0.86
Heavy clay	96.36	Unit weight	98.75	99.11	94.51	95.38	97.68
		Coefficient of variation	0.38	2.57	2.29	1.27	0.54

SOURCE: From Coleman, 1988.

TABLE 4.12   Variation in Density as a Function of Field Density Test Method

Test method	Units	Variation in wet density	Variation in dry density
MIL-STD	lb/ft³	−1.9 to 4.1	−2.2 to 4.0
sand cone	%	−1.3 to 3.0	−1.6 to 3.1
ASTM	lb/ft³	−3.0 to 2.0	−2.5 to 7.6
sand cone	%	−2.1 to 1.6	−1.9 to 6.4
Water balloon	lb/ft³	−5.8 to 1.2	−5.1 to 1.7
	%	−4.0 to 0.9	−3.8 to 0.9
Drive cylinder	lb/ft³	−2.9 to −2.0	−4.8 to −2.2
	%	−2.4 to −1.6	−4.6 to −2.3
Nuclear gauge	lb/ft³	−7.4 to 5.1	−6.3 to 4.9
	%	−5.1 to 4.2	−4.7 to 4.1

SOURCE: From Coleman, 1988.

- Use of method in inappropriate materials
- Inaccurate water content determination
- Inaccurate weight determinations

Sand cone.   The sand-cone test is probably the most widely accepted method of determining in-place densities in geotechnical materials. In this test, a calibrated sand is used to determine the volume of an excavated hole from which a sample has been taken for moisture content determination. The sand-cone test is easily run in any material in which a small excavation made with hand tools will stand open without sloughing or deforming.

To run the sand cone, the surface of the soil to be tested should be level, smooth, and free of loose particles. The base plate of the sand-cone apparatus must be seated snugly to the ground. Through the hole in the base plate, a hole is excavated in the material to be tested. This hole should be as large as practical to minimize the effects of any errors. The minimum size of the hole is specified and depends on the maximum particle size in the material being tested. Table 4.13 provides guidance from ASTM D 1556 on minimum test hole sizes. The sand-cone device, filled with calibrated sand, is then inverted, and the sand is allowed to flow into the hole. The weight of sand required to fill the hole is calculated, and the weight and water content of the excavated material are determined. From this information, the volume of the hole and both the wet and dry densities of the material can be calculated.

The sand cone is a versatile test that can be used to determine in-place densities in compacted or natural soils, aggregates, or soil-

TABLE 4.13   Minimum Test Hole and Moisture Content Samples for the Sand Cone Test

Maximum particle size	Minimum test hole volume		Minimum moisture content sample, g
	cm³	ft³	
No. 4 sieve (4.75 mm)	710	0.025	100
½ in (12.5 mm)	1420	0.050	300
1 in (25 mm)	2120	0.075	500
2 in (50 mm)	2830	0.100	1000

SOURCE: From ASTM D 1556. Copyright ASTM. Reprinted with permission.

aggregate mixes; the material being tested must simply have void spaces small enough that the sand does not enter its natural void spaces. Prior to testing, the container and the sand must be calibrated carefully. When the sand cone is used in hard base courses, the material often must be chiseled out to make a hole. The sand cone should not be run in saturated soils where water seeps into the excavation during testing. It should not be used in soft or friable soils that tend to deform under their own weight or under passing weights (such as a person walking or kneeling nearby).

Potential problems that may cause inaccuracies in the densities determined by the sand-cone method include

- Use of improper calibration sand
- Densification of sand while filling the hole
- Testing of inappropriate soils
- Presence of protrusions, pockets, or overhangs in the hole
- Hole too small
- Improper calibration of sand cone
- Moisture in sand
- Soil surrounding the hole disturbed during excavation

Water balloon. Since the density of water is known and is relatively constant (under normal working temperatures), it can be used to determine the volume of a hole excavated for field density determination. Using a specified watertight device, an elastic (rubber) balloon can be inserted into the hole and filled with water. Use of this device allows the hole's volume to be obtained while maintaining accurate account of the quantity of water required and allowing easy removal of the water.

The water-balloon test, also called the *rubber-balloon test,* is used as follows. The surface of the ground to be tested is leveled, and loose particles are removed. The base plate is firmly attached to the ground surface, and a hole is excavated through the base plate. This hole should be as large as possible to minimize errors in density determination (see Table 4.14). The water-balloon apparatus is placed on the base plate, and the balloon is filled with water, forcing it into the excavated hole. The volume of the hole is read from the calibrated volume indicator on the apparatus. The weight of the excavated material must be determined and a representative sample selected for moisture content determination. From this information, the wet and dry densities can be calculated (see Sec. 4.2.5).

The water-balloon method can be used on a wide variety of materials as long as no sharp or jagged edges protrude into the hole to puncture the balloon. Since a constant-volume material is used to fill the hole, and since volume is read directly from the apparatus, several calculations are eliminated that are necessary in the sand-cone procedure. Disadvantages of this method include the following: the balloon does not conform exactly to the hole, the balloon may be broken easily in sharp materials such as crushed rock, and the method is the least exact of all discussed. The effect of balloon nonconformance is illustrated in gravelly materials where the measured density may be from 1 to 3 lb/ft^3 higher than the actual density.

Possible sources of error in the water-balloon density determination include

- Surface of the hole is too rough for balloon to conform to sides
- Hole is too small
- Device leaks and water is lost out of device or balloon
- Device is used in extreme temperatures where density of water may affect results

Nuclear density. The nuclear density method provides a rapid, nondestructive method of getting in-place wet density and moisture data in

TABLE 4.14   ASTM Guidance on Minimum Test Hole Volume for the Water-Balloon Test

Maximum particle size	Minimum test hole volume, ft^3
No. 4 sieve	0.04
$3/4$ in	0.06
$1\frac{1}{2}$ in	0.10

NOTE: Test hole volume increases incrementally 0.01 ft^3 for each $1/4$ in of maximum particle size up to $1\frac{1}{4}$ in and 0.02 ft^3 for each $1/4$-in increase above $1\frac{1}{4}$ in.
SOURCE: From ASTM D 2167. Copyright ASTM. Reprinted with permission.

soils, soil-aggregate mixes, and compacted aggregates (ASTM D 2922 and D 3017). The nuclear gauge employed in this procedure consists of a single unit that houses a glass bead source of cesium-137 to provide gamma radiation for the density measurement, gamma photon detectors, and a readout device that operates the detectors and displays the results. Densities may be obtained by operating the gauge in one of three modes: backscatter, direct, or air gap. Each mode is based on the principle of emitting gamma photons from the nuclear source, recording the number of photons that arrive at the detector, and relating arriving photon numbers to density of the material traversed.

The nuclear gauge must be "standardized" at the start of each day to offset the effect of aging of the nuclear source and other parts of the system. This is done by taking several readings on a standard material and averaging these results to give a daily reference count that is used to ratio all readings taken that day. When the gauge is used to take readings, the test site location must be prepared by removal of all loose, disturbed material, leveling an area large enough for the gauge to sit on, and filling any voids with fine sand or native fines. When the gauge is used in backscatter mode, it is seated firmly on the ground, the nuclear source remains in the gauge, and one or more readings are taken; the depth of measurement is about 3 to 6 in below the bottom of the gauge. The direct transmission mode requires that a small hole be made perpendicular to the prepared surface before the gauge is seated. This hole should be at least 2 in longer than the probe. The gauge is carefully placed on the prepared surface and over the hole. The probe is extended into the hole to the desired depth, and one or more readings are taken. If the air-gap mode is required to overcome errors caused by calibration on materials of different chemical composition, the gauge should be operated first in the backscatter mode; then it is placed on the cradle, set at the optimum air gap (previously determined) directly over the site used for the backscatter reading, and one or more readings are taken with the source contained inside the gauge. Using the appropriate gauge calibration, the wet density is determined. The moisture content must be determined with the nuclear gauge, by oven drying, or by other means and used with the wet density to calculate the value of dry density.

A major advantage of the nuclear density method is the rapidity with which tests can be run. Not only is the technician freed from digging holes and collecting samples, but the results are obtained almost instantaneously. This is extremely valuable when field density tests are being used to control ongoing construction. Disadvantages of the nuclear density method include

- When holes are not dug, visual examination of the material is not possible.

- Meticulous records of radiation exposure must be kept.

- Equipment must be maintained and batteries charged.

- The volume of soil tested is "indeterminate" and varies with material type.

- The surface 1 in in the backscatter mode and top 3 to 4 in in the direct mode determine over half the total count rate. This does not pose a problem when the material has a uniform density.

In general, the advantages of the nuclear density method far outweigh the disadvantages. This method of density measurement is widely accepted and used in construction.

Possible sources of error in the nuclear density test include

- Gauge not standardized at start of day

- Prepared surface not smooth

- Voids in prepared surface not filled with sand or native fines

- Placement too near (within 9 in) of adjacent vertical wall or side of test pit

4.4.3 Plate load test

The plate load test is conducted to assess the load-carrying abilities of pavement subgrades, bases, and sometimes the entire pavement structure; it also can be used for building foundation design. This test can be used for design of flexible as well as rigid pavements, although it is used most commonly in rigid pavement design. The test procedure involves placing a large circular steel plate on the surface of the geotechnical material to be tested, applying a specified loading sequence, and recording deflections of the plate. Using the load/deflection data, the material's support capabilities can be determined.

To conduct a standard plate load test for rigid pavement design, an area is cleared to expose subgrade at its future use elevation. All loose material and large particles must be removed from the surface. A thin layer of sand or plaster of paris may be used to level the area. A 30-in-diameter, 1-in-thick plate is seated on the prepared surface. Then 24- and 18-in-diameter plates are centered on the larger plate, and a large-capacity (e.g., 25,000 lb), hydraulic jack is centered on the plates. Three dial gauges, accurate to 0.0001 in, are then located on the perimeter of the 30-in plate and are spaced 120 degrees apart. The dial gauges and the reaction beam or device to which the jack is attached must be anchored well outside the loading area so that correct deflections are measured. A small seating load is applied to the plates until deformation ceases; "zero" readings are then taken on the dial gauges. The plate load test is conducted by applying two 3535-lb load increments (5 psi) and monitoring deformation after each load. A preliminary subgrade modulus k is calculated at this point. If k is less than 200 psi/in for a cohesive soil, the test is complete. Otherwise, testing should continue in 3535-lb increments up to a total load of 21,210 lb. Each load should remain until all deformation has ceased; then the three gauges should be read.

The plate load test procedures can be modified to simulate cyclic loading and will provide data on plastic deformation and elastic rebound.

Corrections to the plate load test must be made for plate bending and subgrade saturation. The correction for plate bending is made simply by using a correlation curve that relates calculated k value and corrected k value. This correlation curve is given in Fig. 4.11. A correction to account for future saturation of the subgrade will be necessary if a fine-grained cohesive soil is tested in the field. In this case, an undisturbed soil sample should be taken from the subgrade for laboratory (modified) consolidation testing. Two laboratory tests are usually run to make the saturation correction. One test is run at in situ moisture content, and one is run in a saturated condition; the deformation under a 10-psi load is obtained for each specimen. The ratio of these test results is used as follows to correct the field plate load results for the situation of a saturated subgrade:

$$k = k_u' \left[\frac{d}{d_s} + \frac{b}{75} \left(1 - \frac{d}{d_s} \right) \right]$$

where k = corrected modulus of soil reaction, lb/in^3
k_u' = modulus of soil reaction uncorrected for saturation, lb/in^3
d = deformation of a consolidometer specimen at in situ moisture content under a unit load of 10 psi
d_s = deformation of a saturated consolidometer specimen under a unit load of 10 psi
b = thickness of base course material, in

The value of d/d_s cannot exceed 1.0 in this equation. Also, the equation can be used to calculate a corrected k value for the subgrade whether or not there is a granular base course present. However, if the base course thickness is in excess of 75 in, a correction for saturation of the cohesive subgrade will not usually be made.

When a plate load test is run, additional material samples should be taken for determining the physical characteristics of the various materials tested. Data that are needed include moisture content, gradation (grain-size distribution), Atterberg limits, and density. These data will help in the interpretation of plate load test results.

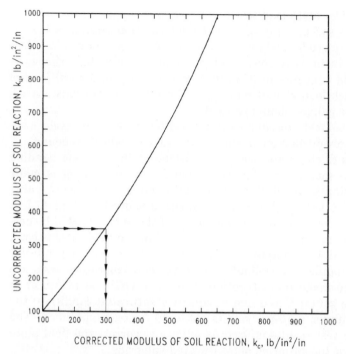

Figure 4.11 Correction curve for bending of steel plate in plate load test (*Department of Defense 1968*).

The plate load test is seldom used with flexible pavements. If used, the plate size is often selected to equal the contact area of the tire that will be using the pavement. For either rigid or flexible pavement design, the maximum plate load used in the test should at least equal the maximum anticipated wheel load. Since test results depend on the plate size and behavior, it is important to use standard-sized plates for each load test.

When the plate load test is run for building foundation design, it is conducted at the elevation of the proposed footing. The plate used in this test is usually 1 ft² and is placed in a 5-ft² (minimum) test pit. The load should be applied in small, equal increments up to 150 percent of the estimated allowable soil pressure. Results are normally presented as load-settlement curves (Terzaghi and Peck 1967). An in-depth discussion of the plate load test can be found in Terzaghi (1955).

Possible errors in running the plate load test include

- Plate not resting on a smooth, level surface
- Hydraulic jack anchored too near the test area

- Dial gauges anchored too near the test area
- Dial gauges not accurate enough
- Insufficient capacity in loading system
- Failure to account for plate size in interpretation of results

4.4.4   Dynamic cone penetrometer

The dynamic cone penetrometer (DCP) was developed originally to determine the strength profile of flexible pavements, but its use has been adapted for soil strength determinations. Although it is used widely overseas, the DCP is used most commonly in the United States by the Corps of Engineers and Air Force for expedient road and airfield soil strength determination; it is normally used in place of field CBR tests in the military theater of operations, where test pits cannot be dug. It is used to determine shear strength on unsurfaced and gravel-surfaced facilities. This device can be used to quantify the shear strength of soils with CBR values ranging from 1 to 100. The DCP is a compact, fieldable device that can be used by relatively inexperienced personnel to obtain strength data which are then correlated with CBR values.

The DCP consists of a 60-degree cone attached to a $\frac{5}{8}$-in-diameter steel rod; a sliding 17.6-lb hammer on top of the rod provides the force to drive the cone into the soil. Two people are needed to run the DCP; one person operates the DCP hammer, and the other measures penetrations and acts as data recorder (see Fig. 4.12). To run the DCP, the device is held vertically, the cone is tapped into the ground until its base (flat side) is flush with the ground surface, and a zero reading is taken. The test is then begun by raising the hammer to the bottom of the handle and dropping it 22.6 in. This is repeated until the number of blows is (generally) 1, 2, 3, 5, 10, 15, or 20 *and* at least 25 mm of penetration has been obtained. At this point, the depth of penetration (to the nearest 5 mm) and the number of blows to cause it are recorded. The DCP index is taken as the average penetration caused by one blow of the hammer. This process is repeated until the penetration depth reaches the desired depth or 39 in, whichever comes first. The DCP index values are converted to CBR values for use in pavement design or evaluation by the correlation equation (Webster et al. 1992)

$$\log CBR = 2.46 - 1.12 \, (\log DCP)$$

where CBR = soil strength used for pavement design/evaluation
DCP = DCP index = average penetration caused by one hammer blow, mm

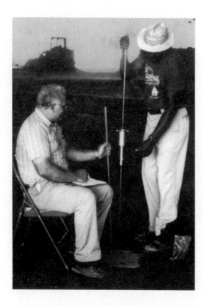

Figure 4.12 Dynamic cone penetrometer being used to obtain soil strength data.

This correlation of DCP index to CBR value was developed by the USAE Waterways Experiment Station; the data from which this equation was derived are shown in Fig. 4.13.

Two hammers can be used with the DCP, the standard 17.6-lb hammer and a 10.1-lb hammer. The lighter hammer causes a penetration of about half that caused by the 17.6-lb hammer; thus the average hammer blow determined with the 10.1-lb hammer must be multiplied by 2 to obtain the DCP index value. The lighter hammer should be used in softer soils, i.e., those with CBR values of 10 or less, since it will provide better test results in these materials. It is suitable for use in materials with CBR values up to 80.

The DCP is an easily conducted, expedient test that is appropriate for use in determining shear strength of underlying soil layers. It can be used to identify stratifications and the presence of soft soil lenses. With the correlation to CBR values, the DCP can readily provide necessary data for pavement design or evaluation when CBR tests cannot be run.

Potential sources of error in conducting the DCP include

- Point of cone dull or deformed
- Apparatus not held vertically
- Hammer drop not appropriate height

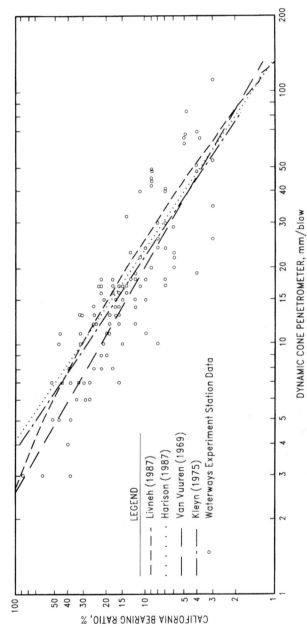

Figure 4.13 Correlation between DCP and CBR (*Webster et al. 1992*).

Legend (rotated with figure):

LEGEND
— — — Livneh (1987)
· · · · · Harison (1987)
———— Van Vuuren (1969)
— · — Kleyn (1975)
○ Waterways Experiment Station Data

Axis labels:
CALIFORNIA BEARING RATIO, %
DYNAMIC CONE PENETROMETER, mm/blow

183

- Hammer hits handle on upward stroke and raises cone from soil surface
- Number of blows not properly counted
- Depth of penetration not accurately measured

4.4.5   Permeability

Permeability, or the rate at which water flows through the voids of a soil mass, can be determined in the field by any one of a number of methods. The particular method selected will depend on the material type, its permeability, and the available equipment. Table 4.15 lists the various methods that are available. Field permeability tests are preferable to laboratory permeability tests in most cases because the small size and discontinuous nature of laboratory specimens will likely preclude identification of discontinuities, thin permeable lenses, and other variations in permeability on the macroscopic scale.

Constant-head test.   This test calculates permeability from the quantity of water that flows out of the bottom of a cased boring under a constant head. A cased boring is advanced to the desired depth, carefully cleaned out, and flushed with clean water. Water is maintained at a constant level in the boring by adding measured amounts of water to counter gravity outflow from the bottom of the casing. A schematic diagram of the test and method of calculating the permeability is shown in Fig. 4.14. Small quantities of silt in suspension in the water used for the test can settle on the bottom of the boring and result in underestimation of the permeability. Consequently, only clean water should be used for testing, and settling tanks and filters should be used to remove sediment from the water, if necessary. Drilling mud should never be used in any boring where an in situ permeability test will be conducted. Ideally, the water used in the test should be warmer than the groundwater to prevent air bubbles from forming and impeding water flow.

Below the water table, the boring must always have a water level higher than the groundwater table. Failure to do this can result in "boiling" or heaving at the bottom of the hole. Any sudden loss of water from the boring should be investigated, since it may indicate the presence of a highly permeable layer such as open work gravel.

Above the water table, it will be difficult to maintain a constant water level in the boring, and water-level fluctuation of a few tenths of a foot over 5 minutes is considered satisfactory (Bureau of Reclamation 1974). If this cannot be done, then a falling-head permeability test that measures the drop of the borehole water level over a period of time can be conducted, and the permeability is calculated as follows (Hvorslev 1951):

TABLE 4.15 Methods of Testing in Situ Permeability in Soils

Method		Technique	Application to				Problems	Method rating	Reference
			Gravel	Sand	Silt	Clay			
A	Augerhole	Shallow uncased hole in unsaturated material above groundwater level	✓	Only where $k > 10^{-3}$ cm/s	?	—	Difficult to maintain water levels in coarse gravels	Poor	Bureau of Reclamation (1968)
	Test pit	Square or rectangular test pit (equivalent to circular hole above)	✓	?	?	—		Poor	Lacroix (1960)
B	Cased borehole (no inserts)	i. Falling/rising head, Δh in casing measured vs. time	✓	✓	?	—	Borehole must be flushed Possible fines clog base (falling head)		Hvorslev (1951)
		ii. Constant head maintained in casing, outflow Q vs. time	✓	✓	?	—	Pumping (rising Δh) where WL lowered excessively	Fair	Bureau of Reclamation (1968)
C	Cased borehole (inserts used)						Single tests only; cannot be used as boring is advanced	Fair	Hvorslev (1951)
	i. Sand filter plug	Generally failing head, Δh measured vs. time only	✓	✓	?	—			
	ii. Perforated/slotted casing in lowest section	Variable heads possible	✓	✓	?	—			
	iii. Well point placed in hole, casing drawn back	As for ii above	✓	✓	✓	—		Fair to good	
D	Piezometers/Permeameters (with or without casing)	i. Suction bellows apparatus (independent of boring); inflow only measured vs. time	—	✓?	✓	—	Restricted to fine sands, coarse silts; variable bellows required, k range 10^{-4} to 10^{-7} cm/s	Good (local zones)	Golder, Gass (1963)
		ii. Short cell (cementation) (independent of boring); outflow only measured vs. time	✓	✓	—	—	Carried out in adit or tunnel		Golder, Gass (1963)
		iii. Piezometer tip pushed into soft deposits/placed in boring, sealed, casing withdrawn/pushed ahead of boring; constant head, outflow measured vs. time; variable heads also possible	—	—	✓	✓	Possible tip "smear" when pushed; Δu set up in pushing tip; danger of hydraulic fracture		Gibson (1956), Wilkes (1974), Hvorslev (1951), Bjerrum et al. (1972)

TABLE 4.15 Methods of Testing in Situ Permeability in Soils (Continued)

Method		Technique	Application to				Problems	Method rating	Reference
			Gravel	Sand	Silt	Clay			
E	Well pumping test	Drawdown in central well; monitored in observation wells on at least two 90-degree radial directions	✓	✓	✓(?)	—	Screened portion should cover complete stratum tested	Excellent (mass permeability of foundation material)	Todd (1959)
F	Test excavation pumping test(s)	Monitoring more extensive than E, during excavation dewatering (initial construction stage)	✓	✓	✓(?)	—	Expensive, but of direct benefit to contractual costing	Excellent (mass permeability of foundation material)	—

SOURCE: "Field Measurement of Permeability in Soil and Rock," V. Milligan, 1975, *In Situ Measurement of Soil Properties*, reproduced by permission of American Society of Civil Engineers.

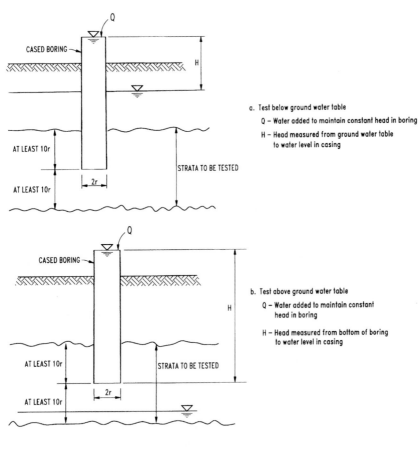

a. Test below ground water table

    Q – Water added to maintain constant head in boring

    H – Head measured from ground water table
       to water level in casing

b. Test above ground water table

    Q – Water added to maintain constant
       head in boring

    H – Head measured from bottom of boring
       to water level in casing

c. Calculation of Coefficient of Permeability

$$k = \frac{Q}{5.5 \, rH}$$

where

k = Coefficient of permeability measured in units of length/time

Q = flow of water required to maintain constant head in units of volume/time

H = head in units of length

r = radius of casing in units of length

Figure 4.14 Constant-head permeability test for field use (*adapted from Cedergren 1989*).

$$k = \frac{\pi r}{5.5\,(t_2 - t_1)} \ln \frac{H_1}{H_2}$$

where r = radius of borehole
  t_1 = time at start of test
  t_2 = time at end of test
  H_1 = head at start of test
  H_2 = head at end of test

Silt settling in the bottom of the hole is always a problem with this test, however.

Borings for permeability determinations should be spaced so that permeabilities across the site are investigated. Permeability tests should be conducted for each strata encountered in the boring, or if the material appears homogeneous throughout, tests should be conducted at several depths in each boring above and below the water table to identify any variation in permeability with depth. The borings should be left open, and the level of the groundwater should be monitored.

Well-pumping test.   If warranted by site conditions, a well-pumping test can be used to determine a more accurate coefficient of permeability for the deposit below the groundwater table. For this test, water is pumped from a central well, and the water table drawdown in nearby observation wells is measured. Figure 4.15 is a schematic diagram of the test and shows the required calculations. Two lines of observation wells perpendicular to each other are needed. At least two and preferably four observation wells are needed in each line. The size of the observation wells is governed only by the diameter necessary to monitor water level. Therefore, boreholes may be used, provided they are not plugged with drilling mud. The pumping well must completely penetrate the pervious stratum and must be pumped until steady-state conditions are reached. The drawdown in the nearest well should be at least 6 in. The well is usually pumped at two different rates of flow to check the permeability calculations (Cedergren 1989).

This test is expensive and is normally conducted only if the constant-head permeability tests find potential seepage problems that require a more accurate determination of the coefficient of permeability. The constant-head test results can be used to locate the well for the pumping test and to estimate the required pump size and observation well spacing. More comprehensive discussions of types and limitations of field permeability testing are available in Cedergren (1989).

Considerations in permeability determinations.   The determination of true permeabilities for geotechnical materials is one of the most (if

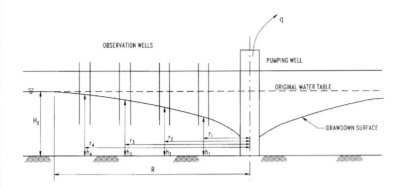

$$k = \frac{2.3q \, \log(\frac{R}{r_L})}{\pi(H_o^2 - h_L)} = \frac{2.3q \, \log(\frac{r_2}{r_1})}{\pi(h_2^2 - h_1^2)}$$

k – Coefficient of permeability in units of length/time

H – Original height to ground water table in units of length

h – Height of drawdown surface in observation well in units of length

R – Radius of influence of pumping well in units of length

r – Distance from pumping well to observation well in units of length

q – Steady state pumping rate in units of volume/time

Figure 4.15 Schematic drawing of well pumping test (*adapted from Bureau of Reclamation 1990*).

not *the* most) difficult problems faced by geotechnical engineers. To further compound the situation, the significance of underestimating permeability can be devastating. Unlike other "loading conditions" (such as consolidation and bearing capacity), where the effect of an external force is applied across some mass of soil, thus minimizing the consequence of small anomalies, "hydraulic loading" takes full advantage of the most minute hole, crack, lense, seam, or other small discontinuity. Water, forced by whatever hydraulic pressure exists in the particular situation, will take the path of least resistance as it moves to reduce its potential head. Not only will the water move through any small permeable space, but it also may cause shear strength reduction, internal erosion (piping), external erosion, under-mining of structures, and eventual failure of structures.

A dramatic example of the underestimation of permeability, lack of proper control on permeating water, and its devastating conse-quences is the failure of Teton Dam in Idaho on June 5, 1976. According to the independent Teton Dam Failure Review Group report, the failure of this 305-ft-high zoned earthfill dam (with a thick central silt core) occurred because of "inadequate protection of the impervious core material from internal erosion" (Department of the Interior 1977). Internal erosion of the dam's core most likely resulted from flow of water through "the open fractures in the abutment foundation rock [which] allowed direct access by reservoir water [under full reservoir head] to the impervious core on the upstream side of the key trench. Any water flowing through the impervious core could exit into open fractures on the downstream side of the key trench. The design failed to provide a defense against flow through the embankment cracks or against erosion of the impervious core at rock surfaces. The rock surface was not adequately sealed under the impervious core upstream and downstream of the key trench" (Department of the Interior 1977).

Permeability is such an important and elusive characteristic of both natural and man-made deposits that some cautions are warranted with regard to its evaluation. Harry Cedergren (1967), a master of dealing with permeability evaluation and control, has given us:

The do nots of permeability

1. Do not design or build hydraulic structures without adequate permeability investigations.

2. Do not depend on novices to perform or interpret permeability studies; give this work to experts who understand earth deposits and their own limitations.

3. Do not underestimate the importance of minor soils and geologic details on the permeability of soil and rock formations. They cause a majority of the failures of dams, reservoirs, and other hydraulic structures.

4. Do not allow low compaction of "permeable" drainage aggregates or embankment materials as a means of obtaining high permeability. Loosely compacted soils and aggregates that become saturated are likely to liquefy under severe shocks.

4.5 References

American Concrete Institute. 1989. Guide for Use of Normal Weight Aggregates in Concrete, ACI 221R-89, Detroit, Mich.
American Society for Testing and Materials (ASTM). 1994. *Annual Book of ASTM Standards*, Part II, ASTM, Philadelphia.
ASTM. 1966. *Testing Techniques for Rock Mechanics*, STP 402, ASTM, Philadelphia.
ASTM. 1970. *Special Procedures for Testing Soil and Rock for Engineering Purposes*, STP 479, ASTM, Philadelphia.
ASTM. 1971. *Sampling of Soil and Rock*, STP 483, ASTM, Philadelphia.
ASTM. 1976. *Living with Marginal Aggregates*, STP 597, ASTM, Philadelphia.
Barksdale, R. D. (ed.). 1991. *The Aggregate Handbook*, National Stone Association, Washington.
Bjerrum, L., J. Nash, R. Kennard, and R. Gibson. 1972. "Hydraulic Fracturing in Field Permeability Testing," *Geotechnique*, 22 (2): 319–332.
Bureau of Reclamation. 1990. *Earth Manual*, part 2, U.S. Department of Interior, Government Printing Office, Washington.

Bureau of Reclamation. 1974. *Earth Manual,* U.S. Department of Interior, Government Printing Office, Washington.

Bureau of Reclamation. 1968. *Earth Manual,* U.S. Department of Interior, Government Printing Office, Washington.

Carter, M., and S. Bentley, 1991. *Correlations of Soil Properties,* Pentech Press, London.

Cedergren, H. R. 1989. *Seepage, Drainage, and Flow Nets,* 3d ed., Wiley, New York.

Cedergren, H. R. 1967. *Seepage, Drainage, and Flow Nets,* Wiley, New York.

Coleman, D. M. 1988. "Variability in Field Density Test Methods Used for Quality Control," master's thesis, Mississippi State University, Starkville, Miss.

Crawford, C. 1969. "Building Damage from Expansive Steel Slag Backfill," *Journal of Soil Mechanics Engineering,* 95 (SM6): 1325–1334.

Department of the Army. 1986. *Laboratory Soils Testing,* Engineer Manual EM 1110-2-1906, Washington.

Department of the Army. 1987a. *Materials Testing,* Field Manual FM 5-530, Washington.

Department of the Army. 1987b. *Confined Disposal of Dredged Material,* Engineer Manual EM 1110-2-5027, Washington.

Department of Defense. 1968. Military Standard: Test Method for Pavement Subgrade, Subbase, and Base Course Materials, MIL-STD-621-A, Washington.

Department of Defense. 1966. Military Standard: Test Methods for Bituminous Paving Materials, MIL-STD-620-A, Washington.

Department of the Interior, Teton Dam Failure Review Group. 1977. *Failure of Teton Dam: A Report of Findings,* U.S. Bureau of Reclamation, Denver, Colo.

Departments of the Army and Air Force. 1989. *Flexible Pavement Design for Airfields (Elastic Layer Method),* Technical Manual TM 5-825-2-1/AFM 88-6, chap. 2, sec. A, Washington.

Dolar-Mantuani, L. 1978. "Soundness and Deleterious Substances," in *Significance of Tests and Properties of Concrete and Concrete-Making Materials,* STP 169B, ASTM, Philadelphia.

Gibson, R. 1956. "A Note on the Constant Head Test to Measure Soil Permeability In Situ," *Geotechnique,* 16 (3): 256–259.

Gilbert, P. A. 1991. "Rapid Water Content by Computer-Controlled Microwave Drying," *Journal of Geotechnical Engineering,* 117 (1): 118–138.

Gnaedinger, J. P. 1987. "Open Hearth Slag: A Problem Waiting to Happen," *ASCE Performance of Constructed Facilities Journal,* 1 (2): 78–83.

Golder, H., and A. Gass. 1963. "Field Tests for Determining Permeability of Soil Strata," in *Field Testing of Soils,* STP 322, ASTM, Philadelphia.

Hvorslev, M. J. 1949. "Subsurface Exploration and Sampling of Soils for Civil Engineering Purposes," report on a Research Project of the American Society of Civil Engineers, sponsored by The Engineering Foundation, Harvard University, and The Waterways Experiment Station, USAE Waterways Experiment Station, Vicksburg, Miss.

Hvorslev, M. J. 1951. "Time Lag and Soil Permeability in Ground-Water Observations," bulletin no. 36, USAE Waterways Experiment Station, Vicksburg, Miss.

Lacroix, Y. 1960. "Notes on the Determination of Coefficients of Permeability in the Laboratory and In Situ," Polytechnic School, Zurich, Switzerland (unpublished).

Lambe, T. W., and R. V. Whitman. 1969. *Soil Mechanics,* Wiley, New York.

Martinez, R. E., D. Bloomquist, M. C. McVay, and F. C. Townsend. 1988. "Consolidation Properties of Slurried Soils," in *Proceedings, Soil Properties Evaluation from Centrifugal Models and Field Performance,* ASCE National Convention, Nashville, Tenn.

Meininger, R. C. 1978. "Abrasion Resistance, Strength, Toughness, and Related Properties," in *Significance of Tests and Properties of Concrete and Concrete-Making Materials,* STP 169B, ASTM, Philadelphia.

Metcalf, J. B. 1988. "The Use of Naturally Occurring Nonstandard Materials for Road Base Construction in Australia," presented at 26th Australian Road Research Board Regional Symposium, Bunburg, Australia.

Milligan, V. 1975. "Field Measurement of Permeability in Soil and Rock," in *In Situ Measurement of Soil Properties,* vol. 2, American Society of Civil Engineers, New York, pp. 3–36.

Morin, W. J., and P. C. Todor. 1975. "Laterite and Lateritic Soils and Other Problem Soils of the Tropics," report AID/CSD 3682, U.S. Agency for International Development, Washington.

Noble, D. F. 1977. "Accelerated Weathering of Tough Shales, Final Report," VHTRC 78-R20, Virginia Highway and Transportation Research Council, Charlottesville, Va.

Organization for Economic Co-operation and Development. 1981. *Use of Marginal Aggregates in Road Construction,* Paris, France.

Peck, R. B., W. E. Hanson, and T. H. Thornburn. 1974. *Foundation Engineering,* Wiley, New York.

Powers, T. C. 1975. "Freezing Effects in Concrete," in *Durability of Concrete,* SP-47, American Concrete Institute, Detroit, Mich.

Reidenour, D. R., E. G. Geiger, and R. H. Howe. 1976. "Suitability of Shale as a Construction Material," in *Living with Marginal Aggregates,* STP 597, ASTM, Philadelphia.

Rollings, M. P. 1994. "Geotechnical Considerations in Dredged Material Management," *Proceedings, First Environmental Geotechnics Congress, Edmonton, Alberta, Canada.* International Society for Soil Mechanics and Foundation Engineering, Edmonton.

Rollings, R. S. 1988. "Substandard Materials for Pavement Construction," in *Proceedings of the 14th Australian Road Research Board Conference,* vol. 14, part 7, Australian Road Research Board, Canberra, Australia, pp. 148–161.

Sanglerat, G. 1972. *The Penetrometer and Soil Exploration,* Elsevier, Amsterdam.

Schiffman, R. L., V. Pane, and R. E. Gibson. 1984. "The Theory of One-Dimensional Consolidation of Saturated Clays: IV. An Overview of Nonlinear Finite Strain Sedimentation and Consolidation," in *Proceedings, ASCE Sedimentation/Consolidation Models,* American Society of Civil Engineers, San Francisco, Calif.

Smith, R. B. 1986. "Evaluation of Sydney Shales for Use in Road Construction," in *Proceedings of the 13th ARRB–5th REAAA Combined Conference,* vol. 13, part 5, Australian Road Research Board, Adelaide, Australia, pp. 133–145.

Strohm, E. E., G. H. Bragg, and T. W. Ziegler. 1978. *Design and Construction of Compacted Shale Embankments,* vol. 5: *Technical Guidelines,* FHWA-RD-78-141, Federal Highway Administration, Washington.

Terzaghi, K. 1936. "Relation Between Soil Mechanics and Foundation Engineering," presidential address to the First International Conference on Soil Mechanics and Foundation Engineering, in *Proceedings,* vol.3, Harvard University, Cambridge, Mass., pp. 13–18.

Terzaghi, K. 1955. "Evaluation of Coefficients of Subgrade Reaction," *Geotechnique,* 5(4), pp. 297–326.

Terzaghi, K., and Peck, R. B. 1967. *Soil Mechanics in Engineering Practice,* Wiley, New York.

Todd, D. 1959. *Groundwater Hydrology,* Wiley, New York.

Townsend, F. C., and M. C. McVay. 1990. "State-of-the-Art: Large Strain Consolidation Predictions," *ASCE Journal of Geotechnical Engineering,* 116 (2): 222–243.

Webster, S. L., R. H. Grau, and T. P. Williams. 1992. "Description and Application of Dual Mass Dynamic Cone Penetrometer," Instruction Report GL-92-3, USAE Waterways Experiment Station, Vicksburg, Miss.

Wilkes, P. 1974. "Permeability Tests in Alluvial Deposits and Determination of K_o," *Geotechnique,* 24 (1): 1–11.

Winterkorn, H. F., and H. Y. Fang (eds.). 1975. *Foundation Engineering Handbook,* Van Nostrand Reinhold, New York.

Wuerpel, C. E. 1944. *Aggregates for Concrete,* National Sand and Gravel Association, Washington.

Basic Geotechnical Construction Techniques

Adversity makes a man wise, though not rich.

Part

2

Basic Geotechnical
Construction Techniques

Chapter

5

Compaction

Among civil engineers wishful thinking amounts
to an occupational disease.

H. J. B. HARDING, 1958

5.1 Introduction

Of all the civil engineering tasks to be accomplished on a project, compaction is one of the most fundamental. Soils for fill or aggregates for pavement bases are placed and compacted to a firm, load-carrying capacity upon which all the remaining structures or pavement surfacings must rest. It is an ancient principle of construction that is as important today as it was in the past. The Roman Marcus Vitruvius describes this fundamental principle of compacting layers in ancient Roman road building as follows (Garrison 1991):

> ...so that account may be taken with special care and great foresight of a solid foundation...it must be rammed carefully with piles...(stones are then added and mixed with lime)....Let it then be laid on, and rammed down with repeated blows by gangs of men using wooden stamps. When the stamping is finished it must be no less than nine inches thick, that is three-quarters of its initial height.

While the equipment may be primitive, the vigor of effort and effectiveness of a 75 percent reduction in loose volume would be laudatory on many modern jobs.

In a review of 32 structural and foundation failures by LePatner and Johnson (1982), inadequately compacted fill under floor slabs was a major contributor to two of the failures. This is significant when one considers that fill compaction is a relatively minor construction detail for buildings, hospitals, storage tanks, and other major structures that were the focus of this study. Consequently, even though we have been compacting materials since ancient times and it may seem to be a minor point compared with the complex structural framing and detailing that may be going on at a site, it behooves us to ensure that it is done correctly.

This chapter concentrates on fundamental "meat and potatoes" compaction with rollers. Some more specialized techniques, such as dynamic compaction using dropped weights, will be covered in Chap. 7.

5.2   Why Compact?

Compaction of geotechnical materials in construction is required for a number of reasons. First, we do not want the materials to densify or decrease in volume under load or in service. One of the first tasks we wish to accomplish with compaction is to densify the material to a degree that future loadings or repetitions of loadings will not cause further reduction in volume with consequent surface distortions. Such surface distortions might obviously adversely affect the overlying pavement or structure's performance.

Next, we want to ensure that the soil's engineering properties such as strength or permeability that the engineer used in design are actually achieved in the field. Consequently, it will be critical to compact the geotechnical materials under conditions and to a degree that ensure that the engineer's design assumptions are met. And finally, we want to compact the materials so that their future non-load-associated volumetric changes are acceptable for the intended purpose.

5.3   Compaction Behavior

The basic principles of soil compaction were understood as far back as the 1930s (Proctor 1933), and a clear understanding of these principles is crucial to understanding soil compaction in the field.

5.3.1   Laboratory behavior

Compaction can be performed in the laboratory with static, kneading, gyratory, vibratory, or impact compactors (Wahls et al. 1966). They each have certain advantages and disadvantages; however, impact compaction tests that use a falling hammer to provide the compaction energy are the standard in practice today. Consequently, only the impact compaction tests will be discussed in this book. Chapter 4 provided a description of the basic versions of the impact-type compaction tests (see Table 4.3) used to define the relationship between molding moisture content and dry density of a soil.

Figure 5.1 shows a typical family of curves developed by laboratory tests on a sandy clay from New Mexico. Compaction procedures follow

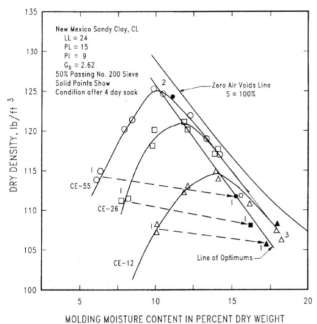

Figure 5.1 Compaction curve for a sandy clay.

Corps of Engineers guidance using 6-in-diameter molds to develop moisture content–dry density–CBR relationships (CBR test results are not shown in Fig. 5.1). Each curve represents a different compaction energy based on the number of blows applied with a compaction hammer to each soil layer in the mold—12,000 ft · lb/ft³ for CE-12 (approximately standard Proctor density), 26,000 ft · lb/ft³ for CE-26, and 55,000 ft · lb/ft³ for CE-55 (approximately modified Proctor density). At each energy level there is an optimum molding moisture content at which point the dry density is a maximum. Above and below this optimum moisture content the dry density drops off. As the compaction energy increases, the optimum molding moisture content decreases, while the corresponding maximum dry density at this moisture content increases. The peaks, or optimums, of the compaction curves can be connected with an approximately straight line known as the *line of optimums.* Any different levels of compaction energy would reasonably be expected to produce optimum molding moisture content and density values that would fall close to this line.

The *zero air voids line* represents the moisture content and dry density where saturation is 100 percent. The compaction test results at moisture contents above the optimum molding moisture content approach but do not reach the zero air voids line. The line of optimums is approximately at a saturation level of 90 percent for this soil. The weight-volume relationships discussed in Chap. 2 are used to calculate the zero air void line.

The solid symbols in Fig. 5.1 represent the final moisture content and density after 4 days of soaking in preparation for soaked CBR tests. For the samples compacted dry of optimum and designated as number 1 in the figure, there is a very substantial increase in moisture content and a decrease in density due to swelling as the clay particles absorb moisture. This loss in density is notable, and the samples are not fully saturated after 4 days of soaking. This result is consistent for all levels of compaction, but higher-energy compaction effort samples have lower moisture contents and higher densities after soaking than do samples at lower compactive effort. For the samples at optimum moisture content and designated as number 2 in the figure, there is little increase in moisture content and only a slight swell resulting in a small drop in density after 4 days of soaking. For the samples compacted wet of optimum and designated as number 3, there is essentially no change in moisture content or density.

Compaction moisture content and energy also significantly affect the particle structure of fine-grained materials (Lambe 1962). At moisture contents wet of optimum, the flat, sheetlike clay particles tend to compact parallel to one another, and this tendency to parallel structure at a given water content above optimum increases as compaction energy increases. Dry of optimum, the particles assume a dispersed, flocculated, or "cardhouse" structure. In Sec. 8.4 this flocculated structure for compaction dry of optimum will be seen to be a contributor to adverse engineering behavior for certain soils, and we will see in a later section that this structure has a major impact on permeability.

Laboratory testing can define an optimum molding moisture content for a soil compacted at a specific energy level with a specific compaction method. When compacted at this optimum moisture content, the soil's dry density will be the maximum achievable with that compaction effort and compaction method. Changes in density and moisture content due to later addition of moisture by soaking will be relatively small compared with other molding moisture contents. If the soil is appreciably drier than optimum, density will be lower for this specific compaction effort, and later soaking will potentially result in large moisture and density changes. Compaction wet of optimum will produce lower densities than compaction at optimum, but future soaking results in small moisture and density changes.

A variety of different terminology has unfortunately grown up around laboratory compaction curves. It is important to remember that this terminology refers to levels of compaction energy used in preparing the laboratory sample, and it has no other significance.

Consequently, *Proctor, standard Proctor, standard density, AASHTO, AASHO,* and *CE-12* all refer to approximately the same thing—a laboratory sample prepared at about 12,300 ft · lb/ft³ of compaction energy using an impact hammer. *Modified Proctor, modified density, modified AASHO, modified AASHTO,* and *CE-55* all refer to samples prepared with about 56,000 ft · lb/ft³ of compaction energy. Each reference to a laboratory density should refer also to the specific ASTM or other test standard used to define the laboratory density. These are all just laboratory reference densities prepared by specified procedures, and they have no other intrinsic or fundamental significance.

5.3.2   Soil behavior

Different soils will have differently shaped compaction curves. Figure 5.2 shows several different soils, identified by location (e.g., Keesler) and USCS soil type, compacted at modified compaction energy and one soil (Goodfellow CH clay) compacted at both modified and standard compaction. The GW, GC, and SC soils all have steep compaction curves that show large changes in density for small changes in moisture content. Compaction of these types of soils in the field requires good control of moisture if optimum compaction is to be achieved. Sharp-peaked compaction curves are fairly typical of silty materials and of granular materials that have a significant silt or clay fraction.

On the other hand, the Goodfellow CH clay at a standard compaction level shows a relatively broad region of molding moisture content where relatively high density can be achieved at this compaction energy. This is probably fortunate because moisture content adjustment (either up or down) is difficult with sticky and low-permeability CH clays.

Materials that behave as completely cohesionless will have a double-peaked compaction curve such as the GW and SW curves in Fig. 5.2. When completely dry, these soils can be compacted to a density about equal to the optimum density at optimum moisture content. At moisture contents between dry and optimum, capillary tension resists compaction, resulting in lower densities for any given compactive effort. This behavior will be encountered in some sands and gravelly materials. Truly cohesionless materials are often best compacted in the field at or near saturation.

The Goodfellow CH curves in Fig. 5.2 also have a zero air voids line plotted alongside. As mentioned previously, this is the line where the soil is completely saturated, and it is the maximum density that can be achieved at a given moisture content. Compaction curves physically cannot plot to the right of the zero air voids line. If this happens, normally the material's specific gravity is in error, or water is running

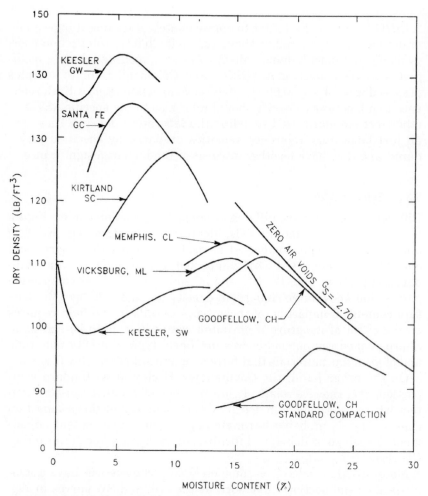

Figure 5.2 Sample compaction curves for different soils.

out of the sample during compaction so that the actual compaction moisture content is lower than the molded moisture content.

If the density of a soil is plotted against the logarithm of the compaction energy for several different compaction energies, the result is approximately a straight line. In Fig. 5.3, the results are plotted for a variety of different soils, normalized to the modified or CE-55 compaction energy. These soils represent subgrade and base course materials from 10 civilian and military airfields in Alabama, Mississippi, Tennessee, Texas, and New Mexico (20 samples) and 18 soils whose characteristics had been reported previously by McRae (1958). The soils are roughly grouped by plasticity index, and the characteristics

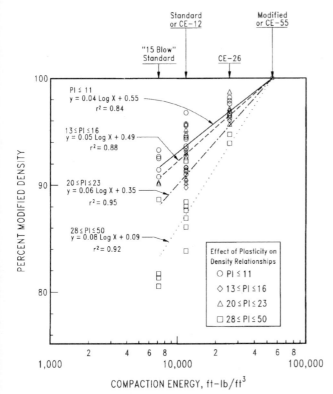

Figure 5.3 Effect of plasticity on density relationships.

of the soils in each group are given in Table 5.1. As a general rule, the more plastic soils have a larger difference between the standard compaction effort and the modified compaction effort. For these specific soils, the maximum standard density averaged 94 percent of maximum modified density for cohesionless and sandy materials, 93 percent for more plastic sandy materials and low plasticity fine-grained materials, 91 percent for plastic soils, and 88 percent for highly plastic soils. These are just rough estimates, and considerable individual variation exists for different soils.

5.3.3 Laboratory versus field behavior

The usefulness of a laboratory soil test depends on how well or poorly it is able to represent or correlate with the actual soil behavior in the field. If a roller compacts a soil in the field at a sequence of different moisture contents, it will produce a compaction moisture content and dry density curve with the same basic shape and characteristics as

TABLE 5.1   Characteristics of Soil Groupings in Fig. 5.3

	Nonplastic and low plasticity	Moderately plastic soils	Plastic soils	Highly plastic soils
Number of soils	15	8	5	7
Plasticity index				
Range	0–11	13–16	20–23	28–50
Mean	5	15	22	38
Number of soils classifying as				
GW	1	—	—	—
GM	1	—	—	—
GC	—	2	—	—
SW	1	—	—	—
SM	1	—	—	—
SC	7	3	2	—
ML	—	—	—	—
CL-ML	1	—	—	—
MH	—	—	1	1
CL	—	3	2	—
CH	—	—	—	7
Ratio of standard to modified density, %				
Mean	93.5	92.9	91.2	87.5
Range	88.2–96.7	90.1–95.6	89.6–93.1	83.7–92.2

was done in the laboratory for the curves in Fig. 5.2. It will show an optimum compaction moisture content and corresponding maximum dry density with decreases in density above and below this optimum moisture content. If the compaction energy produced by the roller is increased by some method such as more passes or increasing the weight of the roller, a new field compaction curve will be produced with, as we saw in the laboratory curves in Fig. 5.2, a higher maximum density and a lower optimum moisture content.

Unfortunately, the roller curve may not match the laboratory curve even though they share a number of common characteristics. Figure 5.4 shows lines of optimums connecting field and laboratory compaction curves for the same soil. There are distinct differences between different rollers and different laboratory procedures. Consequently, it should be clear that there is not a single unique compaction curve and that all compaction curves, whether from the laboratory or the field, are for very specific soil and compaction conditions.

Large earthwork jobs generally will find it worthwhile to develop compaction curves for the specific project soils and compaction equipment. This will potentially avoid controversy and can reduce costs. The Bureau of Reclamation (1974) notes:

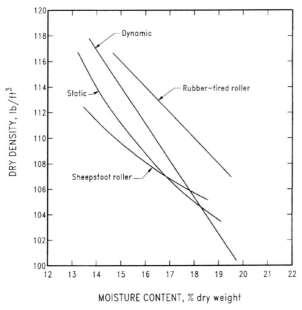

MOISTURE CONTENT, % dry weight

Figure 5.4   Comparison of field and laboratory compaction methods (*Foster 1962*).

In structures where unusual conditions are present, the specifications may include provisions for the contractor to construct one or more test sections of earthfill material. The purpose of a test section is to determine the most practicable excavation, processing, and placing procedures for representative soils under job conditions.

However, as a practical matter, impact laboratory compaction tests have been used very successfully to control compaction of earthwork for better than 50 years and will be adequate for most situations. Nevertheless, as the foregoing discussion has, hopefully, illustrated, the situation is complicated, and judgment is needed to achieve satisfactory results.

5.3.4   Stress distribution and field compaction

The stresses developed in a soil by passage of compaction equipment such as rollers are the agents of compaction. This is true whether stresses are from static, transient, vibratory, or impact forces developed by the equipment. Figure 5.5 shows the stress distribution under individual wheels of four different rubber-tired rollers currently in production in the United States. The sizes in the figure are gross

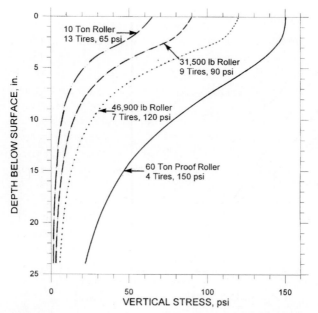

Figure 5.5 Stress distribution under four commercial rubber-tired rollers.

loads, and individual wheel loads are found by dividing the gross loads by the number of tires. In this example, wheel loads vary from 1538 to 30,000 lb, with corresponding tire pressures varying from 65 to 150 psi. At shallow depths, the effect of tire pressure is very significant, with higher stresses providing more compaction energy to the soil.

However, the distribution of stresses with depth suggests that another factor will affect stresses and hence compaction besides tire pressure. The ratio of the stresses from the 65-psi, 10-ton roller to the 150-psi, 60-ton proof roller starts off at 2.3 at the surface but increases to 2.9 at a 2-in depth, 7.2 at 6 in, and 13.3 at 12 in. Clearly, the proof roller is a much more effective roller for achieving compaction at depth than are any of the other rollers in Fig. 5.5. This improved compaction stress is due to more than just higher tire pressure.

Tire pressure is crucial to achieving high density at shallow depth, but at deeper depths, the size of the loaded area becomes more important. The nominal contact area for a tire is the wheel load divided by the tire pressure. The proof roller has a contact area of 200 in^2; the 46,900-lb roller has a contact area of 55.8 in^2; the 31,500-lb roller has a contact area of 38.9 in^2; and the 10-ton roller has a contact area of only 23.7 in^2. The high tire pressure and large contact area are the reasons that the proof roller is effective for reaching high surface density, achieving density in underlying layers, and locating weak spots deep below the surface. Large numbers of repetitions cannot make up for the low stress levels with depth for lightweight equipment. The key is to match the roller to the requirements of the job—sometimes the 10-ton roller will be satisfactory, sometimes the proof roller will be needed, and other times something in between will be appropriate.

In Fig. 5.5 the stress distribution for only one wheel is shown. Some additional stress develops with depth because of stress overlap from adjacent wheels. Techniques for solving for these stresses and using linear superposition to account for stress overlap are readily available in many engineering textbooks (e.g., Yoder and Witczak 1975). The problem of calculating stress distribution under sheepsfoot rollers, vibratory rollers, pad-foot rollers, impact rollers, etc., is appreciably more complex analytically than for the rubber-tired rollers because of the complex geometry involved, variable or uncertain contact area, and dynamic loads. However, the principles remain the same: High surface stresses equate to high compaction near the surface, and large contact areas aid deep compaction. Anything that impedes, dampens, or resists these stresses (e.g., soil plasticity or capillary forces due to low moisture content) will make compaction more difficult.

5.3.5 Compaction and soil properties

Densification. Densification of geotechnical materials when they are loaded after compaction is a common problem. This is particularly noticeable in pavements where thousands or even millions of stress repetitions that can occur under traffic impart considerable energy into the various layers of materials. Considerable compaction is needed during initial construction if densification of these materials is to be avoided under such conditions. This densification under traffic commonly appears as a rut, but it is different in appearance and mechanism from rutting due to shearing in the materials. Densification is caused by a volume decrease with corresponding surface settlement. Ruts from shearing due to inadequate strength typically will show upheaval and material displacement adjacent to the loaded area.

Adequate compaction is equally important for foundations and slabs on grade. The severity of loading is usually less than with pavements, but problems persist. LePatner and Johnson (1982) give an example of a warehouse slab that had been placed directly on fill with no compaction other than spreading and grading. The slab sunk to such an extent that partitions were literally hanging from the ceiling.

Placement of slabs on uncompacted fill is a distressingly common occurrence.

There are some conditions where foundation soils are also vulnerable to densification that may cause serious damage. If the building contains equipment that transmits vibrations to the foundation soils, unexpected densification may occur. This also may be brought about by nearby construction activities such as pile driving. In one circumstance in northwestern Florida, heavy vibratory rollers were placing crushed limestone pads adjacent to a 30-year-old masonry industrial building founded on shallow footings. This work included testing and evaluation of the rollers' effectiveness at different vibratory frequencies and with varying materials and moisture conditions. Consequently, there was relatively intense and repetitive operation of equipment that was somewhat in excess of a normal compaction job. The local foundation soil was a fine silty sand (variable SP to SM), and the vibrations from the roller densified the material under the footings. This resulted in differential settlement and cracking of the masonry walls. Generally, cohesionless soils are particularly susceptible to compaction from vibration, and compaction requirements for such materials used as fill or as foundations should specifically consider future operations or construction that might cause sufficient vibration to densify the soils.

There are also special problems with collapsible soils that densify with dramatic decreases in volume. This can occur in both natural soil deposits and compacted soils. This problem is discussed in more depth in Sec. 8.4.

Airfield pavements are subject to heavy wheel loads and high tire pressures. Compaction is correspondingly very critical to their performance. The heavy Air Force bombers that were introduced in the 1950s forced the Corps of Engineers to examine compaction levels for airfields very critically. They developed a concept of relating compaction requirements to design strength requirements through use of a compaction index C_i. The *compaction index* was defined to be equal to the design CBR that their flexible pavement design method required at any given depth in the pavement. For example, if their design method required a CBR of 10 at a depth of 12 in in the pavement for 10,000 coverages of a B-52, the compaction index would be 10. The change in terminology from CBR to compaction index was to emphasize that although the stress level generated in the material had to affect both the shear resistance (CBR) and the level of compaction (C_i), there is a distinct difference between shear strength and compaction requirements. Figures 5.6 and 5.7 show plots of the compaction index versus the final density measured in pavement materials in full-scale trafficking tests. Early in the analysis of these data, it

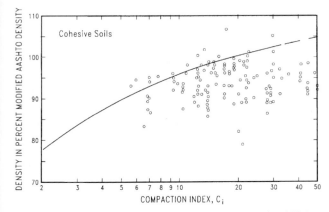

Figure 5.6 Compaction requirements for cohesive soils (*Ahlvin et al. 1959*).

was apparent that there was a distinct difference between cohesive, plastic soils and cohesionless soils, and the difference between the two categories was at a plasticity index of about 2 (Ahlvin 1991). The Corps of Engineers developed the two relationships shown in Figs. 5.6 and 5.7, and these checked against other field results from in-service pavements (Ahlvin et al. 1959). The Corps of Engineers has used these relations since then to set compaction requirements in military airfields, although recent analysis indicates that they may be somewhat conservative (Ahlvin 1991).

The Corps of Engineers approach to specifying required density is the only relatively analytical method of determining required density of which we are aware. Other specifications tend to rely heavily on

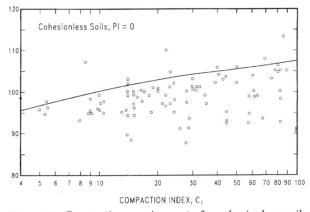

Figure 5.7 Compaction requirements for cohesionless soils (*Ahlvin et al. 1959*).

very specific empirical experience and are often poorly documented. If the strength needed to resist shear can be expressed in terms of CBR, this strength can be related to compaction requirements through the compaction index. Knowing this, one can determine a percentage of modified density that in Corps of Engineers experience with heavy aircraft loads is equal to or greater than the densities that developed under actual traffic. This may be too stringent for lightly loaded pavements, but it is a starting point if there is no other local experience with which to work.

Strength.   Figure 5.8 shows the compaction curves for a Mississippi lean clay compacted at three energy levels. CBR tests were run on these density samples, half on the samples as molded and half on samples that were soaked for 4 days. The moisture content and final dry density of the soaked samples are shown by solid symbols in the figure, and all soaked samples reached or very nearly reached saturation, as indicated by their location along the zero air voids line.

The strength, as measured by the CBR, is plotted above the compaction curves. The unsoaked CBR values are high on the dry side of the optimum moisture content, but there is a dramatic loss in strength as molding moisture content is increased. The soaked CBR curves show a maximum CBR or strength value for each compaction energy level that roughly corresponds to the optimum moisture content for density at that energy level. The reduction of strength due to wetting is a function of the molding moisture content, compaction energy, and soil. Large changes in strength upon wetting are characteristic of fine-grained soils and are less pronounced or even negligible in coarse-grained soils.

Strength tests such as the CBR are too slow and cumbersome to be used for construction control in the field, so normally it will be necessary to control construction by specifying a range of acceptable compaction moisture contents and a minimum percentage of some stipulated standard laboratory compaction test. There are several approaches for estimating the CBR that will be achieved in the field based on these field controls. To obtain an estimate of strength that will be achieved in the field, it is convenient to replot the results of Fig. 5.7 in a form more compatible with the field control methods. Two methods are described in the accompanying illustrative examples.

Illustrative Example 5.1   Determine the likely CBR values for the clay in Fig. 5.8 as placed in the field and later when it is saturated.

solution   This first method follows the standard recommended method of the Corps of Engineers (Department of the Army 1978). The specification limits that are to be used in the field must be selected—for this example, we will use 95 percent of the CE-55 maximum laboratory density with field compaction to

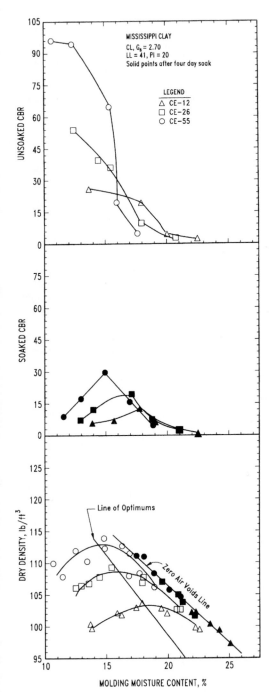

Figure 5.8 Compaction curve for a Mississippi lean clay.

be at CE-55 optimum moisture content $\pm\, 1\frac{1}{2}$ percent. Maximum CE-55 density is 113.0 lb/ft³, and the corresponding optimum moisture content is 16 percent. Consequently, we would expect field placement at 107.4 lb/ft³ or greater (95 percent of CE-55 maximum) and within moisture limits between $14\frac{1}{2}$ and $17\frac{1}{2}$ percent (optimum $\pm\, 1\frac{1}{2}$ percent). Using a constant moisture content, determine corresponding densities and soaked CBR values for each compaction level:

Moisture content	CE-12		CE-26		CE-55	
	γ_d	CBR	γ_d	CBR	γ_d	CBR
14%	100.0	6	107.9	12	112.9	24
15%	101.1	7	108.5	16	113.0	30
etc.						

These are then plotted as shown in Fig. 5.9, and points of equal moisture content are connected. Each curve of constant moisture content will show increasing CBR with increasing dry density up to a point at which there will be a drop in CBR. This drop occurs if the soaked CBR curve for a higher-energy compaction effort (e.g., CE-55 versus CE-26) drops below the CBR curve for the lower energy in plots of soaked CBR versus molding moisture content (see Fig. 5.8). By sequentially plotting each line of constant moisture content from the

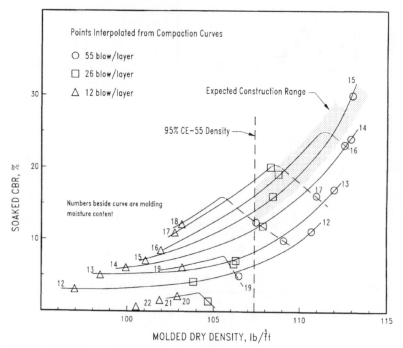

Figure 5.9 Estimate of field CBR values, method 1.

lowest to the highest, the general trends can be identified as they develop. In Fig. 5.9, moisture contents from 12 to 15 percent present easily recognized smooth lines. Then the curves for 16, 17, and 18 percent can be interpolated as parallel to the existing curves up to where the break occurs. The curve for 18 percent has two points on the breakover line, which helps establish the location and shape of the breakover curve shown as a dashed line. This can then be used as a guide to estimate the shape for the 16 and 17 percent curves. The higher moisture content curves above 18 percent show a rapidly dropping CBR value. Once the curves of constant moisture content are sketched in Fig. 5.9, the limits on density and moisture content are shaded. This shaded region provides an estimate of the range of soaked CBR values one might expect for this soil placed at the specified moisture and densities. In this case, CBR values as low as 12 might be encountered.

Soil strength is a complex phenomenon and is a function of soil type, initial molding moisture content, compactive effort, final moisture conditions, and the state and nature of applied stresses. Many soils show a decrease in soaked CBR with increasing density at a constant molding moisture content for some moisture contents. This is seen in the curves for 16, 17, 18, and 19 percent in Fig. 5.9. This behavior is common for fine-grained soils and often occurs for moisture contents above the optimum moisture content from the modified compaction test.

Illustrative Example 5.2   Determine the likely soaked CBR values for the clay in Fig. 5.8 using the same compaction and moisture content controls as in the preceding example.

solution   This example will be based on observations that the CBR values of a soil plotted as dry density versus moisture content will produce parabolic contours of equal CBR with their axes aligned roughly along the line of optimums. CBR and dry density values will be interpolated from Fig. 5.8 at constant moisture content, as was done in the preceding example (or interpolated to determine CBR and moisture content at constant density). The numerical values of CBR are then plotted as shown in Fig. 5.10, and contours of constant CBR are drawn. The specified density and moisture limits are then shaded, and as before, we see that a CBR value somewhat less than 13 could be expected.

Proper selection of the strength value that will be achieved in the field is crucial to design, but it is often poorly done. A large paving project consisting of roadways and very large expanses of parking areas was to be constructed at scattered locations several miles apart. There was considerable pressure to get the project designed and built quickly, and money for engineering was very limited. One surface soil sample was obtained from one of the sites. It had a plasticity index of 12 and classified as a GC soil under the Unified Soil Classification System (USCS). A modified compaction curve and soaked CBR curve were developed, as shown in Fig. 5.11. A design CBR for the pavement was selected as 30 based on these curves, the pavements were designed for this strength soil, the specifications required the soil to be placed at 95 percent of modified laboratory compaction, and the pavements were rapidly built. On the day the pavements were opened

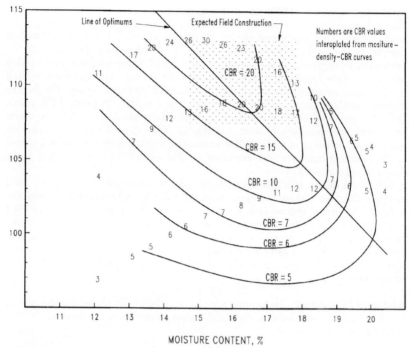

Figure 5.10 Estimate of field CBR values, method 2.

to traffic, they began showing major shear failures. The money that was not available for engineering before was now spent several times over in a failure investigation that revealed that the in situ CBR values were often in the range of 1 to 4 and in better than three-fourths of the cases were less than 10. Estimates for the repair to the pavements to correct the original design deficiency approached two-thirds the cost of the original construction.

The local soil is a residual material derived from cherty, sandy dolomite, and sandstone. It is in a region of rolling hills with active mechanical and chemical weathering forces. From our previous discussions of weathering (Sec. 2.2), residual soils (Sec. 3.4), and soil associations (Sec. 3.3), one could foresee a likely development of a plastic clayey soil with chert gravel remaining unaltered in a sand and clay matrix. The local USDA soil survey shows that several different soil series develop in this residuum depending on their location and steepness of the slope (see Sec. 3.3.2 for a more detailed discussion of why and how this occurs). Consequently, before ever visiting the site, we should expect a plastic soil (residual product of dolomites and limestones), variable properties (common with residual soils and

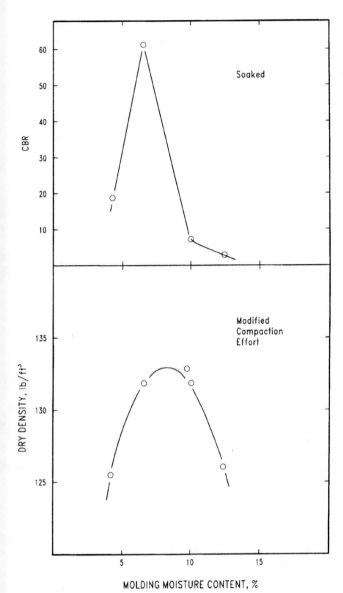

Figure 5.11 Compaction curve for a residual clayey gravel.

residuum from interbedded sedimentary rocks), a gravely component (from the highly weather-resistant chert), and a distinct influence of location and slope on soil properties. These factors should form the basis for the soil exploration program. Under these conditions, it is hopeless to rely on a single sample to determine soil properties. Figure 5.12 compares soil gradations and characteristics of the original sample with samples taken from the constructed sites. The discrepancy between the samples is obvious, and the actual soils used in construction are finer, more plastic, and much weaker than the original sample.

Even if the soil sample had been representative of the actual field conditions, selecting a design CBR value from Fig. 5.11 is problematic at best. The single high 60 CBR value is suspicious. From discussions of the CBR test in Chap. 4, it should be clear that the influence of a single large piece of gravel directly under the 1.95-in-diameter CBR piston could give unrealistically high resistance to the piston's penetration. Because only one curve was run, there are no other data to confirm or deny that such high values are possible. The Corps of Engineers found in the 1950s that there were problems with determining CBR values in the laboratory for granular base and subbase materials simply because of the impact of large aggregate particles on the test and confining effects of the mold (Ahlvin 1991, Department of the Army 1978). To avoid unconservatively high estimates of CBR from laboratory tests, they established the limits in Table 5.2 and required that a material equal or exceed the maximum design CBR in the laboratory tests. In our failure example, the original sample had too many fines to meet the No. 200 recommendation and was at the maximum plasticity index for a *select material* with a design CBR of 20 in Table 5.2. Consequently, there is ample reason to be skeptical of the high CBR value in the laboratory test for this soil.

With only one compaction-CBR curve, there is also no way to determine the interaction of density and moisture content during construction, as was done in the foregoing two illustrative examples. Figure 5.13 shows compaction curves of several of the soils taken from the sites during the failure investigation. These are obviously quite different from the curve in Fig. 5.11, and they also raise the question of how to control compaction on a site where the soils vary significantly (i.e., what laboratory density do you use to calculate the percentage compaction to determine if the soil meets the specification requirement for 95 percent modified density). This problem will be discussed in more detail in Sec. 5.5.3.

Several factors combined together to bring about this unfortunate failure. These included an underfunded design and exploratory effort,

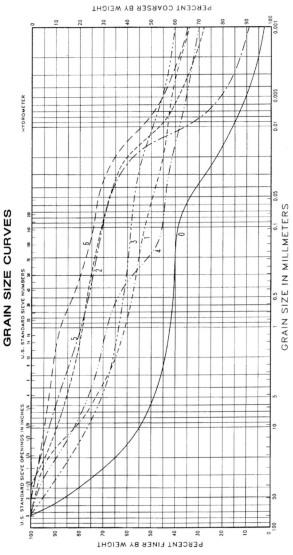

Figure 5.12 Comparison of failure samples and original preconstruction sample gradations.

Curve	Depth, in.	Classification	Coarse LL	Medium PL	Fine PI	Opt. γ, lb/ft^3	Opt. w_o, %	Notes
0	0–10	GC	32	20	12	133.1	8.1	Original Sample
1	8–18	CH	56	15	41	114.3	15.6	Post-Failure Sample
2	12–20	CL–ML	28	21	7	117.7	13.5	Post-Failure Sample
3	6–18	CH	72	17	55	111.3	15.6	Post-Failure Sample
4	6–16	GC	56	16	40	120.0	12.3	Post-Failure Sample
5	16–24	CH	41	15	26	115.5	13.6	Post-Failure Sample
6	18–38	CH	43	17	26	116.3	15.0	Post-Failure Sample

215

TABLE 5.2   Limitations on Design CBR Values

| | | | Maximum permissible value | | | |
| | | | % Passing | | | |
Material	Max. design CBR	Sieve, in	No. 10	No. 200	LL	PI
Subbase	50	3	50	15	25	5
Subbase	40	3	80	15	25	5
Subbase	30	3	100	15	25	5
Select material	20	3	—	25	35*	12*

*Recommended limits, not mandatory.
SOURCE: From Ahlvin, 1991.

undue haste that precluded adequate evaluation of the materials, inadequate appreciation of the geologic factors, unrepresentative and inadequate sampling, inadequate soils testing, lack of appreciation for the limitations of the laboratory testing, and inadequate data upon which to base an engineering judgment of the soil strength following construction.

Permeability.   Compaction has a major impact on soil permeability. As the density of a soil increases, the void space within the soil decreas-

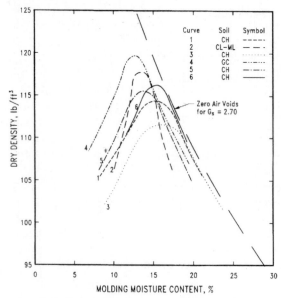

Figure 5.13 Compaction curves for failure samples.

es, with a corresponding reduction in soil permeability. The fabric of the soil also will influence the permeability, with flocculated structures having higher permeability than structures with parallel particle structure. Parallel orientation also indicates that permeability will be anisotropic. As was discussed earlier, compaction on the wet side of optimum and with increasing energy promotes this parallel structure.

Low soil permeability generally will be achieved by high compactive effort and relatively high compaction moisture contents. Unfortunately, these conditions are also associated with low strength. The need for low permeability must be balanced against other needs, such as strength, to define the desired field density and molding moisture content (Daniel and Benson 1990).

As was discussed in Chap. 4, field permeability of an earthen structure will be most influenced by discontinuities in the structure rather than the apparent permeability of intact laboratory specimens. Elsbury et al (1990) found that water penetration in clay liners was a function of macropores and discontinuities rather than the soil permeability. Water preferentially traveled along the interface between lifts in the liner. Also, clods that had not been broken up during construction resulted in higher permeability. The disintegration of clods during construction is encouraged by high water contents and adequate compaction energy (Benson and Daniel 1990).

Specific problems of constructing soil liners and achieving low soil permeability will be examined in more depth in Chap. 10.

Volumetric behavior. As a fine-grained soil changes in moisture content, it undergoes volume changes: swelling when water is added and shrinking when water is removed. This resulting increase and decrease in volume from expansive soils can cause differential movement in structures and pavements and may result in cracking of embankments. One method of controlling these adverse volume changes is to compact the materials with low energy levels (reduce swelling pressure) and at moisture contents above optimum (reduce the future intake of water and lower permeability, which slows the intake of water). The soils that exhibit the largest volume changes are highly plastic clays, and compaction wet of optimum and with relatively low compaction energy means working with a soft, sticky material. As was mentioned earlier, the need for wet of optimum construction to reduce volume change must be balanced with other requirements such as workability and adequate strength. The problems of dealing with expansive soils are covered more fully in Sec. 8.3.

5.4   Compaction Equipment

Early construction used manual methods and horse-drawn equipment for compaction. Steam-driven smooth-wheel rollers were developed in Europe in the middle of the nineteenth century and were later replaced with gasoline-driven equipment in the twentieth century. A horse-drawn sheepsfoot roller was in use in California by 1907 (Porter 1946). Demands for increased compaction in dams and roadways led to a general increase in size for all compaction equipment over the next few decades. In the 1930s, rubber-tired rollers were introduced, and a few huge 200-ton rollers were developed for airfield construction during World War II. These large rollers proved too cumbersome, and today 50- or 60-ton rubber-tired rollers represent the upper size of rubber-tired rollers in use. Since World War II, there have been major advances in developing vibratory rollers, and these now are a major part of the compaction equipment used in construction.

5.4.1   Steel-wheel rollers

Smooth steel-wheel rollers are self-propelled and generally consist of two tandem drums, one in front and one behind. Three-wheel rollers (Fig. 5.14) are another version of the steel-wheel roller, but they are

Figure 5.14   Three-wheel steel roller (*courtesy of Dynapac Mfg., Inc., Schertz, Texas*).

not as common today as they once were. Data on static steel wheel rollers' compaction performance are given by Johnson and Sallberg (1960). Vibratory rollers are more efficient than the static steel-wheel rollers and have largely replaced them for earthwork. The static steel-wheel roller is still widely used for initial breakdown and finish rolling of asphalt concrete. We also have encountered three-wheel rollers being used to compact base-course aggregates in southern Florida and Texas.

5.4.2 Sheepsfoot rollers

Sheepsfoot rollers consist of a drum with protruding feet such as shown in Fig. 5.15. There are a variety of different feet or pad sizes and shapes available today. These variants of the original sheepsfoot roller are variously called *pad-foot, club-foot,* or *tamping-foot compactors.* They typically have shorter feet with more end area and are capable of more speed than the traditional sheepsfoot. This is at the expense of lowered bearing pressure at the tip of the foot. The sheepsfoot roller and variants may be towed or self-propelled and also may be nonvibratory or vibratory. This roller is most widely used to compact cohesive soils. The speed of roller operation has only a minor effect on density (Johnson and Sallberg 1960). As the roller progressively compacts the soil, the feet may penetrate the soil less, and the roller may appear to "walk out" of the soil. This is often taken to be

Figure 5.15 Sheepsfoot roller (*courtesy of Dr. Randy Ahlrich, Vicksburg, Miss.*).

an indication of the degree of compaction, but such an inference may not be correct. For instance, a light roller may rapidly "walk out" of a soil and leave the soil at a lower density than would a heavier roller that did not "walk out" of the soil. If the roller "walks out" of the soil, it simply means that the roller has compacted the soil to the extent it can, but the soil may or may not be compacted to the project requirements.

Figure 5.16 was developed from data reported by the Waterways Experiment Station (1954) and compares the field compaction of a sheepsfoot roller with standard and modified laboratory compaction

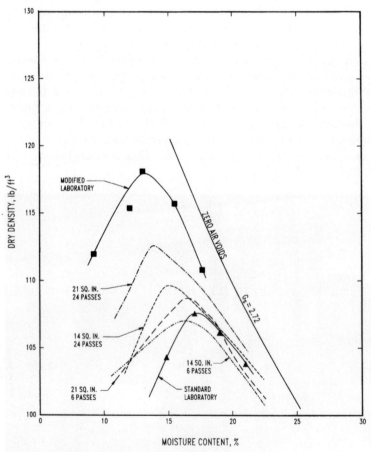

Figure 5.16 Sheepsfoot compaction results on a lean clay.

curves. The soil was a lean clay classifying as a CL. Three differently sized feet varying from 7 to 21 in^2 in area were used, and the roller was ballasted to maintain a constant 250-psi nominal contact pressure per foot. The nominal pressure was calculated as though only one row of feet were in contact with the ground at a time.

The roller developed compaction curves similar to those of the laboratory showing a distinct optimum compaction moisture content. As with the laboratory curves, an increase in roller compaction energy measured either by increased passes or by increased foot size (and hence load) resulted in an increase in maximum density and a decrease in optimum moisture content. Also, the lines of optimums that could be drawn for the laboratory and roller compaction curves in Fig. 5.16 would be similar, although the sheepsfoot roller line of optimums would be to the left and angled more than the laboratory line of optimums. U.S. Bureau of Reclamation experience placing 50 million cubic yards of material with heavy sheepsfoot rollers in dam construction found that they consistently compacted material to their USBR standard laboratory compaction test level, averaging 99.8 percent for gravel-free soils and 103.1 percent for all soils (Hilf 1957a).

The penetration of the individual roller feet into the soil improves the bond between different lifts of materials and helps avoid laminations between lifts. However, compaction of soils with sheepsfoot and pad-foot rollers on the wet side of optimum can result in laminations (e.g., Waterways Experiment Station 1956, Hilf 1957b, and Bureau of Reclamation 1974). Current practice is to compact wet of optimum to try to reduce soil permeability for liners and similar structures, and it is in this moisture regime that laminations develop. As noted in the earlier section on permeability, water will preferentially flow along such interfaces, and interfaces also provide a reduced-strength surface where structural problems may develop. If such laminations develop in the field, it would be prudent to scarify and recompact the materials at a different moisture content or with a different compaction effort.

5.4.3   Rubber-tired rollers

Rubber-tired rollers vary from light two-axle rollers with up to 13 wheels to large 50- and 60-ton rollers with 4 tires on a single axle. Wheel loads vary from 1500 to 30,000 lb for this equipment, and tire pressures range from 25 to 150 psi. The rollers may be towed or self-propelled. An example of a self-propelled rubber-tired roller is shown in Fig. 5.17. Rubber-tired rollers are effective for compacting both cohesive and cohesionless soils. However, laminations between lifts can be a

Figure 5.17   Example of a self-propelled rubber-tired roller.

problem with cohesive soils, particularly if compacted wet of optimum. Scarifying between lifts may be necessary if laminations develop.

Figure 5.18 shows a field compaction curve for four coverages of a rubber-tired roller on a lean clay (CL, $LL = 36$, $PI = 15$). Increased energy is applied by increasing the load and tire pressure, and a typical soil compaction curve develops with a distinct optimum com-

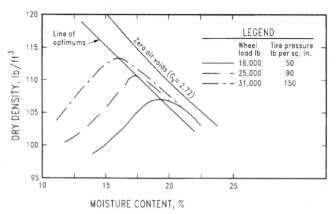

MOISTURE CONTENT, %

Figure 5.18   Field compaction curve for four coverages of a rubber-tired roller on a lean clay (*Turnbull and Foster 1956*). ["Stabilization of Materials by Compaction," W. J. Turnbull and C. R. Foster, 1956, *Journal of the Soil Mechanics and Foundation Division*, reproduced by permission of American Society of Civil Engineers]

paction moisture content that decreases with increasing energy. As was shown previously in Fig. 5.4, the rubber-tired roller line of optimums lies to the right of the laboratory curves. The effect of increasing tire pressure or increasing coverages is shown in Fig. 5.19 for three different moisture contents—one at approximately optimum and one wet and one dry of optimum. Generally, increasing tire pressure proved more effective than increasing coverages, and at higher moisture contents, the effectiveness of either increasing tire pressure or increasing coverages was reduced. Speed of operation also has a minor but measurable effect on density (Johnson and Sallberg 1960).

The depth of effective compaction under 32 coverages of rubber-tired roller is shown in Fig. 5.20. The roller had four wheels ballasted to 25,000 lb each and 90-psi tire pressure. As the lift thickness is increased from 6 to 24 in, the surface density remains approximately

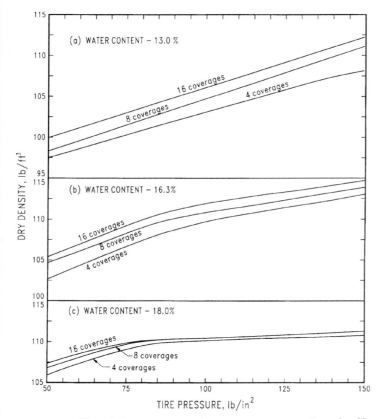

Figure 5.19 Effect of tire pressure and coverages on compaction of a CL clay (*Turnbull and Foster 1956*). ["Stabilization of Materials by Compaction," W. J. Turnbull and C. R. Foster, 1956, *Journal of the Soil Mechanics and Foundation Division*, reproduced by permission of American Society of Civil Engineers]

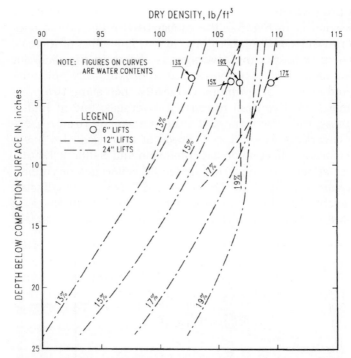

Figure 5.20 Effect of lift thickness on compaction of a CL clay by a 90-psi rubber-tired roller (*Waterways Experiment Station 1957*).

the same, but there is a very distinct density gradient as the lift thickness is increased.

Most common quality-control field density tests such as the sand-cone test (see Chap. 4) determine only the density for a relatively shallow depth. Even the nuclear gauge, which can use a probe to a depth of 12 in on most commercial models, only provides an average density over the depth of measurement. The lower densities at depth may be masked by this averaging. Normal quality-control testing generally will miss the density gradient caused by placement in overly thick lifts unless some excavation to lower lifts is done.

The roller weight, tire pressure, number of passes, and lift thickness are the major factors in determining the compaction that any specific soil at a given moisture content will reach. Roller weight has a greater impact than tire pressure or passes in determining the depth of compaction. The rubber-tired roller is available in a variety of configurations. The light rollers can achieve standard energy levels of compaction if the lifts are sufficiently thin. The heavier rollers can achieve higher levels of density, but there are tradeoffs that must be made between economics of operation, size of roller, tire pressure, number of passes, lift thickness, density gradient, and level of compaction that can be achieved.

5.4.4 Vibratory rollers

Vibratory rollers account for most of the new compaction equipment purchased in the United States today. Smooth steel-wheel rollers consisting of a single drum and two rubber-tired drive wheels such as seen in Fig. 5.21 or tandem drum rollers are probably the most common. However, towed vibratory rollers and vibratory sheepsfoot and pad-foot rollers are also available. Vibratory rollers are available in a wide variety of sizes with static drum weights that exceed 20,000 lb.

Rotating eccentric weights cause the vibration for these rollers. This applies a dynamic load to the soil, and soil pressure measurements under rollers have found 10 to 100 percent increases with vibration as compared with static conditions (Lewis 1954, Bernhard 1952). This vibration also reduces interparticle friction in the soil being compacted and reduces the soil's resistance to densification (O'Reilly 1991, Forssblad 1966). Cohesionless soils will respond best to vibration, while the cohesion of clayey soils impedes the effectiveness of vibration.

Much of the vibratory equipment is marketed based on the size of its dynamic force or the total dynamic and static force. The dynamic force is normally calculated as

Figure 5.21 Typical vibratory roller (*courtesy of Dynapac Mfg., Inc., Schertz, Texas*).

$$F = me\omega^2$$

where $F =$ dynamic force

$m =$ mass of the eccentric weight

$e =$ distance from center of gravity of eccentric mass to the center of rotation

$\omega =$ angular velocity

These claimed dynamic forces are not effective for rating compaction effectiveness, however. Damping within the roller and soil system significantly reduces the effective force that is actually generated in the soil (Yoo and Selig 1979, Rollings 1979, Forssblad 1966, Johnson and Sallberg 1960).

When the frequency of forced vibrations from a vibrator coincides with the natural or resonant frequency of the system, the amplitude of displacements reaches a maximum. This would imply that the optimal compaction of a soil should occur when the soil is vibrated at its resonant frequency by the compaction equipment. However, the value of the resonant frequency depends on the size and weight of the vibrator and the specific test procedures used. Therefore, it is not a unique characteristic of the soil but is a function of the vibratory equipment–soil system.

The problem is further complicated by the fact that the resonant frequency for a specific roller and soil will not remain constant during the compaction process. An increase in a soil's density increases its resonant frequency (Terzaghi and Peck 1967), and a reduction of only 1.2 percent in the moisture content of one granular soil changed the resonant frequency of the soil and vibratory roller system by 120 vibrations per minute (vpm) (LeFlaive and Morel, 1976). Most soils will have a resonant frequency in the range of 800 to 1800 vpm, but the specific value will vary depending on soil type, soil conditions, and the specific vibrator used. Some work has identified two frequencies where peak pressures are developed under vibratory loading (Linger 1963, Yoo and Selig 1979).

It is impractical to specify compaction at the resonant frequency because the frequency is variable and hard to define. Also, at resonance the equipment displacements are a maximum, with a resulting increase in equipment wear and operator discomfort. Fortunately, vibration is effective at a variety of frequencies beside resonance. Johnson and Sallberg (1960) concluded from the data available to them that the dry density achieved by a specific compactor increased with increasing frequency up to a maximum point that may or may not be the resonant frequency. Forssblad (1966) found that increasing the frequency up to the resonant frequency improved compaction, but further increases in frequency above this had relatively little effect on compaction. D'Appolonia et al (1969) found that an increase in frequency from 1140 to 1770 vpm increased density in a fine uniform dune sand by approximately 5 lb/ft^3. Terzaghi and Peck (1967) defined a critical range of frequency between $\frac{1}{2}$ and $1\frac{1}{2}$ times the resonant frequency where settlement, and hence densification, was many times greater than from static force alone. Also, laboratory studies by Brumund and Leonards (1972) found that if the transmitted energy was the same, equal settlements could be achieved for a range of frequencies above and below resonance. Consequently, effective vibratory compaction is possible in the field at a variety of frequencies, and operation at the resonant frequency is not mandatory to achieve good compaction.

The complex interactions of several parameters associated with vibratory rollers are shown in Fig. 5.22. A roller was used to compact a 15-in-thick lift of crushed, well-graded limestone. At different frequencies and amplitude settings, different dynamic forces were generated, and as discussed earlier, the roller and soil system responded differently at different frequencies. Moisture contents were dry, intermediate, and wet (approximately laboratory optimum). At wet or approximately optimum moisture content for this material, higher frequencies and corresponding higher dynamic forces had only a slight effect on the density. However, at lower moisture contents, frequency of operation had a distinct effect, but no consistent pattern was identifiable. This example illustrates that optimizing roller operating parameters requires a trial-and-error approach to determine what is most effective for a specific roller and soil.

Results of a modified energy compaction laboratory curve are plotted in Fig. 5.23 with the field compaction results for several vibratory rollers on a 15-in-thick lift of crushed limestone. The crushed limestone shows the typical double-peaked compaction curve of a cohesionless material, and the vibratory rollers were similarly most effective at high moisture contents and at low moisture contents. Overall, the laboratory and field compaction curves are consistent with one another.

Vibratory rollers are most effective on cohesionless soils, where there is little or no cohesion to impede the rearrangement of soil particles. Similarly, vibratory compaction tends to be most effective at saturation and dry conditions because intermediate moisture contents result in capillary water resistance to particle rearrangement (Forssblad 1966). Other parameters such as increase in fines or increase in plasticity cause further resistance to particle rearrangement and reduce the efficiency of vibratory compaction.

Many extravagant claims are made about the effective compaction depth of vibratory compaction equipment. Figure 5.24 shows the vari-

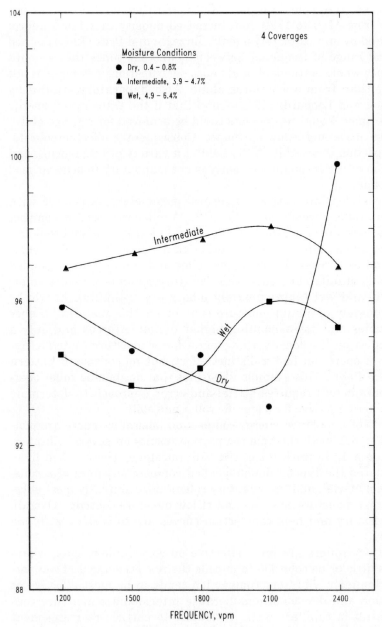

Figure 5.22 Density achieved with four coverages of a vibratory roller on crushed limestone (*from data reported by Rollings 1979*).

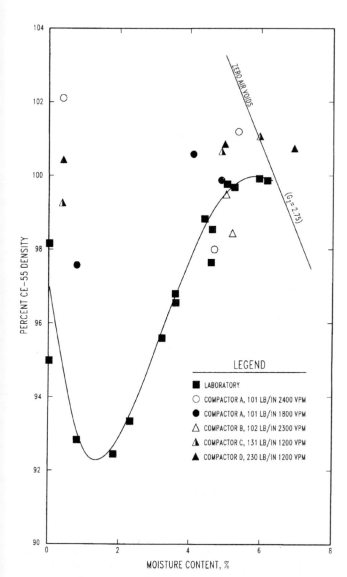

Figure 5.23 Comparison of laboratory and vibratory roller compaction (*Rollings 1979*).

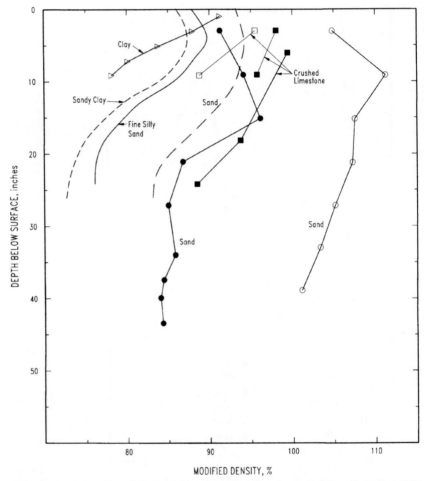

Figure 5.24 Compaction with depth for vibratory rollers (*compiled from Bruzelius 1954, Hall 1968, Rollings 1979*).

ation of density with depth for several vibratory rollers on different soils at optimum moisture content. Only one roller on sand achieved 100 percent of modified AASHTO density throughout its depth, while the others show a very distinct reduction in density with depth. The curves in Fig. 5.24 show a relatively high initial density, an increase in density with depth, and then a distinct reduction in density with depth. Similar patterns of density with depth have been reported by several investigators for vibratory rollers (Johnson and Sallberg 1960, D'Appolonia et al. 1969, Moorehouse and Baker 1969).

Static weight is probably the most reliable single indicator of a vibratory roller's effectiveness (Rollings 1979, Leflaive and Morel

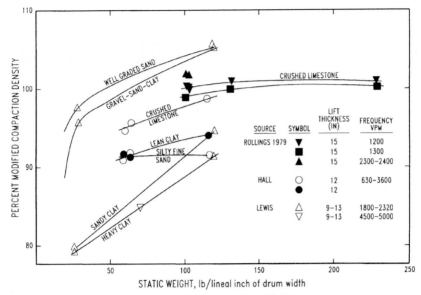

Figure 5.25   Effect of static weight on vibratory roller effectiveness.

1976, Hall 1968, Johnson and Sallberg 1960). Figure 5.25 shows the results from several investigators and in general supports the concept of increasing compaction effectiveness with increasing static weight. However, these results are not conclusive, since there are additional factors such as frequency, lift thickness, and speed of operation besides just the static weight that affect the results shown. The effectiveness of a specific vibratory roller depends on specific roller characteristics (e.g., static weight and ratio of frame to drum weight), dynamic characteristics of the roller (e.g., frequency, dynamic force, and internal damping), roller operation (e.g., passes and speed), soil (e.g., internal damping and resistance to vibration by plastic fines and capillary moisture), and site conditions (e.g., moisture conditions and lift thickness).

Analytical modeling of vibratory rollers has been done (O'Reilly 1985, Yoo and Selig 1979, Leflaive and Morel 1976), but it has not seen much application in practice yet.

5.4.5   Proof rolling

Proof rolling with a heavy rubber-tired roller such as shown in Fig. 5.26 is used to accomplish two objectives: achieve additional compaction above that achieved by normal efforts and identify any undetected soft areas. The roller must be sized to avoid overstressing and

Figure 5.26   Example of a 50-ton proof roller with 150-psi tire pressure.

causing shear failures in the materials that are being proof rolled. Proof rolling on the base course of a heavily loaded military airfield commonly will use 30 coverages of a proof roller with not less than 30,000-lb wheel loads and 150-psi tire pressure. Such large loads would be too severe for a proof-rolling operation directly on the subgrade to identify weak spots.

Proof rolling is not a panacea for poor compaction procedures. Turnbull and Foster (1960) provide an example of unsuccessful proof rolling from construction at Columbus Air Force Base, Mississippi. The base course to be proof rolled dried out between the time that it was compacted and the time proof rolling started. The rubber-tired roller's line of optimums in Fig. 5.4 tends to be to the wet side of the laboratory line of optimums. Consequently, this drying at Columbus AFB would compound the problem of getting density under the proof roller. The proof rolling failed to achieve the desired density, and the 10-in-thick base course later developed a $\frac{3}{4}$-in rut due to densification under traffic. Based on their experience, Turnbull and Foster (1960) concluded:

1.   If the moisture content is in the proper range for compaction, proof rolling will correct compaction deficiencies.

2.   If the moisture content is well on the wet side of optimum, proof rolling will locate this condition and thus permit correction of the condition during construction.

3.   If the moisture content is on the dry side of the proper range for compaction, proof rolling gives a false sense of security, because the layer looks firm and hard; but as moisture increases, the layer will either lose strength drastically or will compact further under traffic.

4. Moderate overrolling during proof rolling will not be detrimental, but excessive overrolling should not be applied, since it may weaken materials that would not be weakened by traffic.

Proof rolling does not replace good compaction procedures and inspection. In fact, during proof rolling, an inspector needs to be present on site to watch for deflection under the roller from any underlying soft areas. Construction equipment such as loaded scrappers or concrete delivery trucks also provide useful tools to detect undiscovered soft spots in a construction site. As they traverse the site, any rutting or large visible deflection under them provides warning of unstable materials that may need to be replaced.

5.4.6 Miscellaneous compaction equipment

A number of riding and walk-behind small rollers of different kinds are manufactured to allow compaction in tight, confined areas where larger equipment cannot operate. There also is a variety of hand-operated equipment such as vibratory plate compactors (Fig. 5.27) and gasoline engine, pneumatic, or hydraulic impact compactors (Fig.

Figure 5.27 Vibratory plate compactor (*courtesy of Wacker Corporation, Menomnee Falls, Wis.*).

Figure 5.28 Impact compactor (*Courtesy of Wacker Corporation, Menomnee Falls, Wis.*).

5.28) for compaction in trenches or adjacent to areas where other larger equipment cannot operate.

Good compaction is possible with this equipment, but lift thicknesses must be small (3 to 4 in), moisture content must be carefully controlled, and an adequate number of coverages of the compactor must be used to achieve good results. Comparative results of several pieces of light compaction equipment on a 6-in lift of crushed limestone are shown in Fig. 5.29. Optimum moisture levels for compaction were not maintained because of hot weather and the free-draining nature of the crushed limestone. Compaction levels that were achieved with all of this equipment under these conditions were modest. In other tests, the small impact compactor was able to consistently achieve 97 percent modified energy compaction levels in 4-in lifts of crushed stone with better moisture control, and the vibratory plate compactor achieved 100 percent modified density in fine dune sands (Rollings 1980).

Small equipment is often needed for placement of material in confined areas. To achieve high compaction levels, thin lifts, good moisture control, and multiple coverages will be needed for this small

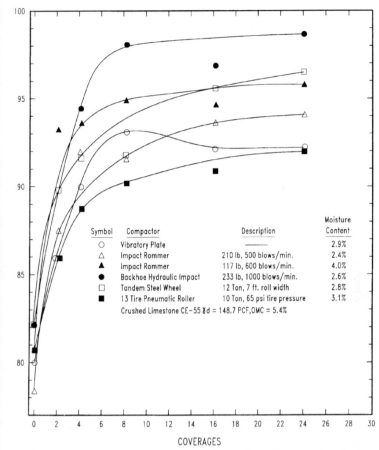

Figure 5.29 Comparison of effectiveness of several small compaction devices on crushed limestone (*Rollings 1980*).

equipment. Backfilled utility trenches have constant problems with settlement and densification under traffic. The primary causes of these problems are an inadequate number of coverages with the small equipment that must be used in the confined space, lifts that are too thick, and poor control of moisture.

Large, heavy vibratory plates, often mounted on equipment such as a dozer chassis, have proven to be effective for compaction. However, they lack the rapid mobility of vibratory rollers and are not common in the United States. They may be encountered overseas, however, and are capable of achieving good compaction.

Despite their low ground pressure, tracked dozers are effective in achieving some compaction. This is apparently due to their vibration, and they are far more effective in cohesionless soils than in cohesive soils. Tracked dozers can compact to 100 percent standard density for many soils and can reach 100 percent of modified density for a few others (Johnson and Sallberg 1960).

Towed impact compactors have been developed and used in Africa (Clifford 1976, Pinard and Ookeditse 1988). These compactors have used three-, four-, and five-sided roller drums. As the roller is towed, the drum rotates, lifting itself up on edge, and then falls back to earth. The impact of the roller striking the ground provides the compactive force. The compactive energy varies depending on the mass of the drum and the height of fall. While a conventional vibratory roller uses relatively high frequencies and low amplitudes for compaction, this compactor uses low-frequency, high-amplitude dynamic energy for compaction. Much of the impetus for developing this equipment came from the need to develop a high-energy compaction device that could be used for compaction at low moisture contents in arid regions and that could help collapse the unstable structure of certain soils often found in arid regions (see Sec. 8.4). The information on the equipment appears promising, but experience is still limited.

Grid rollers consist of towed drums with a meshed surface. This meshed surface typically consists of 1.5-in-diameter steel bars that form a 3.5-in square, and typical roller weights are 12,000 to 30,000 lb (Baryla and Horta 1993). The roller is about as effective as a static steel-wheel roller and has been used in both cohesive and cohesionless soils (Parsons 1959). Its major unique advantage is the ability to crush oversized and larger particles in the material being compacted. It is effective in breaking up material with an LA abrasion greater than approximately 25 or 30 percent (Netterberg 1971). Grid rollers are particularly useful in remote areas and third world countries for in situ processing of relatively soft and oversized materials such as soft limestone, duricrusts (see Secs. 3.2 and 8.2), and laterites (see Sec. 3.4.2).

5.5 Field Compaction

5.5.1 Specification

The engineer preparing the project design and accompanying plans and specifications is responsible for selecting the soil density to be achieved during construction. Only the design engineer can determine what properties are needed in the soil to make it compatible with his or her design concept. The specific behavioral characteristics needed in the soil will form the basis for selecting the project specification compaction requirements.

Too often specifications for compaction are blindly copied from one project to another without considering the actual requirements of the project. Monahan (1994) gives an example of a project that was placing up to 14-ft-thick fill at 95 percent standard Proctor density for a building foundation. Two weeks into the construction, a chance conversation with the building owner's engineer revealed that the building was to have 3000-lb/ft^2 floor loads. Standard Proctor densities are inadequate for this loading, and the contract specifications had to be amended with appropriate extra compensation for the contractor.

Unfortunately, there are not many good guidelines for determining appropriate compaction requirements. Generally, standard Proctor energy compaction is appropriate for light-load-bearing applications (e.g., unloaded utility trench backfill or single-family residential structures and driveways). Modified density compaction is needed for any serious load-bearing applications. The difference between standard and modified energy can be appreciated by Monahan's (1994) wry observation that modified energy corresponds to the "blister density"—the uncalloused hand will blister after just a few hand-dug excavations for sand-cone tests or similar field work. Figures 5.6 and 5.7 provide one method of selecting compaction levels for heavily trafficked pavements, and local and state department of transportation practices may prove helpful in selecting compaction levels for more lightly loaded highway-type pavements. However, any unusual loading conditions or soils for which there is not good established and successful local practice or for cases where specific engineering requirements for the soil are critical, laboratory or field trials may be required to determine appropriate compaction specifications.

The specification for field compaction must define the type of laboratory test to be used as a reference and the percentage of this reference laboratory density that will be required in the field (e.g., 100 percent of modified density or 90 percent of standard density). The specification should clearly state what test method such as ASTM or Military Standard (see Chap. 4) will be used to determine the reference laboratory density, and if the method has alternate methods (e.g., method A, method B, etc.), the one to be used should be explicitly stated.

Compaction moisture contents are usually allowed to be within a range of ±1 or 2 percent of the laboratory optimum to allow for field variability. However, there are instances where specific ranges dry or wet of optimums are needed for engineering reasons, and these must be specified.

The frequency, method of measuring the field density and moisture content, and specific test procedures (e.g., ASTM or AASHTO) to be used to determine compliance with the project specification also should be clearly stated (e.g., sand-cone test in accordance with ASTM D 1557, Method A, for every 500 tons of fill). Similarly, who will conduct the test should be stipulated (owner, contractor, or third-party laboratory).

The maximum lift thickness and whether the lift thickness is measured before compaction (sometimes referred to as *loose lift thickness*) or after compaction also should be specified. The soil type, degree of compaction needed, field compaction energy (type and size of compaction equipment and number of passes), and the contractor's skill in maintaining moisture conditions and handling the material actually determine what lift thickness can be compacted effectively. However, these parameters are not known when the specifications are prepared. Overly thick lifts of materials can develop a serious density gradient (high on the surface where it is tested and low at the bottom), but at the same time, we want the contractor to be able to place the material as efficiently and cost-effectively as possible. A reasonable approach is to specify a maximum lift thickness (e.g., a maximum of 6 in measured after compaction) with a provision that thicker lifts can be approved if testing determines that the contractor's placement methods and equipment can achieve the specified density throughout the lift. The important proviso here is that physical testing in the lower portion of the lift verifies that the specified density is being achieved. This should be monitored throughout the placement and not just at the beginning of the work.

Finally, the specification should address any other special requirements such as proof rolling that will be required by the engineer. Normally, minimum density as a percentage of a specified laboratory standard density, an allowable range of moisture content for compaction, and possibly a maximum lift thickness are adequate to control compaction of most geotechnical materials. However, there are sometimes critical facilities that require additional testing for specification compliance. For example, clay liners often must pass permeability tests before they are acceptable. Overseas, additional requirements may be encountered. For example, in Germany, the foundation soil often must be a minimum strength from a plate load test before it is acceptable.

The specification for compaction should be based on engineering requirements. It must ensure that the geotechnical materials as placed meet the designer's requirements for performance. Since these materials are often the base on which other structures rest, we want to ensure that a due degree of conservatism is used. At the same time, we want to allow maximum flexibility and cost-effectiveness in construction without unnecessary restrictions. Hence requirements for proof rolling or special testing for strength or permeability should be based on the need for the requirement rather than traditional practices or extra conservatism. If it is needed, however, specify it and pay the contractor for it. Preparation of compaction specifications requires a balance of engineering considerations, construction practicality, and economics.

5.5.2 The need for a foundation

It is impossible to adequately compact a material if it is not on a firm foundation. Figure 5.30 shows the stress distribution under two of the rollers from Fig. 5.5 for two different foundations. In one case, the modulus of elasticity of the soil under the roller is assumed to be a constant 25,000 psi, corresponding to a reasonably good granular material. In the second case, the roller is on a 6-in-thick layer of the 25,000-psi modulus soil over a 5000-psi modulus soil. This lower soil corresponds roughly to a CBR of 3 to 4, which would be representative of a soft clay. The stresses available for compacting the 6-in lift of surface material are greatly reduced for the second case because of straining in the soil mass. Surface deflections for each of the four commercial rollers for these two cases are shown in Fig. 5.31, and the weak lower layer causes a major increase in surface deflection. High levels of compaction will not be possible in the thin lift of material over this weak subgrade, and the high stresses on the weaker lower soil may cause shearing and rutting.

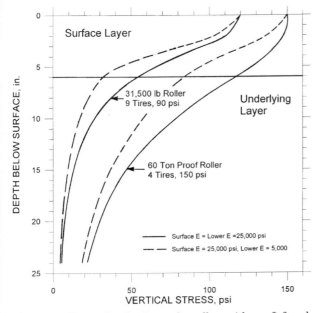

Figure 5.30 Stress distribution under rollers with a soft foundation.

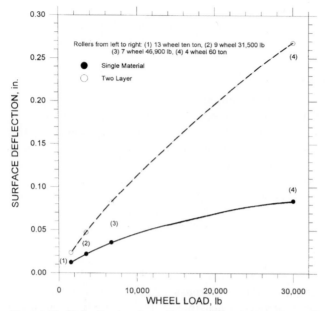

Figure 5.31 Deflections under rollers on different support conditions.

When soil must be placed and compacted over a weak material, the project design must recognize that the initial lifts of materials cannot be compacted to a high degree. In such cases, the designer must recognize that he or she must either remove or improve the existing materials, build a construction platform upon which construction activities such as compaction will be possible, or allow for the lower density with its corresponding impact on engineering properties in the lower lifts (Rollings et al. 1988). These topics are considered in more detail in Chap. 7.

5.5.3 Troubleshooting

When the specified level of compaction is not being achieved in the field, there are a variety of possible causes and remedies. Usually the problem can be traced to one or some combination of the following reasons.

Use the right roller. Certain rollers are more effective in some soils than in others. Rollers using dynamic forces from vibration or impact tend to be more effective in cohesionless than in cohesive soils. Sheepsfoot rollers do best in cohesive soils. General rules like this are simply guidelines, however, and exceptions abound. If the specified

TABLE 5.3   Compaction Equipment* for Different Conditions

Soil	First choice	Second choice	Comment
Rock fill	Vibratory	Pneumatic	—
Plastic soils, CH, MH	Sheepsfoot or pad foot	Pneumatic	Thin lifts usually needed
Low-plasticity soils, CL, ML	Sheepsfoot or pad foot	Pneumatic, vibratory	Moisture control often critical for silty soils
Plastic sands and gravels, GC, SC	Vibratory, pneumatic	Pad foot	—
Silty sands and gravels, SM, GM	Vibratory	Pneumatic, pad foot	Moisture control often critical
Clean sands, SW, SP	Vibratory	Impact, pneumatic	—
Clean gravels, GW, GP	Vibratory	Pneumatic, impact, grid	Grid useful for over-size particles

*Vibratory rollers, Sec. 5.4.4; sheepsfoot, Sec. 5.4.2; pad-foot, Sec. 5.4.2; pneumatic (rubber-tired), Secs. 5.4.3 and 5.4.5; grid roller, Sec. 5.4.6; impact, Sec. 5.4.6; static steel-wheel, Sec. 5.4.1 (usually vibratory rollers will do better on geotechnical materials; can be effective on GP, GW, and GM soils; sometimes useful for smoothing or surface compaction after other rollers).

density is proving elusive, a switch to a different type of roller more suited to the conditions may help. Table 5.3 provides some guidance on matching compaction equipment to different soils.

Probably, one of the most common problems is an inadequately sized roller for the job. If high levels of modified compaction energy laboratory density are being specified, a large roller will be needed. As we saw in Sec. 5.3.4, an undersized roller may never be able to generate enough compaction energy to achieve the desired results.

For sheepsfoot, vibratory, and pneumatic rollers, there has to be enough mass present to carry out the compaction. Variants of the sheepsfoot roller such as the pad-foot roller may encounter problems in achieving density if the bearing area of the foot is too large, and a switch back to a sheepsfoot or another pad-foot roller with a smaller bearing area may be helpful. For pneumatic rollers, an increase in the tire pressure may be helpful in achieving higher surface density, while an increase in wheel load may achieve more density at depth. Also, if several lifts are already in place and the density is found deficient, proof rolling may prove helpful.

Changes in the frequency of vibration may prove helpful for vibratory rollers with adjustable frequencies. Sometimes vibratory rollers loosen the surface of fine, cohesionless soils. Final rolling with the vibrators off or supplementing the vibratory roller with a pneumatic or other roller may densify the surface. Also, when succeeding layers are being placed, the loose surface of one may be densified by the compaction of the overlying lift. This should be checked by testing and not just be assumed to occur, however.

Slowing the speed of roller operation is another technique that sometimes helps improve compaction.

Check site conditions.   Section 5.5.2 emphasized that good compaction cannot be achieved on a weak and yielding material. If a foundation is wet or soft, steps should be taken to drain it, replace it with better material, or stabilize it so that work can proceed. If permeable materials are placed over a fine-grained subgrade, later rain can soak through the permeable surface materials and weaken the underlying fine-grained soil. Also, as construction proceeds, natural drainage channels are often blocked by the ongoing work, and after a rain, water may be dammed up on the construction site, soaking and weakening the soil upon which construction is to proceed. Good attention to site drainage and prompt steps to encourage drainage of soaked areas and removal of ponded water are needed to keep work proceeding smoothly.

On occasion, organic material, vegetation, stumps, trash, etc., are inadvertently covered by fill material during construction. This material is usually weak and soft so that compacting fill material above such material is difficult, and the material is highly undesirable in an engineered structure. Usually these soft spots are evident during compaction, or they can be identified by proof rolling. Such materials need to be excavated and replaced with better materials. This problem seems to be particularly prevalent when filling over old drainage ditches and ravines or at the bottoms of slopes.

Overly thick lifts of material cannot be compacted effectively. Allowable lift thickness is best established by trial and error in the field. Loose lift thickness can be checked manually in the field, but it is good practice to establish several cross sections on the site where profiles can be taken periodically. By knowing the ground elevation changes from these profiles and recording the number of lifts of material that are being placed and compacted, an average compacted lift thickness can be determined easily.

Review test procedures.   In Chap. 4 we saw that a number of errors can creep into both laboratory and field testing. Consequently, if density is proving difficult to achieve in the field, a complete check of the test procedures being used to determine the laboratory density of the soil and to measure the field density of the soil is worthwhile. Weeks and Parker (1987) describe an incident where a contractor placing granular, load-bearing fill was repeatedly directed to remove and replace the material because of inadequate density. After repeated attempts to achieve density failed, an outside review of the problem found that an uncalibrated nuclear gauge had been used in all tests. This resulted in an underestimate of 3 to 5 percent in the density measurement. Also, the gauge was placed directly on the compacted surface without any surface preparation. The minute air gaps between the face of the instrument and the soil resulted in significantly underestimating the soil density. Once these testing deficiencies were rectified, the compaction of the fill was found to meet the specification requirements.

Control the moisture.   Based on previous discussions in this chapter, moisture content is clearly a crucial element during compaction. Despite its importance, it often gets scant attention in the field. If the desired density is not being achieved in the field, checking the moisture content and its consistency throughout the material being compacted is probably the first step to identifying the problem. Often, moisture may be correct at the borrow site or at the start of compaction but will be lost by evaporation before compaction is complete. This can be dramatic in arid regions, where moisture loss can reach 50 percent (Pinard and Ookeditse 1988). However, it can be just as serious a problem in humid climates. We have encountered problems maintaining adequate moisture in a silty sand subbase and a silty clay liner in northern Florida and western Mississippi, respectively. These were both under very humid conditions. Generally, permeable soils, silty soils, high temperatures, low relative humidities, and wind are factors that increase the potential for problems with keeping the desired moisture in the soil during compaction.

One of the basic steps in compaction control is to plot the field moisture contents and dry densities. It is impossible to assimilate columns of numbers, and plotting allows the relations between moisture content and density to be understood much more rapidly and clearly. Control charts that plot each value or averages of moisture content or density measured in the field against date, time, or numerical sequence allow data trends such as a decrease in density or an increase in moisture content to be recognized at a glance. Corrective action can be then be started. Plotting the dry density and moisture content together with the laboratory curves and with details on compaction procedures (e.g., roller weight or number of passes) may help identify the most effective compaction procedures in the field.

The large amount of data often produced during compaction quality-control testing is ideally suited for statistical analysis. An approximately normal distribution of test results occurs in the field, and the

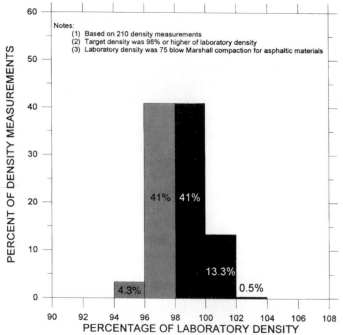

Figure 5.32 Example of a frequency diagram showing field density test results.

range or distribution of results is just as important as the average value. A frequency-distribution diagram such as Fig. 5.32 is a convenient method of visualizing the spread of data results, assessing the average quality of the results, determining the range of results, and quantifying the amount and seriousness of results that fall outside the specified values.

Check the material. The soil compacted in the laboratory must be the same as the soil being compacted in the field, or else the density used to determine specification compliance will be in error. Small variations in gradation or plasticity may cause changes in the value of the laboratory optimum density, and failure to recognize that the soil being placed in the field differs from the one used in the laboratory has caused many construction disputes. This difference in the soils can arise in various ways: poor sampling and handling of the soil used in the laboratory, natural variation of soils in the field, segregation in the field, or breakdown of the soil or aggregates during handling and compaction.

Periodic sampling and testing of the soil in the field should be done routinely to identify any significant variations in the soil being placed in the field. Breakdown of soils and aggregates does occur (see Sec. 3.4.1 for an example), but segregation is more often the cause of the problem (see Sec. 2.5.5). Careful sampling may be needed to find the cause of the problem (i.e., breakdown or segregation).

The residual soil discussed in Sec. 5.3.5 is an example of how natural soils may vary and yet in the field the soils may appear visually identical. In Fig. 5.13 we see that the compaction curves for these soils are significantly different, and if one does not know which curve to use, it is impossible to ascertain specification compliance (i.e., what is the proper maximum laboratory density to determine percent compaction in the field).

One method to overcome this problem is to use a single-point compaction curve. First, a family of compaction curves is prepared at the specified compaction effort for the range of soils found in the construction site. Figure 5.13 is an example of such a family of curves. A soil sample is obtained from the field where compliance is to be checked, its moisture content is determined, and the sample is compacted at the specified compaction effort. This establishes a single point on the soil's compaction curve, and the soil's true compaction curve can then be estimated by interpolating a curve through this point and parallel to adjacent curves. A variant of this is to use two or more points. Points near or dry of optimum moisture are most useful for one- or two-point estimated curves, since the compaction curves of most soils will tend to be very close together wet of optimum.

5.6   References

Ahlvin, R. G. 1991. "Origin of Developments for Structural Design of Pavements," technical report no. GL-91-26, USAE Waterways Experiment Station, Vicksburg, Miss.

Ahlvin, R. G., D. N. Brown, and D. M. Ladd. 1959. "Compaction Requirements for Soil Components of Flexible Airfield Pavements," technical report no. 3-529, USAE Waterways Experiment Station, Vicksburg, Miss.

Baryla, J. M., and J. C de O. S. Horta. 1993. "The Grid Roller: Compactor or Crusher?" *Geotechnical News,* 11 (1): 57–64.

Benson, C. H., and D. E. Daniel. 1990. "Influence of Clods on Hydraulic Conductivity of Compacted Clay," *Journal of Geotechnical Engineering,* 116 (8): 1231–1248.

Bernhard, R. K. 1952. "Static and Dynamic Soil Compaction," *Proceedings,* 31, Highway Research Board, Washington.

Brumund, W. F., and G. F. Leonards. 1972. "Subsidence of Sand Due to Surface Vibration," *Journal of the Soil Mechanics and Foundation Division, ASCE,* SM1: 27–42.

Bruzelius, N. 1954. "On Soil Compaction (Om komprimering av jord)," *Statens Väginstitut Medd.,* 87.

Bureau of Reclamation. 1974. *Earth Manual,* 2d ed., U.S. Department of the Interior, Washington.

Clifford, J. M. 1976. "The Development and Use of Impact Rollers in the Construction of Earthworks in Southern Africa," bulletin no. 16, National Institute for Transportation and Road Research, Pretoria, South Africa.

Daniel, D. E., and C. H. Benson. 1990. "Water Content-Density Criteria for Compacted Soil Liners," *Journal of Geotechnical Engineering,* 116 (12): 1811–1830.

D'Appolonia, D. J., R. V. Whitman, and E. D'Appolonia. 1969. "Sand Compaction with Vibratory Rollers," *Journal of the Soil Mechanics and Foundation Division,* SM1: 263–283.

Department of the Army. 1978. *Flexible Pavement Design for Airfields,* TM 5-825.2, Washington.

Elsbury, B. R., D. D. Daniels, G. A. Sraders, and D. C. Anderson. 1990. "Lessons Learned from Compacted Clay Liner," *Journal of Geotechnical Engineering,* 116 (11): 1641–1660.

Forssblad, L. 1966. "Investigations of Soil Compaction by Vibration," bulletin no. 68, Royal Institute of Technology, Stockholm, Sweden.

Foster, C. R. 1962. "Field Problems: Compaction," in *Foundation Engineering,* G. A. Leonards, ed., McGraw-Hill, New York.

Garrison, E. G. 1991. *A History of Engineering and Technology: Artful Methods,* CRC Press, Boca Raton, Fla.

Hall, J. W. 1968. "Soil Compaction Investigation, Report 10, Evaluation of Vibratory Rollers on Three Types of Soils," Technical Memorandum No. 3-271, USAE Waterways Experiment Station, Vicksburg, Miss.

Harding, H. J. 1958. Discussion on "Consultants, Clients, and Contractors," by K. Terzaghi, *Journal of the Boston Society of Civil Engineers*; reprinted in *Contributions to Soil Mechanics, 1954–1961,* Boston Society of Civil Engineers, Boston, Mass., pp. 289–292.

Hilf, J. W. 1957a. "Compacting Earth Dams with Heavy Tamping Rollers," 83 (SM2), *Journal of the Soil Mechanics and Foundation Division*; pp. 1205-1–1205-28, reprinted in *Award-Winning ASCE Papers in Geotechnical Engineering, 1950–1959,* American Society of Civil Engineers, New York, pp. 507–540.

Hilf, J. W. 1957b. Discussion of "Effect of Tire Pressure and Lift Thickness on Compaction of Soil with Rubber-tired Rollers," by W. J. Turnbull and C. R. Foster, Conference on Soils for Engineering Purposes, Mexico City, STP 232, American Society for Testing and Materials, Philadelphia, pp. 444–445.

Johnson, A. W., and J. R. Sallberg. 1960. "Factors that Influence Field Compaction of Soils," Highway Research Board bulletin 272, Washington.

Lambe, T. W. 1962. "Soil Stabilization," in *Foundation Engineering,* G. A. Leonards, ed., McGraw-Hill, New York.

Leflaive, E., and G. Morel. 1976. "Compaction: Present Trends," translation no. 76-1, USAE Waterways Experiment Station, Vicksburg, Miss.; originally published as "Le Compactage: Orientations Actuelles," *Annales de L'Institut Technique du Batiment et des Travaux Publics,* 325.

LePatner, B. B., and S. M. Johnson. 1982. *Structural and Foundation Failures: A Casebook for Architects, Engineers, and Lawyers,* McGraw-Hill, New York.

Lewis, W. A. 1959. "Investigation of the Pneumatic-Tyred Roller in the Compaction of Soil," Road Research Laboratory technical paper no. 45, Road Research Laboratory, United Kingdom.

Lewis, W. A. 1954. "Further Studies in the Compaction of Soil and the Performance of the Compaction Plant," Road Research Laboratory technical paper no. 33, Department of Scientific and Industrial Research, United Kingdom.

Linger, D. A. 1963. "Effect of Vibration on Soil," Highway Research Record no. 22, Washington.

McRae, J. L. 1958. "Index of Compaction Characteristics," miscellaneous paper no. 4-269, USAE Waterways Experiment Station, Vicksburg, Miss.

Mitchell, J. K. 1993. *Fundamentals of Soil Behavior,* 2d ed., Wiley, New York.

Mitchell, J. K., D. R. Hooper, and R. G. Campanella. 1965. "Permeability of Compacted Clay," *Journal of the Soil Mechanics and Foundation Division,* 91 (4): 41–65.

Monahan, E. J. 1994. *Construction of Fills,* 2d ed., Wiley, New York.

Moorehouse, D. C., and G. L. Baker. 1969. "Sand Densification by Heavy Vibratory Compactor," *Journal of the Soil Mechanics and Foundation Division,* SM4: 985–994.

Netterberg, F. 1971. "Calcrete in Road Construction," bulletin no. 10, National Institute for Transportation and Road Research, Pretoria, South Africa.

O'Reilly, M. 1991. "Analyses of the Performance of Vibrating and Impacting Compacting Plant," research report 316, Transportation Research Laboratory, Crowthorne, U.K.

O'Reilly, M. 1985. "Predictions of the Performance of Compaction Plant," doctoral dissertation, Heriot-Watt University, Edinburgh.

Parsons, A. W. 1959. "An Investigation of the Performance of a 13-$\frac{1}{2}$ ton Grid Roller in the Compaction of Soil," road research note NRN/3563/AWP, Road Research Laboratory, Crowthorne, U.K.

Pinard, M. I., and S. Ookeditse. 1988. "Evaluation of High Energy Impact Compaction Techniques for Minimizing Construction Water Requirements in Semi-Arid Regions," in *Proceedings of the 14th Australian Road Research Board Conference,* part 7, Canberra, Australia, pp. 121–139.

Porter, O. J. 1946. "The Use of Heavy Equipment for Obtaining Maximum Compaction of Soils," American road builders technical bulletin 109.

Proctor, R. R. 1933. "Design and Construction of Rolled Earth Dams," *Engineering News Record*, 31: 245–248, 286–289, 348–351, and 372–376.

Rollings, R. S. 1980. *Interim Report of Field Test of Expedient Pavement Repairs (Test Items 1-15),* ESL-TR-79-08, Air Force Engineering and Services Center, Tyndall AFB, Florida.

Rollings, R. S. 1979. *Tyndall Vibratory Compaction Tests,* Pavement Systems Division letter report, USAE Waterways Experiment Station, Vicksburg, Miss.

Rollings, R. S., M. E. Poindexter, and K. G. Sharp. 1988. "Heavy Load Pavements on Soft Soil," in *Proceedings 14th Australian Road Research Board Conference,* part 5, Canberra, Australia, pp. 219–231.

Terzaghi, K., and R. B. Peck. 1967. *Soil Mechanics in Engineering Practice,* Wiley, New York.

Turnbull, W. J., and C. R. Foster. 1960. "Proof-Rolling of Subgrades," Highway Research Board bulletin no. 254, Washington.

Turnbull, W. J., and C. R. Foster. 1956. "Stabilization of Materials by Compaction," *Journal of the Soil Mechanics and Foundation Division*; 82 (SM2), pp. 934-1–934-23, reprinted in *Award-Winning ASCE Papers in Geotechnical Engineering, 1950–1959,* American Society of Civil Engineers, New York, pp. 409–434.

Wahls, H. E., C. P. Fisher, and L. J. Langfelder. 1966. "The Compaction of Soil and Rock Materials for Highway Purposes," School of Engineering, North Carolina State University, Raleigh, N.C.

Waterways Experiment Station. 1957. "Soil Compaction Investigation: Report No. 8, Effect of Lift Thickness and Tire Pressure," technical memorandum No. 3-271, Vicksburg, Miss.

Waterways Experiment Station. 1956. "Soil Compaction Investigation: Report No. 7, Effect on Soil Compaction of Tire Pressure and Number of Coverages of Rubber-Tired Rollers and Foot-Contact Pressure of Sheepsfoot Rollers," technical memorandum no. 3-271, Vicksburg, Miss.

Waterways Experiment Station. 1954. "Soil Compaction Investigation: Report No. 6, Effect of Size of Feet on Sheepsfoot Roller," technical memorandum no. 3-271, Vicksburg, Miss.

Weeks, A. G., and R. J. Parker. 1987. "Errors in Compaction Site Control," in *Compaction Technology,* Thomas Telford, London.

Yoder, E. J., and M. W. Witczak. 1975. *Principles of Pavement Design,* Wiley, New York.

Yoo, T. S., and E. T. Selig. 1979. "Dynamics of Vibratory-Roller Compaction," *Journal of Geotechnical Engineering,* 105 (GT10): 1211–1231.

6

Stabilization

"Not every soil can bear all things."

VIRGIL

6.1 Introduction

Seldom does nature provide the ideal soil or aggregate for construction. To overcome deficiencies in soil or aggregate properties such as poor grading, excess plasticity, or inadequate strength, we may blend two or more soils together, or we may add stabilizing admixtures such as lime, portland cement, or bituminous materials to the soil or aggregates. These techniques are effective if we can readily mix the materials. Other techniques for improving soil conditions at depth will be covered in Chap. 7.

We often think of stabilization as a method of providing structural strength, but it can have a number of other construction and behavioral effects that are equally beneficial. These might include improved soil workability, an all-weather construction platform, or reduced swelling of expansive materials. Stabilization may improve the properties of an on-site or local material to allow its use rather than incurring the cost of importing a better material from a distant source. In the following sections we will examine the effects of blending and stabilizing with lime, portland cement, bituminous materials, pozzolanic and slag materials, and specialty admixtures.

6.2 Blending

Two or more soils may be blended together to overcome deficiencies in grading. Few natural soils or aggregates will meet commonly specified gradations for base courses and similar high-quality load-bearing geotechnical structures. Also, blends of several different aggregates are used to produce the final aggregate gradation for materials such as asphalt and portland cement concrete.

Blending is best accomplished by feeding the material from aggregate bins or hoppers onto a conveyor and then into a pugmill mixer. Calibrated gates on the aggregate bins control the amount of each material going onto the conveyor, and the pugmill provides vigorous mixing to ensure a homogeneous blend. This setup is highly portable. Attempts to blend materials by winrowing, rototilling, and other similar techniques will face a number of potential problems. These include segregation, contamination, nonuniform mixing, and difficulties in controlling the proportions of different materials.

Blending of soils and aggregates to achieve a desired gradation is usually approached as a trial-and-error procedure. Several combinations of aggregate proportions are tried until a combination that provides the desired gradation is found. This problem is solved very easily and quickly on modern spreadsheet computer programs, and there are also specialized computer programs that will determine aggregate blends. There are graphical approaches also (e.g., Barksdale 1991), but the spreadsheets and computer programs are faster and easy to use.

For any specific sieve size, the final percentage passing that sieve for a blend of aggregates can be calculated as

$$P_i = \sum_{x=1}^{n} A_x \frac{p_x}{100} = A_1 \frac{p_1}{100} + A_2 \frac{p_2}{100} + \cdots + A_n \frac{p_n}{100}$$

where P_i = percentage passing sieve i for the blend
A_x = percentage of aggregate x passing sieve i
p_x = percentage of aggregate A_x in the blend; sum of p_x equals 100 percent

This very simple equation is easily put into readily available computer spreadsheet software so that the calculation of even very complex blends of multiple aggregates can be done very quickly for as many sieve sizes as desired.

Illustrative Example 6.1 Determine a blend of the silty gravel, gravelly sand, eolian sand, crushed limestone, and glacial till from Fig. 2.3 that will meet the gradation requirements for the ASTM D 2940 base course in Table 2.14. For this example, assume that 100 percent of the silty gravel in Fig. 2.3 passes the 2-in sieve.

solution Plotting the soils and the specification requirements on a gradation chart as shown in Fig. 6.1 helps determine if two soils can be blended to meet the specification. If a straight line can be drawn between two soils on this plot and the line passes through the specification zone on the chart, blending of the soils usually will produce a gradation within the specification limits. A quick

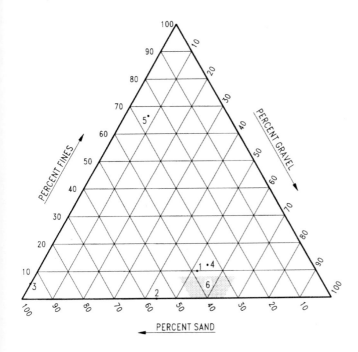

1. Silty Gravel, Oman
2. Gravelly Sand
3. Eolian Sand
4. Crushed Limestone
5. Glacial Till
6. Specification Requirements

Figure 6.1 Aggregate grading chart.

inspection of Fig. 6.1 reveals that the eolian sand (no. 3 in the figure) and glacial till (no. 5 in the figure) will be of no use in solving this problem. A straight line between the gravelly sand (no. 2 in the figure) and either the silty gravel (no. 1 in the figure) or the crushed limestone (no. 4 in the figure) passes through the corner of the specification's shaded area.

Considering a blend of 75 percent silty gravel and 25 percent gravelly sand, the equation for blending gives the following result:

Sieve size	Silty gravel		Gravelly sand		75:25 Blend‡	ASTM D 2940	
	Grad.	Prop.*	Grad.	Prop.†		Min.	Max.
2 in	100	75	100	25	100	100	—
1½ in	93	69.8	100	25	94.8	95	100
¾ in	78	58.5	80	20.0	78.5	70	92

$\frac{3}{8}$ in	62	46.5	67	16.8	63.3	50	70
No. 4	48	56	56	14.0	50.0	35	55
No. 30	22	34	34	8.5	25.0	12	25
No. 200	10	7.5	0	0	7.5	0	8

*Proportion in the blend. Determined as 0.75×percentage passing in the gradation (Grad.) column.
†Proportion in the blend. Determined as 0.75×percentage passing in the gradation column.
‡Sum of the proportion columns for the silty gravel and gravelly sand.

A 70 percent crushed limestone and 30 percent gravelly sand blend would give the following result:

Sieve size	Crushed limestone		Gravelly sand		70:30 Blend	ASTM D 2940	
	Grad.	Prop.	Grad.	Prop.		Min.	Max.
2 in	100	70	100	30	100	100	—
$1\frac{1}{2}$ in	100	70	100	30	100	95	100
$\frac{3}{4}$ in	87	60.9	80	24.0	84.9	70	92
$\frac{3}{8}$ in	64	44.8	67	20.1	64.9	50	70
No. 4	45	31.5	56	16.8	48.3	35	55
No. 30	18	12.6	34	10.2	22.8	12	25
No. 200	11	7.7	0	0	7.7	0	8

A 60 percent crushed limestone and 40 percent gravelly sand blend would give the following result:

Sieve size	Crushed limestone		Gravelly sand		60:40 Blend	ASTM D 2940	
	Grad.	Prop.	Grad.	Prop.		Min.	Max.
2 in	100	60	100	40	100	100	—
$1\frac{1}{2}$ in	100	60	100	40	100	95	100
$\frac{3}{4}$ in	87	52.2	80	32.0	84.2	70	92
$\frac{3}{8}$ in	64	38.4	67	26.8	65.2	50	70
No. 4	45	27.0	56	22.4	49.4	35	55
No. 30	18	10.8	34	13.6	24.4	12	25
No. 200	11	6.6	0	0	6.6	0	8

No blend is ideal. The 75 percent–25 percent blend of silty gravel and gravelly sand in Illustrative Example 6.1 is slightly low on the $1\frac{1}{2}$-in sieve. However, attempts to increase the proportion of silty gravel to raise the amount of $1\frac{1}{2}$-in material cause the No. 30 sieve percentage to fall outside the specification limit. (Setting up readily available personal computer spreadsheet programs to perform the calculations for blending only takes about 10 minutes. Then the proportion of each material can be quickly varied to see the effects on the final blended gradation.) This material would be difficult to produce because the blend is so close to the specification limits at three different points (low on $1\frac{1}{2}$-in material, at the maximum limit for No. 30 sand size, and near the maximum on fines). Small variations in the materials or in the blending proportions would swing the blend out of specification.

The 70 percent–30 percent blend of crushed limestone and gravelly sand is approaching the maximum limit for fines (7.7 percent compared with 8 percent maximum allowed). If the proportion of gravelly sand is increased to 40 percent to reduce the fine content (6.6 percent), the percent passing the No. 30 sieve increases to 24.4 percent, near the 25 percent maximum limit. Quality-control checks of the blended product would have to particularly watch the No. 30 and 200 sieve sizes.

Achieving the specified gradation does not guarantee that the material will be acceptable. Table 2.13 includes Atterberg limits and aggregate quality requirements that also would have to be considered for our blended material in Illustrative Example 6.1.

The aggregate quality criteria for ASTM D 2940 in Table 2.13 requires that 75 percent of the aggregate particles be crushed. Both the silty gravel and gravelly sand are natural products and uncrushed. Therefore, blends of these materials would not meet this requirement. The crushed limestone, if it had 100 percent crushed particles, would require a blend of 75 percent limestone and 25 percent silty gravel to meet the requirement. However, this would raise the fines content to 8.3 percent, which is slightly above the gradation limits. Several of the specifications in Table 2.13 include limitations for LA abrasion, sulfate soundness, and flat and elongated particles that also would have to be checked for the blended aggregate.

Blending is an effective method of upgrading materials. It is used widely to improve gradation of materials, but as was brought out in the discussion accompanying Illustrative Example 6.1, there are a number of other factors that have to be considered in assessing the blend. These include difficulty in staying within specification boundaries during production, plasticity characteristics, strength, durability, and final end use of the material. This assessment also should recognize the variability of natural materials and the inherent difficulties of maintaining tight controls while processing tons of material for construction.

In our example we did not have any ideal solutions to blending the given materials to meet the specification. This is often the case in construction with geotechnical materials. If the material for our example was to be used as the base course of a flexible pavement on a major airport runway, we would probably need to ship in a better material or get the crushing plant that produced the crushed limestone to change the gradation of the crushed material it is producing. If, on the other hand, we want to use low-cost, available materials in a light-vehicle parking lot or other less critical facility, we should be able to accommodate the reduced strength and/or durability of our proposed blends in the design of the structure.

6.3   Lime

Lime is an effective stabilizing agent for plastic soils and can be used to dry soils, improve their workability, limit volumetric changes, and increase strength. Limes that can effectively accomplish this are hydrated lime [$Ca(OH)_2$], quicklime (CaO), and the dolomitic variants of these high-calcium limes [$Ca(OH)_2 \cdot MgO$ and $CaO \cdot MgO$]. Hydrated lime is a fine powder, whereas quicklime is a more granular substance. ASTM C 977, Standard Specification for Quicklime and Hydrated Lime for Soil Stabilization, may be used to specify lime for soil stabilization. Typical lime properties are shown in Table 6.1. Quicklime is more caustic than hydrated lime, so additional safety procedures are required with this material. In the United States, hydrated lime is the predominant form of lime used for stabilization work, but the proportion of quicklime being used in U.S. lime stabilization work has grown from about 15 percent in 1976 (National Lime Association 1976a) to about 25 percent in 1987 (Transportation Research Board 1987). In Europe, quicklime is more common for soil stabilization than is hydrated lime.

TABLE 6.1   Typical Properties of Lime

Properties	Hydrated Lime	Quicklime
Chemical composition*		
Minimum CaO and MgO, %	90.0	90.0
Maximum carbon dioxide, %	5.0	5.0
Maximum free moisture, %	5.0	5.0
Physical properties	≤3% retained on	100% passing
Gradation*	No. 30 sieve	1-in sieve
	≤25% retained on	
	No. 200 sieve	
Slaking rate*†	—	Medium reactive
Specific gravity‡	2.3–2.4, high Ca	3.2–3.4
	2.7–2.9, high Mg	
Bulk density, lb/ft³‡	30–40	55–60

*From ASTM C 977.
†As determined by ASTM C 110.
‡From National Lime Association, 1976a.

Lime is usually produced by calcining limestone or dolomite, but it also may be produced as a by-product of some industrial processes such as production of acetylene. These industrial by-product limes are usually economical, but their composition is often highly variable, which complicates their use in construction. Agricultural limes are calcium or magnesium carbonates ($CaCO_3$ or $MgCO_3$) from grinding limestones and dolomites and are not effective for stabilization. These agricultural limes are sometimes accidentally substituted for hydrated or quicklimes in field stabilization projects. This is a serious blunder, and the soil stabilization will be ineffective.

Additional references on lime stabilization include Transportation Research Board (1987), Little (1987), Terrel et al. (1979), Ingles and Metcalf (1973), National Lime Association (1976a, 1976b), Stocker (1972), Thompson (1966, 1968), and Eades and Grim (1960).

6.3.1 Mechanisms

Table 6.2 shows examples of lime's effects on the properties of several different soils. There is generally a distinct reduction in plasticity and an increase in strength. Several processes are occurring due to the addition of the lime to the soil, and these provide the basis for stabilization using lime.

When lime is added to a soil, hydration of the lime causes an immediate drying of the soil. Anhydrous quicklime will have a more pronounced drying effect than hydrated lime. Consequently, lime can prove to be an effective construction expedient for drying out wet sites.

If lime is added to a plastic soil, plasticity drops, as seen in Table 6.2, and the texture changes. The addition of lime to the soil provides an abundance of calcium ions (Ca^{2+} and magnesium ions (Mg^{2+}). These will tend to displace other common cations such as sodium (Na^+) or potassium (K^+) in the clay particle. This process is known as *cation exchange,* and as we saw in Table 2.2, the replacement of sodium or potassium ions with calcium will significantly reduce the plasticity index of a clay mineral. The addition of lime increases the soil pH, which also increases the cation-exchange capacity. Consequently, even calcium-rich soils may respond to lime treatment with a reduction in the soil's plasticity. A reduction in plasticity is usually accompanied by reduced potential for shrinking and swelling, reduced susceptibility to strength loss in the presence of moisture, and reduced stickiness.

The addition of lime also changes the texture of a soil. A clayey soil becomes more silty or sandy in behavior because of particle agglomeration. Lime and soil mixtures show a significant decrease in clay-sized particles (<2 μm) compared with the original soil even when the lime and soil mixture is treated with a dispersing agent (Verhasselt 1990).

TABLE 6.2   Examples of Lime Effects on Various Soils

Soil	Lime %	Atterberg Limits			Strength	
		LL	PL	PI	$q_u{}^a$	CBR
1. CH, residual clay[b]						
(a) Site 1, Dallas–Ft.	0	63	33	30	76	
Worth Airport,	2	62	48	14	123	
residuum from Eagle	3	60	47	13	202	
Ford shale, Britton member	4	56	46	10	323	
(b) Site 2, Dallas–	0	60	27	33	70	
Ft Worth Airport,	2	48	32	16	171	
residuum from Eagle Ford	3	45	32	13	177	
shale, Tarrant member	5	48	34	14	184	
(c) Site 3, Irving, Texas,	0	76	31	45	64	
residuum from Eagle Ford	2	61	45	16	116	
shale, Britton	3	56	45	11	193	
member	5	57	45	12	302	
2. CH, Bryce silty clay,[c]	0	53	24	29	81	
Illinois, B-horizon	3	48	27	21	201	
	5	NP	NP	NP	212	
3. CH, Appling sandy loam,[d]	0	71	33	38	92	
South Carolina, residuum	3				147	
from granite	6				171	
	8				206	
4. CH, St Ann red bauxite	0	58	25	33	119	
clay loam,[d] Jamaica,	3				127	
limestone residuum	5				334	
5. CL,[e] Pelucia Creek Dam,	0	29	18	11		
Mississippi	1	32	19	13		
	2	31	22	9		
	3	30	21	9		
6. CL, Illinoian till, Illinois,[c]	0	26	15	11	43	
glacial till	3	27	21	6	126	
	5	NP	NP	NP	126	
7. SC, sandy clay, San Lorenzo,	0	54	23	31		8
Honduras[f]	5	61	38	23		20
8. MH, Surinam red earth,[d]	0	60	32	28	72	
Surinam,	3				130	
residuum from acidic	5				136	
metamorphic rock						
9. OH, organic soil with 8.1%	0	63	27	36	4	
organics[g]	2			36	4	
	4			24	8	
	8			25	7	

[a] Unconfined compressive strength in psi at 28 days unless otherwise noted; different compaction efforts used by investigators.

[b] McCallister and Petry, 1990, accelerated curing.

[c] Thompson, 1966.

[d] Harty, 1971, 7-day cure.

[e] McElroy, 1989.

[f] Personal communication, Dr. Newel Brabston, Vicksburg, Mississippi.

[g] Arman and Munfakh, 1972, limits at 48 hours, q_u at 28 days, strength samples prepared with moisture content at the LL.

Verhasselt (1990) experimentally examined various possible bonding mechanisms in lime stabilization. He concluded that hydrogen and hydroxyl bonding by the calcium hydroxide from the lime with the oxygen and hydroxyl functions on the clay particle surface were the most likely mechanisms causing particle agglomeration. These relatively weak bonds link several clay particles together to form stable and larger particles, effectively coarsening the texture of the clay soil.

Provided there is intimate clay and lime mixing, drying, plasticity reduction, and texture change are all very rapid reactions. Tests to determine lime effectiveness and lime contents to accomplish these results are usually conducted 1 hour after mixing the soil and lime.

In Table 6.2 we saw that a long-term pozzolanic strength gain in some soil and lime mixtures can develop. This long-term strength gain is due to formation of calcium silicate and calcium aluminate hydrate cementing compounds. The calcium in these cementing compounds is provided by the lime, whereas the soil clay minerals provide the needed silica and alumina. The pH of the soil will rise as lime is added, and the solubility of the silica and alumina increases as the pH rises. Generally, a soil is considered pozzolanically reactive with lime if its unconfined compressive strength increases 50 psi after 28 days of moist curing at 73°F. The actual long-term pozzolanic strength increase will depend on the specific soil composition, lime characteristics, amount of lime, curing conditions, length of curing, lime-soil mixture density, and interactions between these factors. The strength gain may vary from negligible to in excess of 1500 psi, with values for well-compacted, reactive soils commonly falling in the range of 100 to 500 psi.

When lime is added to a soil, a number of complex reactions may occur. The hydration of the lime will immediately remove moisture and dry the soil. Then cation exchange in the calcium- and magnesium-rich lime-soil mixture will reduce the plasticity of the clay fraction of the soil. This cation exchange and hydrogen and hydroxyl bonding also will link clay particles together, effectively coarsening the soil texture. Up to this point, the reactions are very rapid. The further development of pozzolanic strength from lime stabilization is soil-dependent and is influenced by a number of factors. Detailed discussions of lime and soil reactions can be found in Diamond and Kinter (1965), Stocker (1972), and Verhasselt (1990).

6.3.2 Properties

Strength. Lime stabilization causes both an immediate strength gain and a potential longer-term pozzolanic strength gain. The initial gain in strength of a clayey soil occurs because of the cation exchange and particle linking that are occurring. The amount of strength gain then

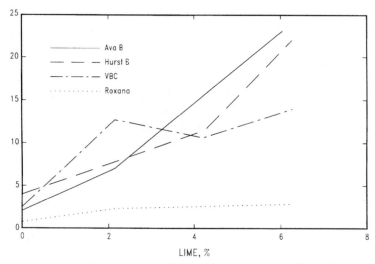

Figure 6.2 Immediate increase of CBR with the addition of lime (*Neubauer and Thompson 1972*).

will be a function of how extensive and complete these two processes are. Figure 6.2 shows examples of increases in CBR for four soils as reported by Neubauer and Thompson (1972).

The pozzolanic strength gain that some clays exhibit with lime requires first that the soil be able to provide the necessary silica and alumina. Pozzolanic strength gain in clayey soils depends on the soil's specific chemical composition. Plasticity is a crude indicator of reactivity, but testing is needed to determine the degree to which any specific soil will exhibit pozzolanic strength gain when mixed with lime. Examples of the pozzolanic strength gain observed in several different soils were shown earlier in Table 6.2.

Pozzolanic strength gain is both time- and temperature-dependent. Temperatures typically need to be above 40°F to maintain the pozzolanic reaction. The rate of pozzolanic reactions is slowed by low temperatures and increased at higher temperatures. The longer a lime-stabilized material cures at any given temperature, the higher its strength will be. Consequently, some care is needed to ensure adequate time and temperature of curing for lime stabilization in the fall and to avoid premature exposure to cold temperatures in the spring.

Durability.   Durability is not a concern for some lime applications (e.g., drying, plasticity reduction, textural change, and immediate uncured strength gain to aid construction). However, more permanent changes such as reduction in the swell potential of an expansive clay or development of pozzolanic strength for structural strength must ensure that lime stabilization effects are not adversely affected by the environment. There are five basic areas of concern that need to be considered:

1. Water
2. Freezing and thawing
3. Leaching
4. Carbonation
5. Sulfate attack

Lime-stabilized soils generally maintain much of their strength (70 to 85 percent) when exposed to water (Thompson 1970, Dumbleton 1962). Many lime-stabilized soils have performed well in structures where they are exposed to continuous or periodic soaking, including canals, levees, and dams (Gutschick 1978). The Friant-Kern Canal in California has had stable lime-stabilized banks that are exposed to fluctuating water levels for more than 12 years with good performance (Gutschick 1978, Transportation Research Board 1987). Also, unsurfaced lime-stabilized airfields built in Honduras have been unaffected by the wet tropical climate except for isolated erosion in areas of concentrated surface runoff flow. These fields have been in operation as long as 7 years. Lime-stabilized soils generally appear to retain their strengths in the presence of water. However, there are examples of poor strength retention for lime-stabilized soils exposed to soaking (Biswas 1972, Kennedy and Tahmoressi 1987). Consequently, testing of stabilized soils for soaked strength retention is prudent.

When lime-stabilized material is exposed to freezing and thawing, the material's volume typically increases, and the strength decreases. Figure 6.3 shows typical results of deteriorating strength with cycles of freezing and thawing for several lime-stabilized soils. The damage for multiple seasons of freezing and thawing may not be cumulative, however, because lime-stabilized materials show an ability to heal autogenously under favorable curing conditions, and the damage of a winter's freezing and thawing cycles may be repaired during the following warm season (Thompson and Dempsey 1969, McDonald 1969). Continued pozzolanic strength gain with continued curing also will help offset the initial winter season's freezing and thawing damage. Some agencies have criteria that assess durability to freezing and thawing by weight loss when subjected to a specified number of laboratory freezing and thawing cycles (e.g., Department of the Army 1983), but specifying an initially high enough compressive strength to

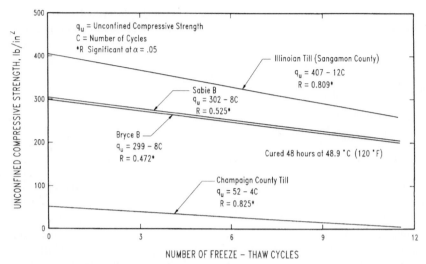

Figure 6.3 Lime-stabilized soil loss of strength when exposed to cycles of freezing and thawing (*Dempsey and Thompson 1968*).

stay above the required residual strength after exposure to freezing and thawing is probably a more common approach (Transportation Research Board 1987). More comprehensive design approaches also have been developed that explicitly examine requirements to maintain a residual strength needed for structural reasons when exposed to the expected number of freezing and thawing cycles (e.g., Allen et al. 1977). Lime-stabilized soils suffer a loss of strength when exposed to freezing and thawing, and this factor must be addressed at the project design stage.

The development of calcium silicate and calcium aluminate hydrates from pozzolanic reactions should be a permanent, nonreversible phenomenon. However, the other beneficial effects of lime stabilization that are due to reversible processes may be affected by extended leaching. For example, the cation exchange that reduces plasticity could be reversed if calcium cations were replaced during percolation of groundwater. The Friant-Kern Canal had its lime content of 4 percent reduced by about 1 percent over 7 years (Byers 1980), and monitoring of 3 percent lime treatment of dispersive clays used in soil conservation dams found a drop in pH and an apparent loss in effectiveness of the lime over the course of 15 years (McElroy 1982, 1987). Laboratory leaching tests on one highly expansive Texas CH clay found that continuous leaching affected the stabilized soil's physical engineering properties (i.e., Atterberg limits, permeability, strength, and swell pressure), but these effects were minimal for samples with 5 percent or more lime (McCallister and Petry 1990). There appears to be a potential problem for leaching to degrade some of the properties of lime-stabilized materials if low lime contents are used, but there is little data on this topic.

Carbonation occurs when atmospheric carbon dioxide (CO_2) combines with lime [$Ca(OH)_2$ or CaO_2] to form calcium carbonate ($CaCO_3$). If the pH of the lime-stabilized system drops sufficiently low, the calcium silicate and calcium aluminate hydrate cementing compounds may become unstable and will react with the carbon dioxide to revert back to silica, alumina, and calcium carbonate. These reactions are obviously detrimental to the lime-stabilized system's long-term strength and durability. Damage of this type has been reported on both lime- and cement-stabilized pavement bases in Africa and seems to be associated with low lime contents (3 to 5 percent), poor-quality soil materials (weathered dolorites, weathered basalts, weathered granites, calcretes, and laterites), intermediate relative humidity conditions around 50 percent (very humid or dry appear less severe), and poor curing (Netterberg and Paige-Green 1984, Bagonza et al. 1987). This is a topic area requiring further investigation, but it has not been reported as a problem in the United States. We cannot say whether this is due to it being unrecognized or to its not occurring. Until more is known, potential problems can be minimized by use of ample lime content, care in selection of materials to be stabilized, placement and compaction of the material to high density to minimize carbon dioxide penetration, prompt placement after mixing the lime with the soil, and good curing.

If sulfates are present in the soil or water where lime stabilization is used, detrimental reactions that result in large volume increases can occur. The calcium from the lime, alumina from clay minerals, and the sulfates combine to form ettringite $\{[Ca_3Al(OH)_6]_2(SO_4)_3 \cdot 26 \, H_2O\}$. The volume of ettringite is over 200 percent that of the original constituents, which can result in massive swelling when it occurs. Detailed discussion of this and related reactions between sulfates and lime-stabilized materials can be found in Hunter (1988) and Mehta and Klein (1966).

Damage due to swelling from sulfate reactions with lime-stabilized soils has occurred in Nevada (Mitchell 1986, Hunter 1988), Kansas (Mitchell 1986), Texas (Perrin 1992, Petry and Little 1992), and Mississippi. Once this reaction starts, it cannot be stopped, and damage is severe. An example of sulfate-induced heaving of lime-stabilized material from a road at Joe Pool Lake, Texas, is shown in Fig. 6.4. Such damage is costly. Repairs to a runway near Del Rio and roads at Joe Pool Dam, Texas, cost over $3 million (Perrin 1992), and

Figure 6.4 Heaving caused by sulfate attack on lime-stabilized material (*courtesy of L. L. Perrin, Ft. Worth, Texas*).

repairs to one street in Las Vegas, Nevada, cost $2.7 million (Hunter 1986). Although the mechanisms of sulfate attack on lime- and cement-stabilized soils had been identified in the laboratory in the 1950s and 1960s (Sherwood 1962), it did not receive much attention until Professor Mitchell's Terzaghi lecture (Mitchell 1986) focused national attention on the problem in 1985. At present, we are unable to predict what combinations of sulfate content, lime content, clay mineralogy, clay content, and environmental conditions will precipitate sulfate attack on lime-stabilized materials with detrimental swelling behavior. Consequently, if sulfates are present where lime stabilization is to be used, it would be wise to test the soil to see if the lime, soil, and sulfate will swell when mixed and exposed to moisture. Recent research into countermeasures such as double lime treatment, raising the system pH to levels where ettringite cannot form, proprietary additives, and barriers have shown promise (Mitchell and Dermatas 1992), but no clearly effective solution has emerged.

6.3.3 Mix design

Lime can accomplish a variety of objectives ranging from reduction of plasticity to long-term pozzolanic strength gain. The purpose of mix design for lime-stabilized materials is to achieve the desired change in soil properties at the minimum lime content that will maintain the desired level of durability. Consequently, all mix design methods should specifically address

- *The objective to be achieved:* PI reduction, texture change, swell reduction, immediate strength gain, or long-term pozzolanic strength gain
- *Durability:* ability to maintain desired change in properties over the time required
- *Economics:* minimum lime content to achieve the objective and maintain the required durability

One basic approach to lime stabilization mix design is to select the minimum lime content that achieves a pH of 12.4 (pH of lime) in a soil-lime mixture of 20 g soil and 100 ml water when measured 1 hour after mixing. A plot of lime content versus pH, such as shown in Fig. 6.5, allows a lime content to be selected above which there is no increase in pH. This procedure is rapid. It floods the pore fluid with

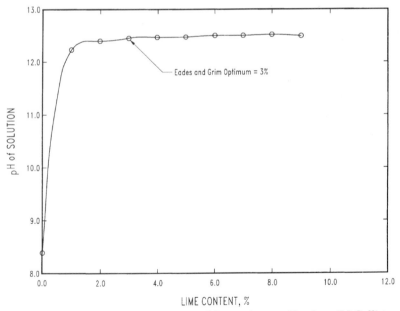

Figure 6.5 Effect of lime content on a soil-lime mixture pH values (*McCallister and Petry 1990*).

calcium to promote cation exchange, flocculation, and particle linkage, and the high pH encourages removal of alumina and silica from the soil to support pozzolanic reactions. This approach was developed by Eades and Grim (1966) as a rapid, simple method of selecting an optimal lime content for soil stabilization. The drawback is that it optimizes the environment for lime stabilization, but since it does not measure any physical soil-lime mixture properties, it provides no information on how or to what degree the soil will respond to lime stabilization. For this reason, the Eades and Grim pH test is often used to obtain an initial estimate of lime content that is then checked further to determine if it achieves the objectives of the stabilization (strength, plasticity reduction, etc.).

An example of the impact of lime content on several different engineering properties of a CH clay is shown in Fig. 6.6. This is the same soil shown in Fig. 6.5 and is from a 15- to 25-ft-thick layer of residual soil developed on the Eagle Ford clay shale formation at the Dallas–Ft. Worth Airport. For this specific example, the Eades and Grim (1966) pH approach did a reasonably good job estimating desirable lime content, but this will not always be the case.

The lime content for mix design should be selected to achieve desired changes in the soil, and testing should establish that these are achieved. Changes in plasticity after lime stabilization can be measured by the Atterberg limits, and textural changes can be measured by change in clay-sized particles with hydrometer tests. Laboratory compaction tests can be used to prepare samples to determine either soaked or unsoaked CBR values for measuring immediate strength gain from lime stabilization. The same procedures and, in some cases, the same samples can be used to determine percentage of swell or swell pressure during soaking. Long-term pozzolanic strength gain is conventionally measured by unconfined compressive tests on samples cured at room temperature (73°F) for 28 days. Accelerated curing at 120°F for 48 hours or some similar shortened period is also often used to ascertain long-term pozzolanic strength gain, but some caution is needed in accelerated curing of these materials because the higher temperatures can promote pozzolanic reactions that do not occur at normal field temperatures. Some current research suggests accelerated curing at 105°F may be superior to higher temperatures (Transportation Research Board 1987).

Long laboratory curing periods for lime-soil mixes to determine pozzolanic strength gain generally have been objectionable to most practitioners. However, other fields of construction such as the concrete industry commonly allow 28 days for cure of concrete specimens for strength determination, and in airfield concrete paving work, the cure time is

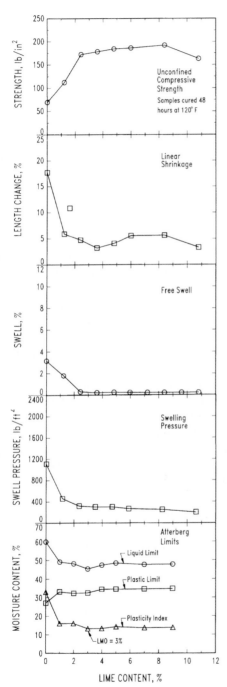

Figure 6.6 Effect of lime content on engineering properties of a CH clay (*compiled from data reported by McCallister and Petry 1990*).

often 90 days. Consequently, for important projects where the degree of pozzolanic strength gain is important, curing for periods and at temperatures more representative of those expected in the field may be more realistic and preferable to the more convenient accelerated curing.

Durability of engineering materials is a very difficult and complex topic. As discussed earlier, lime-stabilized materials generally have a modest reduction in strength when they are tested in a soaked condition—but there are exceptions. Lime-stabilized materials lose strength and increase in volume when exposed to freezing and thawing. Figure 6.3 showed that if a sufficiently high initial strength is achieved, the residual strength after exposure to freezing and thawing will be sufficient for engineering purposes. The specific residual strength required depends on the engineering requirements of the project. Unconfined compressive strength tests of soaked samples and samples exposed to the number of expected first winter freeze-thaw cycles (usually within the range of 3 to 12 cycles) provide specific information on how a specific lime-stabilized soil will perform when exposed to moisture and freezing and thawing cycles.

Leaching and carbonation can result in loss of the effectiveness of lime stabilization with time, but relatively little is known about these topics. Both effects are more pronounced at low lime contents, so avoiding unduly low lime contents coupled with good construction practices seems to offer the best defense against these problems. Neither of these is known to cause major problems, but we do not know whether the damage is simply unrecognized in the field or if the problem is uncommon. Sulfate attack on lime-stabilized materials is a problem area that was recognized only relatively recently. It is not common, but it is terribly destructive when it occurs. If sulfates are present in the soil or water where lime stabilization is to be used, testing to determine the likelihood of sulfate attack and resultant swelling would be very wise.

A number of specific mix design methods have been developed by different organizations, and these are reviewed in Terrel et al. (1979) and Transportation Research Board (1987). These mix design methods are developed for specific purposes and for limited geographic areas and cannot be extrapolated to other uses and geographic regions blindly. A direct comparison of these various methods is very difficult because of their widely varying curing and testing standards.

6.3.4   Suitable soils

The mechanisms for lime stabilization require clay minerals in order to generate flocculation and particle linking with resultant reduction in plasticity and swelling, textural change, and immediate strength increase. It also requires removal of alumina and silica from the clay to support pozzolanic reactions for long-term strength gain. Clearly, then, soils that will be suitable for lime stabilization will be clays or soils that have sufficient clay fines to interact with the lime. A plasticity index (*PI*) of about 10 is a reasonable lower limit for general acceptability for lime stabilization. There are exceptions to this, however.

Organic materials interfere with the reactions between calcium and the clay particles. Soil number 9 in Table 6.2 showed only a modest reduction in plasticity index even at 8 percent lime. Low organic contents may be overcome by increasing the amount of lime used, but if organic contents exceed 1 percent, some alternative to lime stabilization is probably needed. Highly weathered soils also may require higher lime contents than normal for successful stabilization (see, for example, soils 4 and 8 in Table 6.2). The Eades and Grim pH test consistently underestimates the quantity of lime needed for strength gain in weathered subtropical and tropical soils that include laterites (Harty 1971, Morin and Todor 1975).

6.3.5 Construction considerations

The basic steps in lime stabilization construction are

1. Lime delivery and distribution
2. Mixing
3. Compacting
4. Curing

Lime may be applied to a soil either dry or as a slurry. Dry hydrated lime dusts vary severely, which limits the locations where this method is acceptable. Dry quicklime, on the other hand, is granular and poses much less of a dusting problem than does hydrated lime. The dry hydrated or quicklime may be delivered in bags or by trucks equipped for unloading with either augers or pneumatic systems. Spacing bags of lime over the soil to be stabilized, slitting these, and dumping the lime out of the bags requires a great deal of manual labor and is most suitable for small jobs. Pneumatic trucks spreading lime through a spreader bar mounted on the rear of the truck have become a common method of distributing lime on modern construction sites. Dry quicklime may be distributed by dump trucks with tailgate opening controls. This would not be feasible with the lighter and finer hydrated lime because of severe dusting.

A typical hydrated lime slurry mixture would be 1 ton of lime mixed with 500 gallons of water to produce 600 gallons of slurry with 31 percent lime solids (Transportation Research Board 1987). As the contents of solids in the slurry approaches 40 percent, the slurry may become more difficult to pump. The slurry is normally mixed in a central mixing plant, with a portable jet slurry mixer, or by discharging the lime and water separately into a tank truck where mixing is accomplished by a recirculating pump or a compressed air. A relatively new Portabach slaker unit is available for making quicklime slurries. This is a little more complex than preparing hydrated slurries because the initial slaking of the quicklime produces a high exotherm. The slurry is distributed at the site with gravity or preferably pressure spray bars mounted on the rear of tanker trucks.

Lime stabilization will not be successful unless the soil and lime are adequately mixed, and poor mixing is probably the leading cause of unsatisfactory lime stabilization performance. Most mixing for lime stabilization is done on site, although some central plant mixing is used with materials such as base course aggregates. Such aggregates are commonly processed at a central plant for gradation control, and lime may be added and mixed there easily. This central mixing for essentially granular materials is particularly effective for clayey gravels or sands where the clay fraction may prevent the material from meeting strength or plasticity requirements in a specification. Good control of the lime-soil proportions and mixing is possible at a central plant.

Soil and lime may be mixed on site by disking, by repeated blading with a motor patrol, by rotary mixing equipment or traveling plants, or by a combination of these. Repeated passes of a disk harrow turn over the soil and lime, mixing the material in the process, and help break up clods and cohesive soils to allow better effectiveness. When mixing by blading, the motor grader blades the dry materials back and forth until a uniform mixture is obtained. The quality of mixing is directly dependent on the number of blading repetitions and the skill of the equipment operator. Only relatively small amounts of material can be mixed at a time in this manner. Slurried lime is best added in increments to thin lifts (2 in or so) of materials that are then thoroughly mixed together by blading after the last increment of lime is applied. This might be approached by spreading a thin layer of soil, applying an increment of lime slurry, mixing by windrowing to one side, spreading another thin lift of soil, applying an increment of lime slurry, windrowing to one side, etc. Alternatively, one could apply an increment of lime slurry to an in-place soil, blade off a thin lift of material to a side windrow, apply another increment, windrow another thin lift, etc. Once all the lime slurry is applied using either approach, the windrows are mixed by repeated blading operations. In blading operations, the soil is usually relatively dry to allow mixing, and moisture for compaction is added after the lime and soil are mixed.

High-speed rotary mixers and traveling plants are probably the most efficient mixing methods. A number of different models are produced of each; one example of a rotary mixer is shown in Fig. 6.7. The amount and depth of mixing are directly dependent on the type of equipment and number of passes the equipment makes. Water for compaction is added by spraying from trucks during mixing or directly in the rotary mixer's mixing chamber on some equipment models.

When dealing with highly plastic materials, it may be useful to use a two-stage mixing process. The initial mixing and following mellowing of the lime-soil mixture for several days allow the lime to react with the soil to coarsen its texture and reduce the plasticity. The water content for this initial mixing should be on the wet side of opti-

Figure 6.7 Example of a single-shaft rotary mixer used for mixing soils and stabilizers (*courtesy of Seaman-Maxon, Inc., Milwaukee, Wis.*).

mum to encourage destruction of clods. After mixing, the surface should be graded and sealed with light compaction to encourage water runoff in case of rain and to decrease the potential for carbonation reactions. After a day or two, the initial lime reactions should make the soil workable enough to allow final mixing and compaction.

Control of the mixing process may require determining the lime content, lime spread rate, and thoroughness of breakup of the soil and mixing of the soil and lime. ASTM D 3155 provides a quick and easy way to determine lime content of uncured lime-soil mixtures. The spread rate of dry lime can be checked by placing a pan or fabric on the surface where lime is to be spread. Then weighing the amount of lime collected in the pan or on the fabric and knowing the area of the pan or fabric allow calculation of the lime spread rate in pounds per square yard or similar units. The flowmeters on tank trucks are notoriously inaccurate, but it is a simple matter to calibrate them so that a reasonably accurate estimate of slurry discharge can be made. The actual lime content of the lime slurry is easily determined in the laboratory from the slurry's specific gravity determined with a hydrometer. This, in conjunction with the tank truck's discharge rate, speed, and width of placement, will allow a reasonable estimate of the lime application rate. The degree of soil breakup or pulverization during mixing helps determine the speed and quality of the final lime stabilization reactions. This is often evaluated in the field by determining how much material will pass through the 1-in or No. 4 sieve. The contrast between the bright white lime and the soil's natural color allows a quick visual assessment of the depth and thoroughness of lime distribution. Phenolphthalein solution turns red in the presence of lime due to the lime's high pH, and simply spraying it on a soil allows a determination of whether lime is present. This is useful for checking on uniformity and depth of mixing.

Compaction of lime-soil mixtures follows the same principles discussed in Chap. 5. However, addition of lime to a soil will generally tend to decrease the maximum density for any given compaction energy and will increase the optimum moisture content. An example of this effect for the Eagle Ford shale residual clay discussed in Sec. 6.3.3 is shown in Fig. 6.8. This shift in optimum density and moisture content may cause problems determining the percentage of specified density achieved by compaction in the field. If, for instance, the contractor's lime application rate in the field is high (either on purpose to guard against carbonation or to ensure that the minimum specified lime content is met or by accident), then calculating percentage compaction based on the compaction curve for a lower lime content unfairly penalizes the contractor. The higher than specified lime content would generally be a benefit for the owner. If problems are

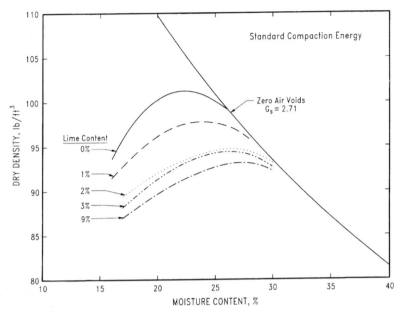

Figure 6.8 Example of lime effects on the compaction curve of a CH clay (*McCallister and Petry 1990*).

encountered achieving the specified density, and the stiffness of underlying materials, compaction equipment, number of passes, moisture content control, and lift thickness are such that one would expect adequate compaction, a one-point compaction curve (see Sec. 5.5.3) of the in-situ lime-soil mixture at the time compaction is taking place may help identify appropriate density values for determining the degree of compaction that is being achieved. Because of the slowness of the pozzolanic reaction in lime, delays between lime mixing and compaction of several days are not detrimental. During this delay period, though, the lime-soil mixture should be kept moist, and steps to minimize carbonation (e.g., cover or seal the surface with light compaction) would be prudent.

Proper curing with adequate moisture and at temperatures above 40°F is necessary for lime-stabilized materials to gain pozzolanic strength. Commonly, a cure period of 3 to 7 days is required; however, if the lime-stabilized material is well compacted, some construction activities such as placement of overlying pavement layers might be allowed sooner if no rutting occurs in the stabilized materials due to construction activities. Loss of water from the lime-stabilized material must be prevented. This is usually accomplished by maintaining moist curing or application of an asphalt seal on the surface of the lime-stabilized material. Moist curing uses trucks, sprinklers, or other methods to keep the exposed surface moist. The greatest danger here is that interruptions in the process will allow the surface to dry out. Alternatively, asphalt cutbacks or emulsions may be sprayed on the surface to form a continuous asphalt seal over the surface. Common application rates are within the range of 0.10 to 0.25 gal/yd^2, but the only important criterion is to form a continuous waterproof seal of the surface. This may require several applications over several days as the emulsions break or cutback solvents evaporate.

6.4   Portland Cement

Portland cement is a *hydraulic cement,* which means it will harden and gain strength through reactions with water. It is composed primarily of calcium oxides, silicates, and aluminates that form a strong cementing paste when hydrated. Basically, five types of portland cement are manufactured:

Type I: Ordinary portland cement

Type II: Moderately sulfate-resistant cement

Type III: High-early-strength cement

Type IV: Low-heat-of-hydration cement

Type V: Highly sulfate-resistant cement

Today, a number of portland cements are manufactured that meet the requirements for both types I and II and are commonly designated type I/II. ASTM C 150 provides the specific requirements for each type of portland cement. Generally, only types I, II, or I/II cements are encountered in stabilization work.

Portland cement is mixed with a soil or aggregate to take advantage of the strength gain that occurs upon hydration of the portland cement. Unfortunately, a wide variety of names have evolved over time to describe products containing cements (e.g., cement-stabilized soils, cement-treated base, soil cement, lean concrete, econocrete, roller-compacted concrete, dry-rolled concrete, rollcrete, etc.). None of these have very precise definitions. More detailed information on various types of soil and aggregate combinations with portland cement is available in American Concrete Institute (1990), Larsen and Armaghani (1989), Rollings (1988a), Kohn and Darter (1981), Terrel et al. (1979), and Portland Cement Association (1979). For the purpose of this book, *cement stabilization* will be used as a generic term to describe all combinations of portland cement and soils or aggregates that are placed by spreading and compacted by rolling as opposed to conventional portland cement concrete, which is essentially a cohesive fluid or semisolid placed between forms or slipformed and consolidated with internal or surface vibrators. Similarly, low-cement-content flowable mixtures such as used for filling utility trenches fall outside the scope of this book.

Cement stabilization can have a variety of uses. It may be used to upgrade a local marginal material to improve its properties so that it may be used in construction without importing more expensive high-quality materials (Rollings 1988b) or to recycle failed asphalt pavements into new pavement base courses (Portland Cement Association 1979, American Concrete Institute 1990). In pavements it may be a structural layer (e.g., a cement-stabilized base course) or a low-grade surfacing for shoulders, low-volume roads, or industrial pavements (e.g, storage yards). Since its original use in a street in Sarasota, Florida, in 1915, cement-stabilized material has been used in pavement structures equivalent to over 100,000 miles of 24-ft-wide pavement (American Concrete Institute 1990). It also has found application for slope protection in over 300 major projects in the United States and Canada, as a liner for water-storage reservoirs (Portland Cement Association 1986, American Concrete Institute 1990), and as a dam embankment (Nussbaum and Colley 1971, Hansen 1986). In addition to water storage, cement-stabilized material has been used to line wastewater-treatment lagoons, sludge-drying beds, ash-settling ponds, and solid-waste landfills (American Concrete Institute 1990). It also may find many applications as a high-quality fill for foundations, backfilling, pipe bedding, and other similar construction activities.

6.4.1 Mechanisms

Portland cements contain excess lime, and this lime will generate the same reactions discussed in the preceding section on lime stabilization. However, the cost of portland cement makes it uneconomical for tasks that lime can do more cheaply (drying soil, reducing plasticity, etc.). The major value of portland cement for stabilization is the strong cementing bonds it forms to tie the soil particles together into a cemented, coherent mass. Unlike lime stabilization, all the chemical components necessary for development of calcium silicate and aluminate hydrate bonds are present in the portland cement, and no chemical contribution is necessary from the soil. Therefore, cement stabilization will not depend on the mineralogy of the soil being stabilized.

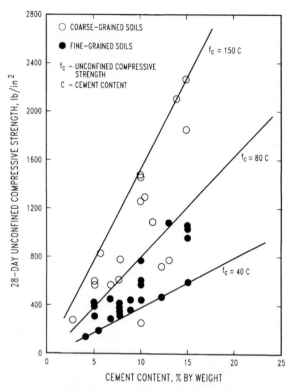

Figure 6.9 Effect of cement content and soil on compressive strength (*American Concrete Institute 1990*).

6.4.2 Properties

Strength.   Unconfined compressive strength (ASTM D 1633) is the most widely used test to evaluate cement-stabilized soil strength. Other tests that have been used for evaluating cement-stabilized materials include flexural beam tests (ASTM D 1635), splitting tensile tests, and direct tension tests. The specific strength that is achieved depends on soil type and amount of cement used. Figure 6.9 shows the relative impact of these factors on strength. Just as with portland cement concrete, longer cure times produce higher strength. Some typical data compiled by Terrel et al. (1979) are shown in Fig. 6.10. Generally, the finer-grained soils require more cement and gain lower strength than do coarse-grained soils. Fine, uniform gradations, such as the eolian sand in Fig. 2.3, also require high cement contents to gain strength.

A wide variation of stabilized soil strength is possible depending on the soil and aggregate characteristics, cement characteristics, cement content, density, curing, and curing time.

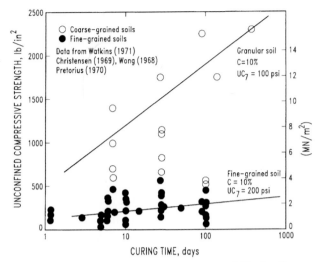

Figure 6.10 Effect of curing time on cement-stabilized soil compressive strength (*Terrel et al. 1979*).

Many compressive strength test results for stabilized soils are run on soil samples from soil compaction tests (e.g., ASTM D 558, D559, and D560) with a height-to-diameter ratio of 1.15. This differs significantly from the ratio of 2.00 commonly used in concrete work and in samples of soil cement prepared in accordance with ASTM D 1632. Correction factors are available in ASTM C 42 to account for the effects of varying height-to-diameter ratios, but they are often neglected in practice.

Durability. Generally, there are four basic problem areas in durability of cement-stabilized soils and aggregates:

1. Water and contaminants
2. Freezing and thawing
3. Carbonation
4. Sulfate attack

Soaked soil samples generally give lower compressive strength results than when tested dry, so normal practice is to test samples after moist curing without allowing drying (ASTM D 1632). When stabilized soils are used as liners or slope protection on channels or dams, they also may be exposed to erosion by flowing water and water-borne debris. To resist stormwater flow velocities of up to 20 ft/s, soil cement slope-protection structures were required to have 7-

day compressive strengths of 750 psi for projects around Tucson, Arizona (Pima County 1986). Tests have indicated that soil cement can resist leachate from municipal waste and hazardous waste such as pesticides, refinery sludges, plastic waste, rubber waste, and pharmaceutical waste, but tests of resistance of a specific soil cement and a particular waste always should be done for compatibility (American Concrete Institute 1990). Portland cement–stabilized materials will not be resistant to acids, however, because of the basic chemical nature of cement hydration products.

The durability of cement-stabilized materials is usually assessed by ASTM D 559 for resistance to wetting and drying and ASTM D 560 for resistance to freezing and thawing. Under ASTM D 559, the weight loss of cement-stabilized material under wire brushing is determined after 12 cycles of wetting and drying, and a similar technique is used for 12 cycles of freezing and thawing. Typical durability requirements for cement-stabilized materials are shown in Table 6.3.

As with lime-stabilized materials, there is a relation between compressive strength and durability. Work by the Portland Cement Association (1971) found that only about 20 percent of the samples with a compressive strength of 300 psi would pass the laboratory

TABLE 6.3   Typical Durability Requirements

Portland Cement Association (1971)			Corps of Engineers (Department of the Army 1983)	
AASHTO soil	USCS soil	Maximum weight loss, %*	Soil	Maximum weight loss, %*
A-1-a	GW, GP, GM, SW, SP, SM	14	Granular, $PI<10$	11
A-1-b	GM, GP, SM, SP	14	Granular, $PI>10$	8
A-2	GM, GC, SM, SC	14†	Silt	8
A-3	SP	14	Clay	6
A-4	CL, ML	10		
A-5	ML, MH, CH	10		
A-6	CL, CH	7		
A-7	OH, MH, CH	7		

*After testing under wetting and drying (ASTM D 559) or freezing and thawing (ASTM D 560).

†Maximum loss of 10 percent for A-2-6 and A-2-7.

Note: Additional PCA criteria:

a. Maximum volume change < 2 percent of initial volume.

b. Maximum moisture content during the test should be less than the quantity required to saturate the sample at the time of molding.

c. Compressive strength should increase with age of specimen.

d. The cement content is for pavements and will be adequate for slope protection that is 5 ft or more below minimum water elevation. For higher elevations, increase the cement content 2 percent.

freeze and thawing test, whereas about 70 percent of the samples would pass with a compressive strength of 500 psi. The association developed a relation between compressive strength and percentage of samples passing laboratory freezing and thawing tests that can be used to select a minimum compressive strength to achieve some degree of protection against freezing and thawing (Portland Cement Association 1971). However, freezing and thawing damage is a function of pore structure and saturation level, so the relation to strength is coincidental. Certain factors that provide higher strength, such as higher cement contents and higher density, also improve pore structure and reduce permeability, which in turn make a sample more difficult to critically saturate. Thus, while higher strength also may improve durability, strength alone is no guarantee of durability.

Carbonation is potentially a durability problem for cement-stabilized materials, as it was for lime-stabilized materials, as discussed earlier. However, relatively little is known of this phenomenon, and the earlier discussion on carbonation for lime-stabilized materials should be equally valid for cement-stabilized materials.

There are conflicting statements in the technical literature about whether cement-stabilized materials are susceptible to sulfate attack in the same manner as conventional concrete (e.g., Terrel et al. 1979, American Concrete Institute 1990). In sulfate attack on conventional concrete (Neville 1981, American Concrete Institute 1977), the destruction of the concrete is caused by the formation of the same expansive ettringite compound that caused sulfate attack on lime-stabilized materials discussed earlier. In this case, however, the calcium, silica, and alumina needed to form ettringite are all present in conventional portland cement concrete, and no contribution from clay minerals is needed, as was the case with lime-stabilized materials.

There appear to be two potential mechanisms for sulfate attack on cement-stabilized materials:

Conventional sulfate attack, where calcium, silica, and alumina are provided by the portland cement hydration products

Sulfate attack where the calcium comes from the portland cement and the silica and alumina come from clay minerals in the stabilized material

Sulfate attack on cement-stabilized soils was demonstrated in the laboratory over 30 years ago (Sherwood 1962), but the only field problems with sulfate attack on cement-stabilized materials of which the authors are aware are in the United Kingdom (Thomas et al. 1989) and in Georgia. In the first case, sulfates from the oxidation of pyrite in cement-stabilized mine waste used as pavement base courses caused disruptive swelling. In the second case, the sulfates were present in the mixing water for a cement-stabilized clayey sand base course for a secondary road, and reactions between the sulfates, cement, and clay resulted in extensive swelling.

Problems with conventional sulfate attack on cement-stabilized materials might be helped by use of sulfate-resistant cements (type II or type V depending on the severity of the sulfate problem) that limit the tricalcium aluminate content in the cement. This reduces the alumina available to form ettringite. However, sulfate-resistant cements will do nothing to help in the second type of sulfate attack, where the alumina comes from the clay minerals present in the material being stabilized.

Sulfate attack does not appear to be a major problem in the field. In the case of conventional sulfate attack, the cement contents used in most stabilization work are about one-third to one-half that used in conventional concrete, so the amount of ettringite formed may be small enough not to cause widespread damage. Also, the cement-stabilized materials normally have a higher void content than conventional concrete. This may provide empty volume to accommodate the ettringite growth. Most cement stabilization uses coarse-grained materials with little clay, which further minimizes the likelihood of the second form of sulfate attack.

However, our understanding of when such sulfate attacks occur is limited. For example, the sulfate attack in Georgia occurred with only 10 percent of the soil being silt- or clay-sized material. Since problems with sulfate attack of lime- or cement-stabilized materials have only recently been fairly widely recognized, it is hard to decide whether the problem is relatively rare or if it has simply gone unrecognized in the past. Consequently, if sulfates are present in material to be stabilized, it would be advisable to conduct laboratory tests to evaluate the likelihood of expansion in the field.

Permeability.   Generally, the permeability of a soil decreases as cement is added, and this reduction can be several orders of magnitude or more depending on the specific soil and the amount of cement added. However, discontinuities in the cement-stabilized material due to shrinkage cracking or multiple lift construction will determine the actual water flow in the field. For example, Nussbaum and Colley (1971) found that permeability of cement-stabilized material was 2 to 20 times greater parallel to the compaction plane than perpendicular to the plane. Consequently, water flow through cement-stabilized materials in the field must be assessed based on discontinuities in the field and not on laboratory-derived permeability values from intact samples.

Volume change. During hydration of portland cement and as a consequence of temperature and moisture content changes, cement-stabilized materials undergo volume changes in the same manner as conventional portland cement concrete. Restraint due to friction or bonding to adjacent materials results in differential volume changes within the material, and cracking occurs. This cracking is commonly referred to as *shrinkage cracking* and is inherent in all portland cement–based materials. Factors that encourage higher differential volume changes within the material result in increased shrinkage cracking. Higher cement contents, increased fines or plasticity in the soil, and high temperatures are all examples of factors that will increase the occurrence of shrinkage cracking. Cement-stabilized clays tend to develop relatively narrow cracks spaced at 2- to 10-ft intervals, whereas cement-stabilized coarse-grained materials tend to have fewer, somewhat wider cracks at approximately 10- to 20-ft intervals (American Concrete Institute 1990).

This shrinkage cracking has serious engineering ramifications. If an asphalt concrete surfacing is placed on a cement-stabilized base course, the shrinkage cracks normally will reflect through the asphalt surface, resulting in an unsightly and ongoing maintenance problem. Similarly, if the cement-stabilized material is to retain liquids, the crack is a major potential leakage source. Problems with shrinkage cracking have led to some proposals to limit the maximum strength of cement-stabilized materials (e.g., 800 psi by Ingles and Metcalf 1973) or maximum cement content (e.g., 5 percent by Department of Defense 1978) to avoid reflective cracking problems in pavement structures. If high-strength cement-stabilized materials are used and cracking is a concern, it may be useful to cut contraction joints in the same manner as for conventional concrete.

6.4.3 Mix design

The cement content for a stabilized material must be as low as possible for economy but also must be high enough to be durable and to achieve desired engineering properties such as strength or permeability. Different organizations have developed specific requirements for mix design procedures (e.g., Portland Cement Association 1971 or Department of the Army 1983), and there are significant differences between different methods. The paramount objective is to achieve the desired level of performance in the field for the minimum cost, and one should expect methods oriented toward high-quality, stabilized bases for heavily loaded pavements to differ significantly from those intended for slope protection. The user must select the approach that is compatible with his or her intended use of the material.

A general conceptual approach to mix design for cement-stabilized materials is shown in Fig. 6.11. The initial checks of soil pH and organic content are to identify soils that tend to react poorly to cement stabilization (American Concrete Institute 1990). The sulfate content guidance is based on recommendations by Dunlap et al. (1975) and should be viewed as tentative until more is known about this mechanism. The initial trial cement contents are for a relatively good-quality stabilized material that would be suitable for a structural layer in a pavement. If the application for the stabilized material does not require high strength, the trial cement content can be reduced 20 percent.

The recommended trial cement contents increase as the gradation becomes poorer and the quantity of fines increases (Table 6.4). Fine, poorly graded sands such as eolian sands are particularly troublesome, and often an inordinately large quantity of cement is needed to achieve stabilization. Poorly graded uniform gravels may be impractical to stabilize unless they are blended with smaller-sized materials such as sand to help fill the void space. Organic material in the soil interferes with the cement hydration, and such soils may prove difficult to cement stabilize. Consequently, if the A horizon of a soil deposit is being stabilized, it may be necessary to increase the trial cement contents to overcome the effects of organic material.

Once the initial trial cement content is selected, it must be checked in the laboratory to determine if it will produce the engineering characteristics needed (e.g., flexural strength, compressive strength, fatigue life, or modulus value) and if it will be durable. The design engineer is responsible for identifying the engineering characteristics that are needed, and Table 6.3 provides some guidance on durability. The drawback to Table 6.3 is that it does not differentiate well between applications (e.g., stabilized pavement base course versus stabilized reservoir liner) or locations (e.g., New Mexico versus New Hampshire). Often the durability requirements of Table 6.3 will require a higher cement content than many project requirements for strength.

The final step in the mix design is to optimize the cement content. If the trial cement content fails to meet the engineering or durability requirements, then another trial at a higher cement content is warranted. If it does meet the requirement, then the possibility of achieving the project objectives at a lower, more economical cement content needs to be explored.

6.4.4 Suitable soils

The most economical materials for cement stabilization will generally be sandy and gravelly materials with between 10 and 35 percent

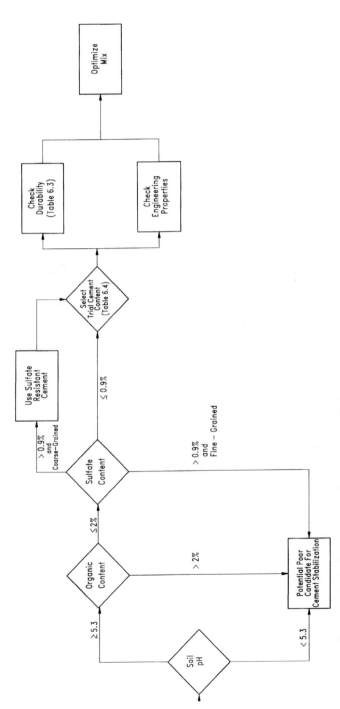

Figure 6.11 Conceptual mix design approach for cement-stabilized materials.

TABLE 6.4   Suggested Trial Cement Contents

Soil type	Trial cement contents, percent by weight
Granular, well graded, GW, SW	4–5
Granular, poorly graded or silty, GP, GM, SP, SM	5–6
Granular, clayey fines, GC, SC	7–8
Uniform fine sands, SP, SM	9–10
Clays and silts, low plasticity, CL, ML	9–10
Clays and silts, high plasticity, CH, MH	11–12
Organic soils, OL, OH	13–14

passing the No. 200 sieve, 55 percent or more passing the No. 4 sieve, 37 percent or more passing the No. 10 sieve, and no material larger than 2 in (Portland Cement Association 1979). Material coarser or finer than this typically will require higher cement contents. Clay balls may tend to form when the plasticity index is greater than 8. Difficulty in achieving adequate mixing often limits use of cement stabilization with CH and the more plastic CL clays. Organic, acidic sands often do not react well to cement stabilization and can be particularly difficult to deal with (Robbins and Mueller 1960, American Concrete Institute 1990). Additions of calcium chloride, friable clayey soil, or calcareous material such as limestone screenings or limerock to these soils may be helpful (Portland Cement Association 1979).

Portland cement stabilization is best suited for cementing individual soil and aggregate particles into a coherent mass. It will reduce soil plasticity with resultant effects on swelling and similar behavior, but these effects can be achieved more economically with lime. Almost any soil can be cement stabilized, but the cement content may be uneconomically high for clays and uniformly graded materials. The physical process of blending and mixing cement with highly plastic soils also may prove too difficult to actually achieve economically. Organic and acidic soils may require special treatments before they can be reasonably stabilized. Materials containing sulfates pose the risk of sulfate attack, but no firm criteria are available on this problem.

6.4.5   Construction considerations

Cement-stabilized materials may be mixed in place, or they may be centrally mixed in a plant. More uniform mixing and better control of materials are achieved with a central plant, and central mixing is best when high-strength or high-quality cement-stabilized materials are needed. In-place mixing may be accomplished using transverse rotary mixers with either single or multiple shafts or a windrow-type traveling mixer (e.g., Fig. 6.7). The continuous-flow, twin-shaft pugmill is the most common central plant, but batch plants and drum mixers also have been used successfully to mix cement-stabilized materials.

The nature of the soil and aggregates being stabilized has a major impact on the mixing operations. Difficulties and mixing effort increase as the fines and plasticity of the material increase. Plastic soils may require pulverization prior to addition of the cement and subsequent mixing. Additions of lime may make a very plastic soil more workable and may ease pulverization and mixing difficulties. Windrow-type traveling mixers work best with nonplastic or slightly plastic materials. The transverse rotary mixers can mix almost any material, but multiple passes may be needed in plastic soils. Rotary-drum central mixing plants do a poor mixing job if the material has plastic fines. Revolving-blade central mixing plants can handle somewhat plastic material. Twin-shaft pugmills provide a very vigorous mixing action and can handle the widest range of materials.

Accurate control of the moisture is important during mixing. Mixing clayey soils below the optimum moisture content encourages formation of clay balls and clods, whereas mixing at or slightly above optimum eases pulverization and helps break down clods. Above optimum moisture contents, clayey soils can become sticky and difficult to mix. Mixing granular soils at low moisture contents may result in cement balls forming.

Placement of cement-stabilized materials is controlled with moisture-density relationships as with conventional soils and lime-stabilized materials. Test procedures for developing compaction curves for soil and cement mixtures are given in ASTM D 558. The addition of portland cement causes particle flocculation in fine-grained soils (same as lime) with a resulting increase in optimum moisture content and decrease in maximum dry density for any given compaction effort. Portland cement has a specific gravity of 3.15, which is appreciably higher than the normal soil range of 2.6 to 2.7. The addition of this high-specific-gravity material consequently will tend to increase density. The compaction curve of a soil or aggregate will shift upon addition of cement, but the direction of the shift cannot be predicted because of the counteracting processes that are occurring. An example of the effect of cement on a compaction curve is shown in Fig. 6.12.

The basic steps in construction of cement-stabilized materials are

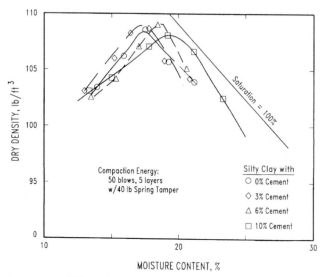

Figure 6.12   Effect of cement on a silty clay compaction curve (*based on data reported by Mitchell et al. 1972*).

- Mixing the soil and cement (in situ or central mix)
- Hauling and spreading (central mix)
- Compacting
- Final grading
- Curing

Most in-situ mixing jobs deliver and spread the cement mechanically, and only the smallest jobs use hand-placed bags of cement. The spread rate of dry cement by mechanical spreaders can be checked by placing a pan or fabric on the surface where cement is to be spread. Then weighing the amount of cement collected in the pan or on the fabric and knowing the area of the pan or fabric allow calculation of the cement spread rate in pounds per square yard or similar units. After spreading the cement, it is mixed using the in-situ mixers discussed earlier. The soil and cement should be mixed to a uniform color, and the presence of cement in a soil can be determined with phenolphthalein solution, just as with lime. In the field it is difficult to obtain the uniformity of blending and exact proportions of cement and soil that were produced in the laboratory mix studies. Consequently, the cement content is often increased 1 or 2 percent (by total mass) above the laboratory-determined value as a hedge against the uncertainties of in-situ mixing in the field. Central mixing plants generally have tighter control over proportions and can achieve better mixing. The cement content of freshly mixed material can be determined by chemical analysis (ASTM D 2901), but it is not used widely in practice.

The cement and soil or aggregate mixture is typically transported from the plant to the construction site, where it may be spread by motor grader, spreader box on a dozer, or a paving machine. Sheepsfoot, tamping or pad-foot rollers, pneumatic rollers, and vibratory rollers have all been used to compact cement-stabilized materials. The same principles of compaction that were discussed in Chap. 5 still apply to compaction of cement-stabilized materials. The final grading can be done with a motor grader or an electronic automatic profiler. If necessary, the surface may be roughened with various drags, a hydraulic scratcher, a spiketooth harrow, or similar equipment. This removes the planes formed during compaction and provides a better surface to which succeeding layers can bond. In multiple-lift construction, bonding between lifts can be enhanced by roughening the surface, keeping times between lift placement short, keeping the surface clean, or mixing a retarder with the cement-stabilized material to slow the setting of the cement. The use of bonding grouts to improve bonds between lifts has given mixed results. When construction joints must be placed in the cement-stabilized material due to work delays, at the end of the day's construction, adverse weather, or similar causes, the exposed edge should be trimmed to a vertical face using the blade of a motor grader, concrete saw, special cutting blades attached to compaction equipment, or similar devices. It is impossible to obtain good compaction adjacent to free edges because of the material's tendency to displace laterally, so it is good practice to remove the outer 3 to 6 in of poorly compacted material along the joint when it is trimmed. The vertical face will allow the succeeding material to be placed and compacted properly. If the joints are not trimmed vertically, the new material placed on the initial rounded or sloped joint edge will tend to develop raveling problems.

Proper curing with adequate moisture and at temperatures above 40°F is necessary for cement-stabilized materials to gain strength. Commonly, a cure period of 3 to 7 days is required; however, if the cement-stabilized material is well compacted, some light traffic may be allowed sooner if no rutting occurs in the stabilized materials due to construction activities. Loss of water from the cement-stabilized material must be prevented. This may be accomplished by maintaining moist curing, covering the surface, spraying the surface with a membrane curing compound, or application of an asphalt seal on the surface of the cement-stabilized material. Moist curing uses trucks, sprinklers, or other methods to keep the exposed surface moist. The greatest danger here is that interruptions in the process will allow the surface to dry out. The surface also may be covered with wet burlap, plastic, or similar material that will prevent loss of moisture. Alternatively, asphalt cutbacks or emulsions may be sprayed on the surface to form a continuous asphalt seal over the surface. Common application rates vary from 0.15 to 0.30 gal/yd^2, but the only important criterion is to form a continuous waterproof seal of the surface. This may require several applications over several days as the emulsions break or cutback solvents evaporate. Cement-stabilized material should not be allowed to freeze within 7 days of placement. During spring and fall construction this may require covering newly placed material with insulating materials such as straw, soil, or blankets.

A crucial difference between placement of cement-stabilized material and either lime-stabilized material or conventional soils is that cement-stabilized material must be placed and compacted within a relatively short time after mixing the water, cement, and soil. Failure to place and compact the material before the cement begins to set will result in lower density and lower strength. Commonly specified time periods for placement of cement-stabilized materials are

Haul time: not more than 30 minutes

Start of compaction: not more than 1 hour after mixing

End of compaction: not more than 2 hours after mixing

Start of curing: when moisture loss begins, as evidenced by drying or graying of the surface; water sprinkling may be needed during compaction and finishing operations

Cement hydration is a function of temperature, so these times may be too long on hot days, while on cool days a little more time may be available. Retarders have been used with cement-stabilized materials to try to extend construction times, improve the quality of finish grading, and enhance bonding between lifts; however, they are not in common use (Murphy 1988, American Concrete Institute 1990).

6.5   Bituminous Materials

Bitumens include a wide variety of semisolid, predominately hydrocarbon products that occur naturally or that form from the distillation of petroleum, coal, wood, or peat. Natural bitumens have been used in construction since ancient time, and asphalts from distillation of petroleum are of major importance in modern construction.

Asphalts encountered in soil stabilization are generally classified as shown in Table 6.5. Asphalt cements usually require heat to lower their viscosity to a point where they can be mixed with other materi-

TABLE 6.5   Classification of Asphalts

| Product | Description | Relevant standards | |
		ASTM	AASHTO
Asphalt cement	Product of distillation of petroleum		
Penetration grade	Classified on basis of the penetration test	D 946	M20
Viscosity grade	Classified on basis of viscosity	D 3381	M226
Cutback asphalt	Asphalt cement mixed with a solvent		
Rapid curing	Naptha-type solvent	D 2028	M81
Medium curing	Kerosine-type solvent	D 2027	M82
Slow curing	Low-volatility solvent	D 2026	M141
Emulsified asphalt	Asphalt cement mixed with an emulsifying agent and water		
Anionic	Negatively charged emulsifying agent	D 977	M140
Cationic	Positively charged emulsifying agent	D 2397	M208

als. Cutback asphalts are liquids that harden as the solvents evaporate. Environmental restrictions limit the use of cutbacks today. Emulsified asphalts are also liquid at room temperature and harden as the emulsion breaks, allowing the asphalt particles to coalesce and the water to evaporate and drain. The amount and type of emulsifying agent determine whether the asphalt will be a rapid-setting, medium-setting, or slow-setting emulsion. The emulsified asphalt droplets also carry an electrical charge. Consequently, negatively charged, anionic emulsions would be most compatible with positively charged aggregates such as limestone, and positively charged, cationic emulsions would be particularly compatible with negatively charged acidic aggregates such as siliceous sandstone. Guidance on selection and use of emulsified asphalt is given in ASTM D 3628 and other general references (e.g., Asphalt Institute 1993). Asphalt cements tend to be used for higher-quality structural applications, whereas the liquid asphalts (cutbacks and emulsified asphalts) are more common in less severe structural applications.

Most asphalt cements in the United States today are specified using the viscosity grading system (ASTM D 3381), and the asphalt grades are identified by a two-letter designation (AC or AR) followed by a numerical indicator of the asphalt viscosity. The older penetration grading system (ASTM D 946) uses the depth of penetration of a standard needle into the asphalt at 77°F as described in ASTM D 5 to designate grades of asphalt. The penetration of the needle (or pen) is measured in tenths of a millimeter, and the asphalt is graded as a 40 to 50 pen asphalt which is harder than a 60 to 70 pen asphalt, etc. The recently completed Strategic Highway Research Program (SHRP) developed a new asphalt binder specification that will be coming into use during the 1990s. Cutback asphalts have a designation consisting of two letters identifying the curing time (RC, MC, and SC, for rapid, medium, and slow curing, respectively) followed by a numerical value related to the kinematic viscosity (ASTM D 2170) in centistokes at 140°F. Anionic emulsified asphalts are designated with a two-letter prefix indicating the speed of setting (RS, MS, and SS, for rapid, medium, and slow setting, respectively) and the numeral 1 or 2 reflecting the amount of asphalt in the emulsion (typically 55 percent residue from distillation for 1 compared with 65 percent for 2). A suffix h after the designation indicates that a harder base asphalt was used in the emulsion. A prefix of C (e.g., CMS-2) indicates a cationic emulsion. Cationic and anionic asphalts of the same setting grade (e.g., MS-2 and CMS-2) differ, so their respective ASTM standards should be consulted for specific requirements.

6.5.1 Mechanisms

Asphalt does not react chemically with the soil or aggregate to be stabilized but instead coats the individual particles or agglomerates of particles. This provides a waterproof covering and cohesive binder for the particle. Since there are no chemical reactions involved, the stabilization is complete as soon as hot asphalt cement cools, solvents in cutback asphalt evaporate, or emulsions break and the water evaporates. Consequently, stabilization with asphalt should reduce water susceptibility of the original material and should increase its strength. Asphalts are viscoelastic materials, so relatively large changes in their rheologic properties occur as a function of temperature and of rate of loading. Asphalts lose stiffness as temperatures rise, and their ability to resist deformation under load decreases as the time of loading increases (creep).

6.5.2 Properties

The Marshall mix design procedure (ASTM D 1559) is still the most widely used method of examining the effect of asphalt content on the properties of an asphalt and aggregates. The compressive load

(Marshall stability) on a compacted cylindrical specimen is measured along with deformation (flow). The void structure is also commonly analyzed as part of the test. Detailed description of this approach may be found in Asphalt Institute (1994), Department of the Army (1987), and Roberts et al. (1991).

Figure 6.13 shows sample Marshall mix design results for four natural aggregates that might be considered as candidates for stabilization for a pavement base or subbase or low-grade surfacing. The Marshall density (usually determined at either 50 blows per sample side for lightly trafficked facilities or 75 for heavily trafficked facilities) is often used to control field placement of asphaltic materials in a manner analogous to laboratory compaction for soil compaction. The Marshall stability (resistance to compressive load applied to the sample) is a useful indicator of the mix's ability to resist load, but it is not a reliable measure of actual strength or stiffness of the mix. Flow is the vertical deformation of the Marshall specimen under compressive load. High values of flow indicate a plastic mix that will deform under traffic, while low values of flow indicate a lean mix that may not be durable, may ravel, and may be brittle. High voids in a mix suggest that permeability will be high and that oxidation of the asphalt cement will cause premature hardening and deterioration of the asphalt with resulting raveling and cracking. Low voids in the mix are associated with rutting and instability of the mix. Low voids filled with asphalt are similarly associated with durability problems, and high voids filled with asphalt are associated with field stability problems.

Any bituminous stabilization must balance the competing demands of having sufficient asphalt to provide low permeability and durability, sufficient compaction to avoid densification under use, and the right combination of mineral aggregate and asphalt to provide strength. The strength of asphalt-stabilized materials depends more on the strength of the soil or aggregate being stabilized than lime- and cement-stabilized materials. The selection of an optimum asphalt content and interactions of the various parameters of asphalt mix design, whether for conventional hot-mix asphalt concrete or for stabilization, are quite complex. More detailed information than can be provided here is readily available (e.g., Asphalt Institute 1994, Department of the Army 1987, and Roberts et al. 1991).

A number of different tests are available to examine asphalt properties besides the Marshall series mentioned before. These includes the Hveem tests, indirect tensile test, gyratory tests, beam fatigue, creep tests, complex modulus, and resilient modulus, among others. These are described in detail in a number of references, including Roberts et al. (1991).

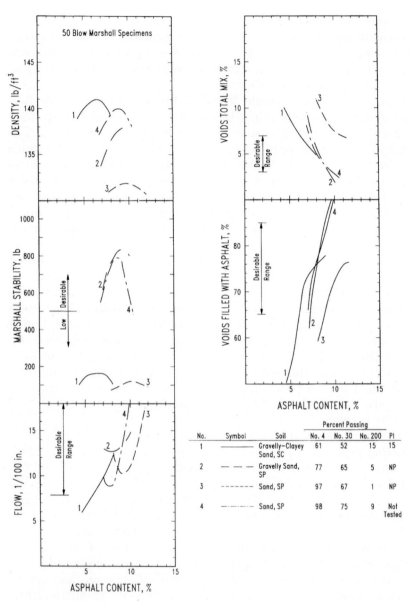

Figure 6.13 Marshall mix design results for stabilization of four natural aggregates with asphalt cement (*based on data reported by Grau 1979*).

The available tests for asphalt and aggregate mixtures have been oriented primarily toward asphalt concrete used as pavement surfacings. Most approaches for analyzing and evaluating asphalt-stabilized materials simply have been adapted from criteria for conventional dense-graded asphalt concrete used as a pavement surfacing. Highly empirical systems such as the Marshall methods that give reasonably good results for conventional dense-graded asphalt concrete often do a poor job of handling the less standard materials (poorly graded, gap-graded, plastic fines, etc.) that we often encounter in stabilization work (Rollings 1988b). Therefore, some caution is needed in applying conventional asphalt concrete criteria to asphalt stabilization, where the materials may depart significantly from our experience base and for which performance requirements may be significantly different.

Water is the primary durability problem of asphalt-stabilized materials. Often the Marshall stability or splitting tensile strength after soaking in water is required to be 70 or 75 percent of the unsoaked value as a crude check on water susceptibility of the mix. Under certain conditions, water also can strip asphalt off the aggregates, and this was discussed at greater length in Chap. 2. This is a potentially serious problem for asphalt-stabilized materials that may be used in pavements or as water barriers where exposure to water is likely.

6.5.3 Mix design

Several mix design approaches to asphalt stabilization have been developed, and most of these are adaptations of methods used for conventional asphalt concrete surfacings. A survey of state highway departments reported by Terrel et al. (1979) found that 13 states used a Marshall-type approach to mix design of asphalt-stabilized base courses, 8 used the Hveem approach, 2 used unconfined compressive strength, and others reported use of the Hubbard-Field, triaxial compression, repeated-load triaxial, and various penetration tests. Comparisons and descriptions of different approaches may be found in Asphalt Institute (1993), Department of the Army (1983), Terrel et al. (1979), and Epps et al. (1971).

There is no consensus on mix design practice for stabilization with bituminous materials. Some use an initial estimate of asphalt content based on past experience or empirical formulas with the intent of adjusting this estimate during construction as needed. Others may use a more rigorous (although not necessarily better) approach based on laboratory testing such as the Marshall or Hveem method. This discussion will examine these approaches for asphalt cements and for liquid asphalts separately.

Asphalt cement. Asphalt cement and the material with which it is mixed must be heated to allow proper mixing and coating of particles. This normally will be done at a central plant. Generally, the grade of asphalt cement being used in local paving will be indicative of the grade that would be suitable for stabilization work under the local environmental conditions. Typically, this might be an AC-20 or 60 to 70 pen material in the hot southern states of the United States and an AC-10 or 85 to 100 pen asphalt in the cooler northern states. If the material to be stabilized is predominately sand and relatively fine material, it may be helpful to use a lower-viscosity-grade asphalt to ease mixing. If, on the other hand, there are concerns about the stability of the aggregate, or if it must withstand heavy traffic, a higher-viscosity-grade asphalt may be helpful.

Often asphalt-stabilized materials will be within a pavement or other structure where they will not be exposed directly to the environment. This allows a little more latitude in selection of asphalt grades than for asphalt concrete surfacings. The surfacings must use a grade stiff enough to withstand softening in the summer heat and also retain flexibility to prevent cracking due to cold weather. Like many aspects of work with asphalts, competing requirements have to be balanced to reach a practical solution. Table 6.6 shows some suggested grades of asphalt for stabilization.

An initial estimate of asphalt content suggested by the Asphalt Institute (1994) is

$$P = 0.035a + 0.045b + Kc + F$$

where P = approximate asphalt content of the mix, percent by weight of mix

a = percent retained on No. 8 sieve

TABLE 6.6   Suggested Asphalt Cement Grades for Soil Stabilization

Climate	Asphalt grading system	Sand asphalt	Crushed stone or gravel-sand mix
Hot	Viscosity	AC-20	AC-40
	Penetration	60–70	40–50
	Viscosity of aged residue	AR-8000	AR-16000
Moderate	Viscosity	AC-10	AC-20
	Penetration	85–100	60–70
	Viscosity of aged residue	AR-4000	AR-8000
Cold	Viscosity	AC-5	AC-10
	Penetration	120–150	85–100
	Viscosity of aged residue	AR-2000	AR-4000

Based partly on Epps et al., 1971.

b = percent passing the No. 8 and retained on the No. 200 sieve

c = percent passing the No. 200 sieve

K = 0.15 for 11 to 15 percent passing the No. 200 sieve

   = 0.18 for 6 to 10 percent passing the No. 200 sieve

   = 0.20 for 5 percent or less passing the No. 200 sieve

F = 0 to 2.0 percent based on light or heavy aggregate absorbtion; in lieu of other data, 0.7 may be used

This may be used as an initial starting point with adjustments in the field during construction, or it may be used as a starting point for more careful laboratory testing.

Some variant of the Marshall mix design approach is probably the most common approach to selecting optimum asphalt content and evaluating the quality of the mix. Typical criteria for evaluating conventional asphalt paving mixes with this procedure are shown in Tables 6.7 and 6.8. A common current practice is to select an initial design asphalt content at an air void content of 4 percent and then check the other Marshall results (such as in Fig. 6.13) against criteria such as shown in Table 6.7.

Asphalt-stabilized materials are not exposed to the levels of stress or to the temperatures and oxidation conditions that asphalt paving surfaces commonly are. Consequently, it is probably very conservative to require asphalt-stabilized materials to meet the same requirements as asphalt pavement surfaces. This conservativeness can be redressed to some degree by making adjustments to the test temperatures at which the Marshall tests are conducted in recognition that the buried underlying asphalt-stabilized layers will never be exposed to the same temperatures as the asphalt pavement surface. Epps et al. (1975) found in tests of one asphalt cement and four different aggregates that decreasing the Marshall test temperature from the standard value of 140°F to 77°F increased the Marshall stability between 4.4 and 17.6 times. Other investigators have reported similarly dramatic effects of temperature on asphalt properties.

An approach that has been suggested by a number of investigators is to conduct the Marshall tests at some temperature lower than the standard 140°F that would be more representative of the temperatures to which the material would be exposed in the field. Then criteria such as those in Table 6.7 could be used to evaluate the suitability of the mix.

The air voids and percentage voids filled with asphalt in Table 6.7 are probably too restrictive also. For many applications, the air voids content could be allowed to increase and the percentage voids filled with asphalt could be allowed to decrease.

TABLE 6.7 Typical Marshall Mix Design Criteria

Property	Asphalt Institute (1994)			Department of the Army (1987)		
	Light traffic*	Medium traffic*	Heavy traffic*	Light load†	Heavy load†	Sand asphalt†
Compaction energy‡	35	50	75	50	75	50
Marshall stability, lb	≥750	≥1200	≥1800	≥500	≥1800	≥500
Flow (1/100 in)	8–18	8–16	8–14	≤20	≤16	≤20
Percentage air voids	3–5	3–5	3–5	3–5§	4–6§	4–6
Percentage voids filled with asphalt	70–80	65–78	65–75	70–80¶	55–75¶	70–80
Percentage voids in mineral aggregate	Table 6.8	Table 6.8	Table 6.8	Not used	Not used	Not used

*Light < 10⁴, medium = 10⁴ to 10⁶, heavy >10⁶ design equivalent axle loads.

†Light ≤ 100 psi tires (roads and streets), heavy >100 psi tires (airfields, tracked vehicles, solid tires), sand asphalts for ≤100-psi tires.

‡Number of blows on each end of the sample with a Marshall hammer (10-lb weight, 3⅞-in diameter hammer face, 18-in drop.

§For intermediate course, surface course has 2 to 4 percent air voids.

¶For intermediate course, surface course has 80 to 90 and 75 to 85 percent filled for the 50-blow light and heavy load categories.

TABLE 6.8   Voids in Mineral Aggregate Criteria

Nominal maximum particle size*	Minimum voids mineral aggregate, %		
	Design air voids, %†		
	3.0	4.0	5.0
No. 16	21.5	22.5	23.5
No. 8	19.0	20.0	21.0
No. 4	16.0	17.0	18.0
3/8-in	14.0	15.0	16.0
1/3-in	13.0	14.0	15.0
3/4-in	12.0	13.0	14.0
1.0-in	11.0	12.0	13.0
1.5-in	10.0	11.0	12.0
2.0-in	9.5	10.5	11.5
2.5-in	9.0	10.0	11.0

*Defined as one size larger than the first sieve to retain more than 10 percent of the aggregate.
†Interpolate minimum voids in the mineral aggregate between the listed air void values listed.
SOURCE: From Asphalt Institute of America, 1994.

There are analytical techniques available to estimate the temperature of underlying stabilized layers (e.g., Dempsey and Thompson 1969, Dormon and Metcalf 1965). The Asphalt Institute (1994) suggests that asphalt-stabilized bases that are 4 in or more below the surface should be evaluated with Marshall tests run at 100°F. This is considered representative of conditions in the United States. Epps et al. (1975) suggest testing for asphalt-stabilized base evaluation at 100°F for the cooler areas in the northern hemisphere and 120°F for the warmer climates in the northern hemisphere.

At present, a reasonable laboratory approach to determining the amount of asphalt cement to use in asphalt-stabilized materials appears to be as follows:

- Select a trial asphalt content based on experience or empirical formulas.

- Run Marshall tests (or Hveem method or other established hot mix asphalt concrete mix design approach) at a reduced temperature representative of the expected field conditions and at several asphalt contents.

- Select an initial optimum asphalt content based on conventional criteria such as Table 6.7.

- If the project warrants, run any additional significant engineering property test (resilient modulus, splitting tensile, beam fatigue, etc.) at the initial optimum asphalt content to confirm desired structural capacity and conformance to design requirements.

- Evaluate water effects (see Sec. 2.5.4).
- Adjust asphalt content as needed.

Although the Marshall or similar conventional mix design methods may provide a useful and simple approach to selecting an asphalt content for stabilization, these conventional mix design procedures are empirical and were developed for well-graded aggregates with a low content of nonplastic fines. Consequently, they may not give the best solution when working with nonstandard materials that might be poorly graded, have excess fines, contain plastic fines, etc. (Rollings 1988b, Grau 1979). Therefore, field adjustments to the asphalt content may be necessary, or more comprehensive testing based on actual engineering properties needed in the field (e.g., resilient modulus, splitting tensile, beam fatigue, etc.) may provide a different and perhaps superior mix design. The drawback, of course, is the cost of such testing and the designer's difficulty in defining exactly what properties he or she needs.

Liquid asphalts.   Cutback and emulsified asphalts may be used for either central plant mixing or field mixing in stabilization work. Table 6.9 shows suitable liquid asphalts for stabilization of different soils. Generally, the highest-viscosity liquid asphalt that can be mixed into the soil will provide the best results. Consequently, the proper grade of liquid asphalt will vary depending on the available mixing equipment and the soil. Uniformly graded material requires higher-viscosity liquid asphalts, and increasing fine content in a material will require lower-viscosity liquid asphalts. Other factors that influence selection of asphalt include properties of the asphalt cement from which it was made, rate of curing, ambient temperature conditions, and aggregate charge characteristics (for use of cationic emulsified asphalts).

Several empirical equations have been developed to determine liquid asphalt contents in a mix. Like all empirical design systems, these equations are only reliable when used with materials similar to those from which the equations were derived and the intended use is similar to the original use. However, these equations can be used for simple low-cost jobs or as an initial estimate of liquid asphalt content to be field adjusted later or as a starting point for a more comprehensive laboratory mix design. These equations are:

McKesson's equation for sands (McKesson 1950):

$$P_e = 0.75(0.05B + 0.01\beta + 0.50f)$$

TABLE 6.9 Liquid Asphalts for Soil Stabilization

Mixing	Soil/aggregate	Cutback asphalts	Emulsified asphalts
Central plant	Open graded	MC 800	MS-1, MS-2, MS-2h; CMS-2, CMS-2h
	Well graded	RC 250; MC 250, 800, 3000	SS-1, SS-1h; CMS-2; CSS-1, CSS-1h
Field Mixed	Open graded	RC 250, 800, 3000; MC 800, 3000; SC 250, 800	MS-1, MS-2, MS-2h; CMS-2, CMS-2h
	Well graded	RC 250; MC 250, 800; SC 250, 800	SS-1, SS-1h; CMS-2; CSS-1, CSS-1h
	Sands	RC 250, 800; MC 70, 250, 800	SS-1, SS-1h; CSS-1, CSS-1h
	Sandy soils	RC 70, 250; MC 250, 800	SS-1, SS-1h; CMS-2; CSS-1, CSS-1h

SOURCE: Adapted from Asphalt Institute, 1991.

Bird equation for material with 50 percent or more passing a No. 10 sieve (Bird 1959):

$$P_e' = 0.02b + 0.1f \qquad \text{(5 to 12 percent passing No. 200)}$$

$$P_e' = 0.2b + 0.1\delta + 4 \qquad \text{(>12 percent passing No. 200)}$$

Oklahoma equation (Oklahoma Highway Department 1967):

$$P_c = k + 0.005b + 0.01c + 0.06f$$

The Asphalt Institute's equations (Asphalt Institute 1991):

$$P_c = 0.02D + 0.07\gamma + 0.15\epsilon + 0.20f$$

$$P_e = 0.05A + 0.1\alpha + 0.5f$$

where P_e = percent of emulsified asphalt by weight of dry aggregate
P_e' = pounds of emulsified asphalt per cubic foot of loose, dry aggregate
P_c = percent of cutback asphalt residue by weight of dry aggregate; percent cutback can be calculated as $[P_c/(100-\text{percent solvent})]\times 100$
a = percent passing No. 8 sieve

A = percent retained on No. 8 sieve

α = percent of material passing No. 8 sieve and retained on No. 200 sieve

b = percent passing No. 10 sieve

B = percent retained on No. 10 sieve

β = percent of material passing No. 10 sieve and retained on No. 200 sieve

c = percent passing No. 40 sieve

D = percent retained on No. 50 sieve

γ = percent of material passing No. 50 sieve and retained on No. 100 sieve

ϵ = percent of material passing No. 100 sieve and retained on No. 200 sieve

f = percent passing No. 200 sieve

$\delta = 24 - f$

k = 1.5 if $PI \leq 8$

   = 2.0 if $PI > 8$

These empirical methods of selecting liquid asphalt content may be adequate for small jobs or to provide a starting point with field adjustment later. However, a sounder approach based on testing may be justified on larger jobs.

The earlier discussions concerning stabilization tests with asphalt cement apply also for liquid asphalts. However, an extra complexity is added because the asphalt-stabilized material also must undergo some curing (solvent evaporation for cutbacks and emulsion breaking for emulsified asphalts). ASTM D 4223 provides a procedure for preparing samples made with liquid asphalts for further evaluation using Marshall tests (ASTM D 1559), Hveem tests (ASTM D 1560), or resilient modulus (ASTM D 4123). The Asphalt Institute (1991) also provides detailed procedures for preparing samples and evaluating mixes with emulsified and cutback asphalts.

6.5.4 Suitable soils

Asphalt stabilization will work best with granular materials. Excess fines make it difficult to achieve a good particle coating, require high asphalt contents, and pose construction and mixing difficulties. Generally, plasticity index values above 6 and more than 12 percent passing the No. 200 sieve indicate potential problems for asphalt stabilization. Limiting the product of the plasticity index and the percentage passing the No. 200 sieve to 72 or less also has been suggested as a useful measure of soil suitability for asphalt stabilization (Asphalt Institute 1991). The sand-equivalent test (ASTM D 2419) is a rapid method of indicating the relative proportion of clay fines in sand-sized material. Generally, materials with a sand equivalent above 30 can be stabilized with asphalt, those with a sand equivalent below 20 probably cannot, and between these values results depend on the ability of the asphalt to waterproof the particles (Asphalt Institute 1991). Table 6.10 shows one suggested rating of materials for asphalt stabilization. Of course, the intended use of the stabilized material and its overall gradation are major factors that will determine the suitability of the material for stabilization. In general, asphalt cements are more suited for materials with low contents of fines, whereas liquid asphalts can tolerate higher contents of fines.

Some aggregates have a high affinity for water (hydrophilic aggregates; see Sec. 2.5.4) and are very difficult to coat with asphalt. Attempts to use liquid asphalts with these aggregates are usually futile until the electrical charge of the asphalt or the aggregate is changed (Asphalt Institute 1991). Addition of lime or a proprietary antistrip compound or use of the properly charged emulsified asphalt (anionic or cationic) for the aggregate charge may be helpful in accomplishing this.

TABLE 6.10 Materials Suitable for Asphalt Stabilization

Requirement	Sand-asphalt	Soil-asphalt	Sand-gravel-asphalt
Gradation, % passing			
1½ in	100		100
1 in			
¾ in			60–100
No. 4	50–100	50–100	35–100
No. 10	40–100		
No. 40		35–100	13–50
No. 100			8–35
No. 200	5–12	Good, 3–20	0–12
		Fair, 0–3 and 20–30	
		Poor, >30	
Atterberg limits			
Liquid limit		Good, <20	
		Fair, 20–30	
		Poor, 30–40	
		Unusable, >40	
Plasticity index	<10	Good, <5	<10
		Fair, 5–9	
		Poor, 9–15	
		Unusable, >12–15	

Note: Includes slight modifications later made by Herrin (Robnett and Thompson 1969)
SOURCE: Herrin, 1960.

6.5.5 Construction considerations

Stabilization with asphalt cement is usually carried out at a central plant where all the materials can be appropriately heated and thoroughly mixed. These would be the same drum or batch plants used to produce asphalt concrete. In general, production and placement of asphalt cement–stabilized materials will follow the same procedures as for conventional hot-mix asphalt concrete (Roberts et al. 1991, Transportation Research Board 1991, Asphalt Institute 1978, 1983). The materials are heated and mixed to produce a mix typically between 275 and 300°F. The mix is delivered to the site by dump trucks and placed with an asphalt paving machine. Good workmanship greatly improves the quality of the construction joints between lanes [see Roberts et al. (1991), Transportation Research Board (1991), and Asphalt Institute (1978) for details].

Rolling consists of three steps: breakdown, intermediate, and finish rolling. Breakdown rolling is crucial and should be applied as soon as possible. If it is applied too late, the desired density may not be achieved, but if, on the other hand, it is applied too soon, the hot mix may push, shove, and crack under the roller. The timing of rolling depends on the mix temperature and the quality of materials in the mix. For high-quality crushed stone mixes, breakdown rolling may be applied at temperatures between 250 and 275°F. For mixes containing uncrushed, rounded gravel and sand particles, the rolling temperature will likely be around 225°F. Sand asphalts or asphalts with much fine material may require cooling to 200°F or less. These last two cases are most likely to be encountered with stabilized materials.

A tandem steel-wheel static roller is the most common breakdown roller. If low-stability mixes crack under steel-wheel rollers, pneumatic rollers may prove useful for breakdown rolling. Intermediate rolling is carried out after the breakdown rolling with vibratory steel-wheel rollers or pneumatic rollers to achieve the specified density. Vibratory rollers on asphalt materials generally do better at higher frequencies than those commonly used with soils. Finish rolling with a static steel-wheel roller is the final step to remove any surface marks or blemishes.

Liquid asphalts may be used to produce stabilized materials either at a central plant or on site. At a central plant, better control of materials and more uniform mixing are possible. In addition, the materials may be heated, which allows a higher-viscosity liquid asphalt to be used. This is useful because generally the higher-viscosity liquid asphalts will produce the most stable mixes. The mixed material is hauled to the site and placed with asphalt paving machines, Jersey spreaders, towed spreaders, or motor graders. Cutback materials may need to be aerated before compaction to allow some solvent evaporation.

On-site stabilization with liquid asphalts consists of mixing, aerating, and compacting the materials. The mixing may be accomplished by rotary mixers, motor graders, or traveling plants in much the same way as described for other stabilizers. The liquid asphalt is sprayed from an asphalt distributor just ahead of the mixing equipment, or some traveling plants and rotary mixers are equipped to allow discharge of the liquid asphalt directly into mixing chamber. Some of the diluent (water for emulsified asphalts and solvent for cutbacks) that makes the liquid asphalt flowable must be lost before compaction. Otherwise, the mix usually will not support the compaction rollers. This loss of diluent often will occur during mixing, but additional aeration may be needed depending on the specific mix (e.g., fine mixes lose diluents more slowly than coarse ones), the nature of the liquid asphalt (e.g., SC or SS liquid asphalts will need more time than RC or RS liquid asphalts), and the environment (e.g., warm, dry weather encourages evaporation). Usually rolling needs to start for emulsified asphalt mixes as soon as the emulsion breaks. This point occurs when the material changes from brown to black. Compaction for cutback asphalts normally should start when about half the solvent has evaporated.

Compaction can use one or some combination of static steel-wheel rollers, vibratory rollers, or pneumatic rollers to achieve the specified density.

6.6   Pozzolanic and Slag Materials

Pozzolanic materials are natural materials or by-products of industry. When they are mixed with water and alkali or alkaline earth hydroxide activators, they will react to form cementitious compounds. The most common source of these activators is calcium hydroxide in lime or portland cement.

Natural pozzolans were used in ancient construction, and the Romans in particular developed the process to a high degree. They found that the sandy volcanic ash from near Vesuvius and elsewhere in Italy could be mixed with lime, water, and aggregates to make a strong waterproof concrete. Today, the most common pozzolan encountered in construction is fly ash recovered from the burning of ground or powdered coal. Natural pozzolans and fly ash are specified in ASTM C 618 in one of three classes: N, raw or calcined natural pozzolans; C, by-product of burning lignite or subbituminous coal; and F, by-product of burning anthracite or bituminous coal.

Granulated blast furnace slag, a by-product of the production of iron, can be ground to produce another material that forms cementitious compounds when combined with water and calcium hydroxide. This material is described in ASTM C 989 and is classified in one of three grades—80, 100, and 120—based on the results of the slag activity index at 7 and 28 days. The slag activity index is the ratio of the strength of a mortar cube composed of 50 percent slag and 50 percent portland cement to the strength of a cube composed entirely of portland cement.

Table 6.11 compares the nominal chemical composition of portland cement, ground granulated blast furnace slag, and fly ash. The pozzolanic fly ashes are rich in silica and alumina, portland cement has a high calcium oxide content, and the slag falls between the two. Some class C fly ashes and ground granulated blast furnace slags contain some self-cementing capability, but generally these materials and class F fly ashes will require the addition of some source calcium hydroxide to develop appreciable cementitious reactions. Additions of lime or portland cement would be obvious ways to provide sources of calcium hydroxide to start these cementitious reactions.

For stabilization work, combinations of fly ash or slag with an activator offer several potential advantages over the conventional portland cement stabilization of soils and aggregates that was discussed earlier. These fly ash- or slag-stabilized materials are often cheaper than portland cement, and their setting times often are appreciably slower. Consequently, there is more time to mix, spread, and compact the materials than when conventional portland cement is used. In addition, shrinkage is often less, so shrinkage cracking with resulting permeability problems and reflective cracking can be less severe than with portland cement stabilization. These materials are often more resistant to sulfate attack and related durability problems than are portland cements. Strength gain is slower than with portland cement,

TABLE 6.11   Typical Chemical Composition of Portland Cements, Blast Furnace Slag, and Fly Ash

Material	Chemical composition		
	CaO	SiO_2	Al_2O_3
Portland cement*	60–70	17–25	3–8
Blast furnace slags†	29–42	32–40	7–17
Fly ash, class C‡	12–29	44–64	20–30
Fly ash, class F‡	1–7	45–64	20–30

*Neville, 1981.
†American Concrete Institute, 1987a.
‡American Concrete Institute, 1987b.

however. If early strength is needed, combinations of fly ash or slag with portland cement to provide both the activator and some early strength will provide an intermediate composite stabilizing medium.

The effects of fly ash and lime on the soil or aggregate to be stabilized include a complex mixture of reactions. These include reactions between the fly ash and lime and fine fraction of the soil similar to those with lime and clay alone, acting as a filler for voids to increase strength and density, and developing cementing reactions to bind particles together. Many clays are naturally pozzolanic, so the use of fly ash and lime for their stabilization would not appear to be generally as effective as simple lime stabilization alone. However, for silty materials that do not contain appreciable quantities of reactive clay, the fly ash and lime combination can be highly effective. However, it is probably predominately granular materials in which the void-filling role and the pozzolanic cementing can both be brought into play that offer the best promise for use of fly ash stabilization. In this role they have been effective as stabilizers for a variety of gradations of sand, gravel, and crushed stone (American Coal Ash Association 1991, Terrel et al. 1979).

Adequate curing time and temperature are needed for the reactions to occur and for the stabilization to become effective. Generally, the reactions cease around 40°F, but freezing does not seem to cause any long-term damage. The reactions will begin again when the temperature rises (American Coal Ash Association 1991, Terrel et al. 1979). The material also exhibits autogenous healing with time, since the reactions apparently can cement across existing cracks if sufficient fly ash and lime are available for the reaction to occur. Adequately cured lime–fly ash mixtures often will have compressive strengths in the range of 500 to 1000 psi, with ultimate long-term strength gain above 3000 psi being possible (Terrel et al. 1979). ASTM C 593 provides some general requirements for lime–fly ash mixes and includes a minimum compressive strength requirement of 400 psi for the stabilized material and a durability requirement for a vacuum-saturated compressive strength of 400 psi also.

The mix design for these materials requires determining the combination of fly ash, activator (usually lime), and water that will provide the most economical stabilization. Past experience suggests that the best strength and durability for the stabilized material occur when the coarse aggregate particles are floating in a matrix of the finer components of the blend (American Coal Ash Association 1991). A conceptual approach to mix design for fly ash stabilization is (1) determine the amount of fly ash that develops maximum density using an appropriate modified or standard energy laboratory soil compaction test; (2) the target fly ash content will be 2 percent higher than the maximum density fly ash content from (1) [this floats the coarse aggregate in a matrix of finer material (American Coal Ash Association 1991)]; (3) select activator content and type (lime or portland cement) to achieve the desired amount and rate of strength gain; (4) determine optimum compaction moisture content using an appropriate laboratory compaction curve based on density of samples with the target fly ash content and selected activator type and content compacted at various moisture contents; (5) prepare and cure compacted samples at this optimum moisture content; (6) evaluate for strength (e.g., ASTM C 593 requires a minimum 400 psi compressive strength after a 7-day cure at 100°F); (7) evaluate for durability (e.g., ASTM C 593 requires a minimum 400 psi compressive strength after vacuum saturation of cured samples); and (8) if warranted, optimize the mix by adjusting the fly ash content or the activator type or quantity to achieve the desired strength, durability, and economics. American Coal Ash Association (1991) and Terrel et al. (1979) provide additional discussion of mixture proportioning. As with lime- and cement-stabilized materials, the compressive strength of the fly ash–stabilized materials can be used as an indicator of resistance to freezing and thawing. The American Coal Ash Association (1991) suggests that when exposed to the first cycle of freezing, the stabilized material should have achieved 1000 psi compressive strength for severe conditions (more than 10 freezing and thawing cycles), 800 psi for moderate conditions (6 to 10 freezing and thawing cycles), or 600 psi for mild conditions (less than 6 freezing and thawing cycles).

A wide variety of materials has been stabilized with fly ash. Generally, granular materials respond best to fly ash stabilization, and maximum content of the fines in the material to be stabilized is often specified to be 15 to 30 percent. Mixtures have been used with 2 to 8 percent lime and 8 to 36 percent fly ash, but the normal range is $2\frac{1}{2}$ to 4 percent lime and 10 to 15 percent fly ash (Terrel et al. 1979). The American Coal Ash Association (1991) reports that coarse aggregates typically require 10 to 20 percent fly ash, whereas sandy aggregates typically require 15 to 30 percent fly ash. Initial strength gain can be improved with additions of 0.5 to 1.5 percent cement.

Stabilization with ground granulated blast furnace slag is not common in the United States, but it has proven useful overseas. Typical proportions have been 8 to 20 percent ground slag mixed with 1 percent lime (Ray 1986). At present, there is relatively little U.S. experience with this material as a stabilizing agent, but successful use overseas suggests that it should be viable in the United States also.

6.7 Miscellaneous Materials and Methods

A number of specialized stabilizers are sold under a variety of trade names. There is often little technical information or test data for some of these products. Identification of the mechanisms of stabilization is necessary in order to evaluate these commercial stabilizers. Once the mechanism of stabilization is identified, appropriate tests can be selected to allow evaluation of the material's effectiveness in accomplishing its stabilization objective.

ASTM D 4609 provides a general guide for evaluating the effectiveness of chemicals in stabilizing fine-grained soils. The effectiveness of the chemical is ascertained by measuring the unconfined compressive strength, Atterberg limits, particle size, moisture-density relation, and volume change of treated and untreated samples.

A wide variety of potential stabilizers exists. Some of these include

1. *Phosphoric acid and phosphates.* These have given some good results in acid soils and appear to react well with the troublesome mineral chlorite and to flocculate clays effectively to aid in compaction. However, they are too costly for most applications (Ingles and Metcalf 1973).

2. *Salts.* Salts such as sodium and calcium chloride have long been used as dust palliatives. They generally are not durable and require repeated applications to maintain their benefit.

3. *Lignin.* Lignin is a waste from paper production that can form bonds with silica surfaces and thereby act as a cementing agent. A mixture of potassium dichromate and lignin has been found to be particularly effective because the chromium ion enhances the lignin-soil bond. This mixture is expensive and may pose an environmental hazard to wildlife. It may prove particularly effective for treatment of troublesome volcanic soils and soils containing chlorite (Ingles and Metcalf 1973).

4. *Cation exchange.* Chemicals that promote favorable cation exchange in the soil are potential stabilizers in much the same way as lime's substitution of calcium for sodium and potassium in clay (see Sec. 6.3.1).

5. *Electrolytes.* Materials that serve as strong electrolytes decrease the ability of a clay particle to hold water on its surface. This would reduce the volume-change characteristics of expansive soils.

6. *Enzymes and polymers.* Proprietary enzyme and polymeric systems are also available that may serve as cementing agents, waterproofing agents, or to change the internal structure of clay particles.

7. *Combinations of stabilizers.* Research in France has developed a sophisticated emulsion of 22 to 40 percent asphalt, 33 to 44 percent cement, and 27 to 37 percent water to make a material combining the flexibility of asphalt-based materials with the strength-

ening effect of cementitious materials while avoiding the undesirable effects of temperature on asphalt and shrinkage cracking in cementitious materials (Godard 1991).

A number of commercial materials are marketed as stabilizers, but little data are available on most of them. Scholen (1992) has collected performance data on 60 projects that used proprietary stabilizers including three pozzolans, four bioenzymes, two sulfonated oils, ammonium chloride, a mineral pitch, and two clay fillers. All appeared to be doing well when used properly. Careful laboratory testing and well-designed and well-monitored test sections will be needed to carry out fair evaluations of the continuously evolving area of specialty stabilizers.

6.8   Selection of Stabilizers

The first step in selecting a stabilizer is to identify what end result is desired from the stabilization. If one wishes to reduce swelling in an expansive soil, a stabilizer that affects the soil's ability to gain and hold water is needed (e.g., lime or a strong electrolyte). If, on the other hand, a high-strength base course is needed under a major airport pavement and the pavement will be opened to traffic fairly quickly, a strong cementing material such as portland cement will be needed. The objective of the stabilization should determine the spectrum of stabilizers to consider.

Inherent in this first step is a clear understanding of how the stabilizer under consideration works. At a recent international conference, an engineer presented the results of his work trying stabilize an industrial waste that was composed on nonplastic silt- and fine sand–sized particles. Part of his large matrix of laboratory testing included in-depth assessments of the effects of adding various percentages of lime to the waste material. As borne out by his results, this work was wasted because there was no clay mineral present to give up the necessary silica and alumina to support a pozzolanic reaction with the lime. Combinations of lime and fly ash or lime and ground granulated slag offered more economical and effective approaches, and portland cement stabilization offered another, though more costly, alternative.

There are generally bands of plasticity and gradation within which each stabilizer is most effective. Combinations of stabilizing techniques may extend the range of soil properties for which stabilization can be effective. For instance, lime may be used to reduce plasticity and to coarsen the gradation of clayey materials so that it may be possible to then treat them with other cement, pozzolanic, or bituminous stabilizers to achieve the desired result.

Next, special project considerations need to be examined to determine if they will influence the selection of the stabilizer. This might include some of the following considerations: Are the stabilization effects to be permanent, or are they a temporary construction expedient? What type of equipment is available for mixing? How rapidly will the strength gain be needed? And very important, what are the weather implications on strength gain, curing, and durability?

Often massive amounts of materials are involved in stabilization efforts, so the economics of stabilization must be evaluated as part of the selection process. The benefits achieved by stabilization must be commensurate with the cost of the stabilization effort.

The foregoing steps should identify what stabilizers can reasonably achieve the desired results. Laboratory testing can then verify that the stabilizers will achieve the objectives needed and determine which stabilizer and what application rate will achieve the most economical and effective results. There is often a tendency not to test stabilizers in the laboratory except for large government projects. However, it is prudent to run at least minimal verification tests in the laboratory with the actual soils, stabilizers, and water that will be used on the project to check that the intended results will be achieved. Geotechnical materials are complex and variable, and the stabilization reactions are often very intricate and poorly understood. It is a small investment of money, time, and energy to check things out in the laboratory beforehand compared with the heartache of discovering an unexpected problem in the midst of construction. Full-blown laboratory investigations to optimize stabilizer selection and content are probably only economically justified for larger projects.

A number of guides for selecting stabilizers have been published, and these are summarized by Terrel et al. (1979). However, these selection guides are strongly oriented toward strengthening materials for use in highway and airfields. These guidelines may not prove useful for other project conditions and stabilization objectives. A general approach to stabilization may be summarized as follows:

1. Determine the objective of stabilization, and identify those stabilizers which will accomplish the objective.

2. Determine limitations imposed by the material to be stabilized, and determine if combinations of stabilizers can overcome these limitations.

3. Evaluate the impact of specific project limitations on the stabilizers under consideration.

4. Compare the cost and benefits of stabilization.

5. Optimize stabilizer selection and application rate in the laboratory

TABLE 6.12   Stabilization Methods

Method	Soil	Effect	Remarks
Blending	Moderately plastic	None	Too difficult to mix
	Others	Improve gradation Reduce plasticity Reduce breakage	—
Lime	Plastic	Drying	Rapid
		Immediate strength gain	Rapid
		Reduce plasticity	Rapid
		Coarsen texture	Rapid
		Long-term pozzolanic cementing	Slow
	Coarse with fines	Same as with plastic soils	Dependent on quantity of plastic fines
	Nonplastic	None	No reactive material
Cement	Plastic	Similar to lime	Less pronounced
		Cementing of grains	Hydration of cement
	Coarse	Cementing of grains	Hydration of cement
Bituminous	Coarse	Strengthen/bind waterproof	Asphalt cement or liquid asphalt
	Some fines	Same as coarse	Liquid asphalt
	Fine	None	Can't mix
Pozzolanic and slags	Silts and coarse	Acts as a filler Cementing of grains	Denser and stronger Slower than cement
Misc. methods	Variable	Variable	Depends on mechanism

using actual materials to be used in the project, or at least verify in the laboratory that the anticipated results of stabilization will be achieved with the actual project materials.

Table 6.12 provides a brief summary of some of the selection factors for different stabilization techniques, and more detailed discussion on these aspects was provided in the foregoing sections on each of the stabilizers.

6.9   References

Allen, J. J., D. D. Currin, and D. N. Little. 1977. "Mix Design, Durability, and Strength Requirements for Lime-Stabilized Layers in Airfield Pavements," Transportation Research Record 641, Transportation Research Board, Washington.

American Coal Ash Association. 1991. *Flexible Pavement Manual,* Washington.
American Concrete Institute. 1990. "State-of-the-Art Report on Soil Cement," ACI 230.1R-90, Detroit, Mich.
American Concrete Institute. 1987a. "Ground Granulated Blast-Furnace Slag as a Cementitious Constituent in Concrete," ACI 226.1R-87, Detroit, Mich.
American Concrete Institute. 1987b. "Use of Fly Ash in Concrete," ACI 226.3R-87, Detroit, Mich.
American Concrete Institute. 1977. "Guide to Durable Concrete," ACI 201.2R-77 (reapproved 1982), Detroit, Mich.
Arman, A., and G. Munfakh. 1972. "Lime Stabilization of Organic Soils," paper presented at 51st Annual Meeting of the Highway Research Board, Washington.
Asphalt Institute. 1994. *Mix Design Methods for Asphalt Concrete and Other Hot-Mix Types,* MS-2, 6th ed., Lexington, Ky.
Asphalt Institute. 1993. *A Basic Asphalt Emulsion Manual,* MS-19, 2d ed., Lexington, Ky.
Asphalt Institute. 1991. *Asphalt Cold Mix Manual,* MS-14, 3d ed., Lexington, Ky.
Asphalt Institute. 1983. *Asphalt Plant Manual,* MS-3, 5th ed., Lexington, Ky.
Asphalt Institute. 1978. *Asphalt Paving Manual,* MS-8, 3d ed., Lexington, Ky.
Bagonza, S., J. M. Peete, D. Newill, and R. Freer-Hewish. 1987. "Carbonation of Stabilized Soil-Cement and Soil-Lime Mixtures," *Proceedings*, PTRC Transport and Planning Meeting, University of Bath, pp. 29–48.
Barksdale, R. D., 1991. *The Aggregate Handbook,* National Stone Association, Washington.
Bird, G. C. 1959. "Stabilization Using Emulsified Asphalt," *Proceedings,* Canadian Good Roads Association.
Biswas, B. R. 1972. "Study of Accelerated Curing and Other Factors Influencing Soil Stabilization," doctoral dissertation, Texas A&M University, College Station, Texas.
Byers, J. G. 1980. "Treatment of Expansive Clay Canal Lining," *Proceedings of the Fourth International Conference on Expansive Soil,* American Society of Civil Engineers, New York.
Dempsey, B. J., and M. R. Thompson. 1969. "A Heat-Transfer Model for Evaluating Frost Action and Temperature Related Effects in Multilayered Pavement Systems," Illinois Cooperative Highway Research Program, Project IHR-401, University of Illinois, Urbana, Ill.
Dempsey, B. J., and M. R. Thompson. 1968. "Durability Properties of Lime-Soil Mixtures," Highway Research Record 235, Transportation Research Board, Washington.
Department of the Army. 1987. "Bituminous Pavements Standard Practice," TM5-822-8, Washington.
Department of the Army. 1983. "Soil Stabilization for Pavements," TM5-822-4, Washington.
Department of Defense. 1978. "Flexible Pavement Design for Airfields," DM21.3/TM5-825.2/AFM88-6, chap. 2, Washington.
Diamond, S., and E. B. Kinter. 1965. "Mechanisms of Soil-Lime Stabilization, an Interpretive Review," Highway Research Record 92, Highway Research Board, Washington.
Dorman, G. M., and W. D. Metcalf. 1965. "Design Curves for Flexible Pavements Based on Layered System Theory," Highway Research Record 71, Highway Research Board, Washington.
Dumbleton, M. J. 1962. "Investigations to Assess the Potentialities of Lime for Soil Stabilization in the United Kingdom," Technical Paper 64, Road Research Laboratory, Crowthorne, U.K.
Dunlap, W. A., J. A. Epps, B. R. Biswas, and B. M. Gallaway. 1975. "United States Air Force Soil Stabilization Index System: A Validation," AFWL-TR-73-150, Air Force Weapons Laboratory, Kirtland AFB, New Mexico.
Eades, J. L., and R. E. Grim. 1960. "Reaction of Hydrated Lime with Pure Clay Minerals in Soil Stabilization," Highway Research Record 262, Highway Research Board, Washington.
Eades, J. L., and R. E. Grim. 1966. "A Quick Test to Determine Lime Requirements for

Lime Stabilization," Highway Research Record 139, Highway Research Board, Washington.

Epps, J. A., W. A. Dunlap, and B. M. Gallaway. 1971. "Basis for the Development of a Soil Stabilization Index System," AFWL-TR-70-176, Air Force Weapons Laboratory, Kirtland AFB, New Mexico.

Epps, J. A., B. M. Gallaway, and W. A. Dunlap. 1975. "Realistic Mixture Design Requirements for Asphalt Stabilized Base Courses," AFWL-TR-73-35, Air Force Weapons Laboratory, Kirtland AFB, New Mexico.

Godard, E. 1991. "Stabicol: A Composite Material for Tomorrow's Roads," *Revue Générale des Routes et des Aerodromes,* no. 691:117–121.

Grau, R. E. 1979. "Utilization of Marginal Construction Materials for LOC," Technical Report GL-79-21, USAE Waterways Experiment Station, Vicksburg, Miss.

Gutschick, K. A. 1978. "Lime Stabilization Under Hydraulic Conditions," Fourth International Lime Congress, National Lime Association, Washington.

Hansen, K. D. 1986. "Soil-Cement for Embankment Dams," Bulletin no. 54, U.S. Committee on Large Dams, Denver, Colo.

Harty, J. R. 1971. "Factors Influencing the Lime Reactivity of Tropically and Subtropically Weathered Soils," AFWL-TR-71-46, Air Force Weapons Laboratory, Kirtland AFB, New Mexico.

Herrin, M. 1960. "Bituminous-Aggregate and Soil Stabilization," in *Highway Engineering Handbook,* K. B. Woods, ed., McGraw-Hill, New York.

Highway Research Board. 1961. "Soil Stabilization with Portland Cement," Bulletin 292, Washington.

Hunter, D. 1988. "Lime-Induced Heave in Sulfate-Bearing Clay Soils," *Journal of the American Society of Civil Engineers,* 114 (2): 98–107.

Ingles, O. G., and J. B. Metcalf. 1973. *Soil Stabilization, Principles and Practice,* Wiley, New York.

Kennedy, T. W., and M. Tahmoressi. 1987. "Lime and Cement Stabilization," in *Lime Notes, Updates on Lime Applications in Construction,* National Lime Association, Washington.

Kohn, S. D., and M. I. Darter. 1981. "Structural Design of Composite Pavement," *Proceedings of the 2nd International Conference on Concrete Pavement Design and Rehabilitation,* Purdue University, West Lafayette, Ind., pp. 107–117.

Larsen, T. J., and J. M. Armaghani. 1989. "Florida Econocrete Test Road, 10 Year Progress Report," *Proceedings of the 4th International Conference on Concrete Pavement Design and Rehabilitation,* Purdue University, West Lafayette, Ind., pp. 547–560.

Little, D. N. 1987. "Fundamentals of the Stabilization of Soil with Lime," Bulletin no. 332, National Lime Association, Arlington, Va.

McCallister, L. D., and T. M. Petry. 1990. "Property Changes in Lime Treated Expansive Clays Under Continuous Leaching," Technical Report GL-90-17, USAE Waterways Experiment Station, Vicksburg, Miss.

McElroy, C. H. 1989. "Soil Stabilization Experiences in the Soil Conservation Service," *Proceedings of the Soil Stabilization Technology Workshop,* Army Corps of Engineers, Department of Agriculture, Department of the Interior, Department of Transportation, and Environmental Protection Agency, Denver, Colo.

McElroy, C. H. 1987. Using Hydrated Lime to Control Erosion of Dispersive Clays, ASTM STP 931, American Society for Testing and Materials, Philadelphia.

McElroy, C. H. 1982. "Effectiveness of Dispersed Clay Treatment, Progress Report," Soil Conservation Service, Ft. Worth, Texas.

McDonald, E. B. 1969. "Experimental Stabilization Expansive Shale Clay—Lyman County. Four Year Report," South Dakota Department of Highways, Pierre, S.D.

McKesson, C. L. 1950. "Suggested Method of Test for Bearing Value of Sand-Asphalt Mixtures," in *Procedure for Testing Soils,* American Society for Testing and Materials, Philadelphia.

Mehta, P. K., and A. Klein. 1966. "Investigation on the Hydration Products in the System $4CaO \cdot 3Al_2O_3 \cdot SO_3$-$CaSO_4$-CaO-$H_2O$," Highway Research Board Separate Report 90, Washington.

Mitchell, J. K. 1986. "Practical Problems from Surprising Soil Behavior," *Journal of Geotechnical Engineering,* 112 (3): 259–289.

Mitchell, J. K., and D. Dermatas. 1992. "Clay Soil Heave Caused by Lime-Sulfate Reactions," in *Innovations and Uses for Lime,* STP1135, D. C. Hoffman and D. D. Stanley, eds., American Society for Testing and Materials, Philadelphia.

Morin, W. J., and P. C. Todor. 1975. "Laterite and Lateritic Soils and Other Problem Soils of the Tropics," U.S. Agency for International Development, Report AID/CSD3682, Washington.

Murphy, H. W. 1988. "Cement Treated Pavements: From the Unpredictable to the Dependable," *Proceedings of the 14th Australian Road Research Board Conference,* vol. 14, part 7, Canberra, Australia, pp. 140–147.

National Lime Association. 1976a. "Lime Stabilization Construction Manual," Bulletin 326, Washington.

National Lime Association. 1976b. "Lime Handling, Application, and Storage," Bulletin 213, Washington.

Netterberg, F., and P. Paige-Green. 1984. "Carbonation of Lime and Cement Stabilized Layers in Road Construction," Technical Report RS/3/84, National Institute for Transport and Road Research, Pretoria, South Africa.

Neubauer, C. H., and M. R. Thompson. 1972. "Factors Influencing the Plasticity and Strength of Lime-Soil Mixtures," Highway Research Record 381, Highway Research Board, Washington.

Neville, A. M. 1981. *Properties of Concrete,* 3d ed., Pitman Books, London.

Nussbaum, P. J., and B. E. Colley. 1971. "Dam Construction and Facing with Soil-Cement," Research and Development Bulletin RD010W, Portland Cement Association, Skokie, Ill.

Oklahoma Highway Department. 1967. "Engineering Classification of Geologic Materials," Oklahoma Highway Department Research and Development Division.

Perrin, L. L. 1992. "Expansion of Lime-Treated Clays Containing Sulfates," *Proceedings of the Seventh International Conference on Expansive Soils,* vol. 1, International Society of Soil Mechanics and Foundation Engineering, Dallas, Texas, pp. 409–414.

Petry, T. M., and D. N. Little. 1992. "Update on Sulfate-Induced Heave in Lime and Portland Cement Treated Clays: Determination of Potentially Problematic Sulfate Levels," presented at 1992 Annual Meeting of the Transportation Research Board, Washington.

Pima County. 1986. "Soil-Cement Applications for Use in Pima County for Flood Control Projects," Pima County Department of Transportation and Flood Control District, Tucson, Ariz.

Portland Cement Association. 1986. "Soil-Cement for Facing Slopes and Lining Channels, Reservoirs, and Lagoons," Information Sheet no. IS126W, Skokie, Ill.

Portland Cement Association. 1979. "Soil-Cement Construction Handbook," Engineering Bulletin EB003.09S, Skokie, Ill.

Portland Cement Association. 1971. "Soil-Cement Laboratory Handbook," Engineering Bulletin EB052S, Skokie, Ill.

Ray, M. 1986. "French Practice in the Stabilization of Road Pavements with Hydraulic Binders and in Surface Dressings," *Proceedings 13th Australian Road Research Board Conference,* vol. 13, part 1, Adelaide, Australia, pp. 65–89.

Robbins, E. G., and P. E. Mueller. 1960. "Development of Test for Identifying Poorly Reacting Sandy Soils Encountered in Soil-Cement Construction," Bulletin 267, Highway Research Board, Washington.

Roberts, F. L., P. S. Kandhal, E. R. Brown, D. Y. Lee, and T. W. Kennedy. 1991. *Hot Mix Asphalt Materials, Mixture Design, and Construction,* National Asphalt Pavement Association Education Foundation, Lanham, Md.

Robnett, Q. L., and M. R. Thompson. 1969. "Stabilization Recommendations for Illinois Soils and Materials," Illinois Cooperative Highway Research Program, Project IHR-94, University of Illinois, Urbana.

Rollings, R. S. 1988a. "Design and Construction of Roller Compacted Concrete Pavements," *Proceedings of the 14th Australian Road Research Board Conference,* vol. 14, part 8, Canberra, Australia, pp. 149–163.

Rollings, R. S. 1988b. "Substandard Materials for Pavement Construction," *Proceedings of the 14th Australian Road Research Board Conference,* vol. 14, part 7, Canberra, Australia, pp. 148–161.

Scholen, D. E. 1992. "Non-Standard Stabilizers," FHWA-FLP-92-011, Federal Highway Administration, Washington.

Sherwood, P. T. 1962 "Effect of Sulfates on Cement and Lime-Stabilized Soils," Highway Research Bulletin 353, Highway Research Board, Washington.

Stocker, P. T. 1972. "Diffusion and Diffuse Cementation in Lime and Cement Stabilized Clayey Soils," Special Report 8, Australian Road Research Board, Vermont South, Australia.

Terrel, R. L., J. A. Epps, E. J. Barenberg, J. K. Mitchell, and M. R. Thompson. 1979. *Soil Stabilization in Pavement Structures: A User's Manual,* vol. 1: *Pavement Design and Construction Considerations*; vol. 2: *Mixture Design Considerations,* FHWA-IP-80-2, Federal Highway Administration, Washington.

Thomas, M. D., R. J. Kettle, and J. A. Morton. 1989. "Expansion of Cement Stabilized Minestone Due to Oxidation of Pyrite," Transportation Research Record 1219, Washington.

Thompson, M. R. 1970. "Soil Stabilization for Pavement Systems: State of the Art," Technical Report, U.S. Army Construction Engineering Research Laboratory, Champaign, Ill.

Thompson, M. R. 1968. "Lime-Treated Soils for Pavement Construction," *Journal of the Highway Division,* vol. 94 (HW2): 430–440.

Thompson, M. R. 1966. "Lime Reactivity of Illinois Soils," *Journal of the Soil Mechanics and Foundation Engineering Division,* vol. 92 (SM5): 67–92.

Thompson, M. R., and B. J. Dempsey. 1969. "Autogenous Healing of Lime Soil Mixtures," Highway Research Record 263, Highway Research Board, Washington.

Transportation Research Board. 1991. *Hot-Mix Asphalt Paving Handbook,* Washington.

Transportation Research Board. 1987. "Lime Stabilization," State of the Art Report no. 5, Washington.

Verhasselt, A. F. 1990. "The Nature of the Immediate Reaction of Lime in Treating Soils for Road Construction," in *Physico-Chemical Aspects of Soil and Related Materials,* STP1095, K. B. Hoddinott and R. O. Lamb, eds., American Society for Testing and Materials, Philadelphia.

Construction-Site Improvement

"Surely the earth, that's wise being very old,
Needs not our help."   D. G. ROSSETTI

7.1 Introduction

Construction sites are normally selected not only for their location
but also for their suitability for the intended end use. In some
instances, and increasingly so, nonoptimal land must be used as a
construction site. This occurs in highly urbanized, developed areas; in
regions where the entire landmass is low, soft, and wet; and in circumstances where project requirements dictate the exact location of a
facility (regardless of site characteristics), among others. Nonoptimal
land may be described alternately as marginal, derelict, soft, marshy,
soggy, wet, mucky, deltaic, floodplain, old dump, and other equally
descriptive and forbidding terms. To illustrate the nature and relative
condition of various soils, Table 7.1 compares engineering properties
of several soft soils that might be encountered on construction projects (and which would probably require improvement prior to *any*
use) with some of the better known "bad actors" of geotechnical engineering.

These poor construction sites historically have been avoided if at all
possible. When other considerations (economic, social, political, environmental, location, etc.) dictate that a site be used despite poor site
conditions, then one of the many available methods of site improvement will likely be used to change the undesirable characteristics of
the site (see Table 7.2). The method selected for site improvement will
depend on the condition of the potential construction site; geotechnical material type(s) present and their locations (depth, areal extent,

TABLE 7.1 Engineering Properties of Selected Soft Soils and Waste Slurries

Material type	Engineering Properties				
	w_c, %	G_s	LL	PI	Shear strength (kPa)
Dredged material	50–400	2.4–2.8	0–270	0–185	0–50
Marsh/organic soil	100–500	1.7–2.5	50–200	100–165	<4
Peat	100–1800	1.5–1.75	250–500	150–400	<4
Phosphatic clay	300–1000	2.5–2.9	76–245	45–175	<1
London clay	20–40	2.71	65–95	35–65	50–500
Norwegian marine clay	18–45	2.77	20–44	4–23	3–50
Boston blue clay	32–42	2.78	40–52	18–32	40–80
Mexico City clay	100–550	2.35	150–500	100–400	25–175
Coode Island silt, Australia	23–130	2.4	37–90	20–53	7–120
Alumina red muds	55–65	2.8–3.3	42–46	7–39	NA
Pulp and paper wastes	210–265	1.9–2.3	70–413	40–380	<7
Fly ash, type F	≈50	2.1–2.5	–	NP	Pozzolanic

TABLE 7.2   Improvement Methods for Soft Foundation Soils

Method	Soil type	Structure type
Removal by excavation	Cohesive and noncohesive	All structures
Removal by displacement	Cohesive	Embankments
Preloading	Cohesive and noncohesive	All structures
Lightweight fills	Cohesive	Embankments
Self-supporting fills	Cohesive	Embankments
Grouting	Cohesive and noncohesive	All structures
Electro-osmosis	Cohesive	All structures
Blasting	Cohesive and noncohesive	All structures
Earth reinforcement	Cohesive and noncohesive	All structures
Vibrodensification	Noncohesive	All structures
Vibroreplacement	Cohesive	All structures
Dynamic consolidation	Cohesive and noncohesive	All structures

SOURCE: Modified from ASCE, 1978.

and possibly vertical location relative to other layers of different properties); ultimate site use requirements; materials, equipment, and local experience available for use in site improvement; time available in which to conduct the improvement; and cost. At very soft sites, some type of improvement may be necessary before access to the site can be gained. Otherwise, site-improvement techniques should improve the soil characteristics to a sufficient depth to adequately distribute and reduce the stresses from the intended construction project to levels that can be borne by the underlying weak soils.

The construction-site improvement techniques presented in this chapter are organized by their mechanism of improving the site. Actual removal of poor-quality material by excavation or displacement is discussed in Sec. 7.2. Section 7.3 presents methods of (often temporarily) reinforcing the surface of the soil deposit. Methods of reducing soil water content are addressed in Sec. 7.4, which includes the diverse topics of surface trenching, precompression, and electro-osmosis. Site strengthening is the subject of Sec. 7.5 and includes sections on dynamic compaction, vibratory methods, stabilization and grouting, thermal methods, and geosynthetics. Section 7.6 summarizes the available techniques for site improvement and gives a basis for selecting a particular improvement technique.

7.2   Removal of Poor-Quality Soils

When the in situ soils at a construction site have inadequate shear strength, unacceptable compressibility characteristics (total and/or differential compressibility), and less than optimal foundation stability, and generally do not exhibit the needed load-carrying capacity for the proposed project or construction activities, the material may be removed and replaced with a more suitable material. This method is most often used in embankment construction. It also can be used on other construction projects, although for structures the foundation design is more commonly modified, e.g., to point-bearing piles, to transfer the load to a stronger underlying stratum. Removal and replacement of poor-quality soil are usually feasible when the deposit to be removed is less than 20 ft thick and is underlain by stronger material (see Fig. 7.1). The feasibility also depends on the availability of suitable backfill material. Either select fill can be used to replace the poor material or, if appropriate, the poor material may be dried and/or stabilized for reuse.

Several approaches to removal and replacement can be used. Complete removal (to full depth of the deposit and full width of the project) is the most positive approach to dealing with weak and/or compressible soils. In some instances, partial excavation may be used, as shown in Fig. 7.1. The inadequate soil may be removed to partial depth, leaving a thinner layer of in situ material that can consolidate within an acceptable time period (usually prior to construction). Alternatively, the poor-quality material may be removed to full depth but only under a portion of the width of the project; this method usually involves excavation and replacement of material under the toes of the embankment, although it also can be used under the central portion of the fill.

Prior to initiation of any field work, the final geometric design of the excavation should be developed by analyzing the various possibilities for stability, compressibility, and/or differential settlement using appropriate material properties. Factors that will influence design of the excavation include size, shape, and loading of the proposed structure because these affect the stresses that must be supported at various depths; engineering properties of the in situ soil and the potential backfill materials; extent of the unacceptable soil deposit; position of the groundwater table; duration of the construction project; and the maximum allowable settlement and/or differential settlement.

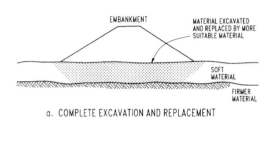

a. COMPLETE EXCAVATION AND REPLACEMENT

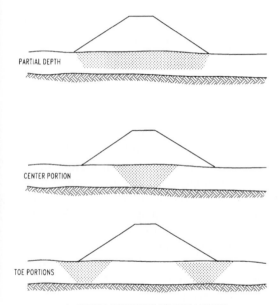

b. PARTIAL EXCAVATION AND REPLACEMENT

Figure 7.1 Typical use of excavation and replacement methods for improving site conditions.

The method of excavation and type of replacement material to be used will depend on the location of the water table relative to the bottom of the excavation. When the water table is below the area to be excavated, conventional construction equipment can be used effectively to remove the poor-quality material. The excavation can then be backfilled with any soil (either fine- or coarse-grained) that has adequate strength. If necessary, the backfill material can be stabilized with an admixture such as lime, portland cement, fly ash, or other appropriate stabilizers (see Chap. 6). If stabilization is required, the mixing can be accomplished in place in the excavation if there is sufficient space, or it can be done nearby or at a central mixing plant. After procurement of proper backfill material, it should be placed in the excavation in lifts and compacted to the required density for the specific project.

In areas of high water tables (e.g., above the base of the excavation), removal of material can best be accomplished with dredges and draglines. Barge-mounted draglines and dredges are suitable for use on inundated construction sites, whereas draglines working from mats or other construction platforms (see Sec. 7.3) can be used effectively on low-strength soils where the water table is at or below the ground surface. When excavations are made in wet areas where a dewatering system is not used, relatively flat side slopes should be used, and it may be advantageous to maintain the water level in the excavation at the phreatic surface. These precautions will minimize the possibility of slope failures that could lead to entrapment of soft soil below the select backfill. The backfill used in wet excavations should be a granular, non-water-susceptible material such as sand, shell, gravel, or slag. This material often can be pumped hydraulically into the excavation. If compaction is necessary, various vibratory or dynamic methods can be used (see Secs. 7.5.1 and 7.5.2).

Construction of the excavation normally should proceed by a continuous operation of excavation and replacement with higher-quality material, since this minimizes side-slope slumping and bottom heave in soft, weak soils. Equipment generally should work at a distance from the sides of the excavation to prevent slope failures caused by overloading the top of the slope. Close monitoring of the excavating operation is necessary to ensure proper removal of soft material so as to prevent postconstruction problems with instability, differential settlement, and lateral displacement. Consideration must be given to the location and method for disposing of the waste material. Identification of easily accessible waste disposal areas near the excavation will minimize costs while maximizing production. In environmentally sensitive areas, waste placement must not adversely affect the existing water regime, e.g., interrupt the sheet water flow and periodic tidal inundation. If pollutants are present, proper methods must be used to prevent contaminant migration (Rollings 1994, USACE/USEPA 1992).

In some instances, the excavation of undesirable, soft, cohesive material may be unfeasible or too expensive for embankment construction. This poor-quality material can then be removed by displacement. The displacement method involves end dumping fill material and pushing it onto the soft soil with dozers. Sufficient quantities of fill must be used to cause shear failure of the in situ soil; the backfill then replaces this soil and rises above it to form both the belowground base and the embankment itself (Fig. 7.2). The displacement method requires shearing of the underlying soft soils. Consequently, the side slopes of the fill should be as steep and as high as possible to

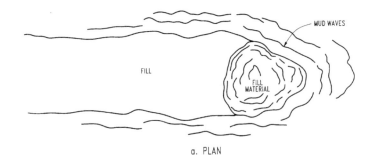

a. PLAN

b. PROFILE

Figure 7.2   Advancement of fill using displacement technique (*Hammer and Blackburn 1977*).

induce failure. The fill should be advanced with a V-shaped leading edge (with the center of the fill always in the most advanced position) to force displacement of underlying material to the sides. This method has been used in soft deposits up to 65 ft deep (ASCE 1978). It has been applied successfully in situations where a firm bottom existed as well as where no definite firm bottom existed (Hammer and Blackburn 1977). Figure 7.3 illustrates both these situations.

Factors affecting the selection and use of the displacement method include in situ material properties, extent of the poor-quality deposit, cost of excavation, influence of mudwaves on adjacent structures, availability of suitable backfill material, required construction time, and potentially higher maintenance costs caused by long-term deformations (OECD 1979).

The displacement method is most applicable in soft, cohesive soils that exhibit some sensitivity and have an average in situ shear strength of less than 150 to 200 lb/ft^2. In general, the more sensitive the soil and the lower its shear strength, the easier the material is to displace. If a harder, desiccated surface crust or a mat of fibrous roots

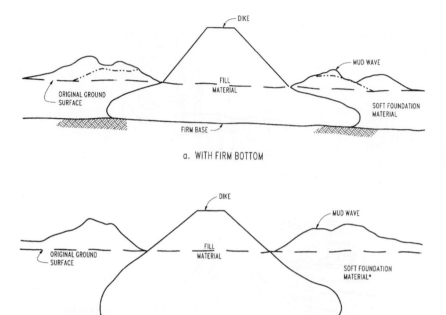

Figure 7.3 Final dike cross sections after displacement of soft foundation soil (*Hammer and Blackburn 1977*).

exists above the soft soil, this harder layer may need to be excavated or at least broken up prior to fill placement. Blasting reportedly has been used to break up this surface crust (OECD 1979).

When the displacement technique is used, mudwaves are typically formed by the displaced material. Mudwaves normally will be seen both in front of the advancing fill and along its sides. These mudwaves, not uncommonly, reach heights equivalent to that of the embankment. The lateral extent of such mudwaves can be significant, and structures at distances of five times the depth of displacement have reportedly been affected (OECD 1979). Mudwaves developing in front of the advancing fill should be excavated at about the same rate as the rate of fill advancement; this will minimize entrapment of soft soils under the embankment. Lateral mudwaves should be left in place to act as stability berms, if possible.

7.3   Construction Platform

Construction can only proceed on a site if there is a firm platform upon which to work. If a firm platform is not available, equipment

Figure 7.4   Excavation equipment lost in a soft soil.

may be immobilized (Fig. 7.4), placement of material may prove impossible, good-quality granular fill placed on weak materials may be fouled or may sink into the soft underlying materials, and it will prove impossible to achieve high levels of compaction in overlying materials placed on weak foundations. Under such conditions, some form of construction platform may be needed to simply gain access to the site to allow site improvement or construction to proceed. Poor site conditions also add to the cost of construction. For example, Fig. 7.5 illustrates that as rutting increases from poor support conditions, the cost of earthwork increases dramatically. The quality, speed, and economics of construction generally will be enhanced by a construction surface that provides a reasonable degree of support under the expected construction weather conditions.

Timber mats (Fig. 7.6) are commonly employed to allow construction equipment to operate on soft sites. This construction equipment then can begin trenching to obtain drying of the surface, excavation, or other needed construction activities. Such activities may promote surface drainage and an improvement of on-site mobility. If equipment can operate on the surface, stabilization techniques discussed in Chap. 5 or placement of granular fill materials can then be used to progressively build a firm, relatively weatherproof surface upon which further construction can proceed.

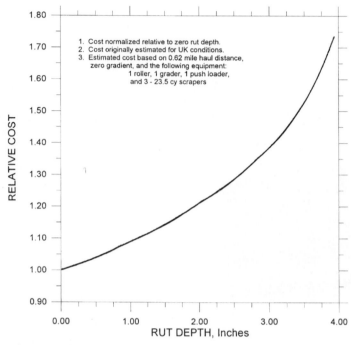

Figure 7.5 Effect of rutting on cost of earthworks (*Organization for Economic Co-Operation and Development 1979*).

Figure 7.6 Equipment operating on a timber mat over soft soil.

On very soft sites, however, some structural system may be needed to allow construction and provide foundation support. Brush and logs have been used widely as an expedient method of building lightweight structural platforms on marshy and swampy soils. "Corduroy roads" made of transversely laid logs are a common and historically old method of construction on soft soils throughout the world. Along the Gulf of Mexico and Atlantic Ocean coasts in the United States, dredged oyster shells loosely dumped in 4- to 7-ft-thick lifts have been used widely to build a lightweight, flexible construction platform for embankments, industrial yards, roads, and warehouses on soft coastal soils (ASCE 1978). Recently, some companies in the north central United States have developed equipment to cut trees as large as 3 in in diameter into chunks that are then used as fill material. Lightweight materials such as brush and logs, wood chunks, sawdust, oyster shells, coral, expanded shale or clay, and fly ash have been used to simultaneously reduce the load to be carried by underlying soft materials and provide a construction platform for further work.

Geosynthetics offer a very effective and economical method of providing a structural system over soft materials and are used widely in that role today. More details on different types and uses of geosynthetics can be found in Chap. 9, but for this discussion it will suffice that there are geosynthetics that can be used as separators, reinforcing elements, or both when building a construction platform over soft soils. If granular materials are simply placed directly on fine-grained materials, the fines will be extruded into the coarser-grained granular material. This results in reduced permeability and strength in the granular materials, and their desirable engineering characteristics are badly degraded. If an appropriately selected geosynthetic is placed between the fine material and the granular material, then the intrusion of fine material can be prevented, and the engineering properties of the granular fill will be maintained. If the fine-grained material is also soft, then the geosynthetic also will provide significant reinforcement support to the overlying materials. Generally, it is hypothesized that this support develops due to tension in the geosynthetic from vertical deflection in the underlying material. Consequently, this reinforcing action will develop only on soils that deflect sufficiently to mobilize the tensile resistance of the geosynthetic. If the CBR of the soil is 3 or less, this tensile reinforcing capacity of geosynthetic is quite pronounced, but on higher-strength materials, the geosynthetic will probably serve more as a separator than as a structural reinforcement. Detailed guidance on selection of geosynthetics for different roles and appropriate design criteria and methods for use with geosynthetics in these roles may be found in Koerner (1990), Richardson and Koerner (1990), Rollings et al (1988), Broms (1987), and Hausmann (1987), as well as in Chap. 9.

Generally, construction of a geosynthetic reinforced construction platform will follow these steps, but specific site conditions and design requirements will dictate the actual procedures needed:

1. The site should be cleared of all material such as rocks, logs, brush, and similar objects that may puncture the geosynthetic. On occasion, trees and stumps simply have been cut off at ground level, but if possible, stump removal is desirable.

2. The site is rough-graded if it is trafficable by dozers or similar equipment.

3. The geosynthetic is spread over the site by equipment or by hand. Commonly, equipment such as a front-end loader can be used to lift and deploy the geosynthetic roll. The equipment should not operate on the unprotected geosynthetic unless absolutely necessary. In such cases, the geosynthetic should be inspected carefully for damage.

4. Subsequent rolls of geosynthetic should be overlapped, and if soil conditions are particularly bad, the overlapped materials may need to be joined by sewing, with adhesives, or by seaming.

5. Granular fill or aggregate is dumped adjacent to the geosynthetic or on previously placed fill, not directly on the geosynthetic. The delivered fill or aggregate is then spread into place on the geosynthetic with a bulldozer or similar equipment (Fig. 7.7).

Figure 7.7 Spreading granular fill on a geosynthetic.

6. Grading and compaction follow fill or aggregate placement to provide a construction platform capable of providing support for subsequent site-improvement efforts such as described in the following sections or for direct construction of embankments, roadways, or the like.

Many problems, claims, and delays in construction can be minimized by ensuring that there is adequate support for the construction activities to proceed smoothly and with minimal delay under the expected weather conditions. The designer can help this problem by recognizing and incorporating stabilization of difficult soils or reinforcing systems such as geosynthetics within his or her design. However, most of the responsibility for establishing and maintaining effective site conditions rests with the contractor. It will be up to the contractor to establish and maintain drainage at the site to minimize rain effects, select support systems to allow site access for equipment, establish and maintain haul roads, and generally make the decisions on how the work will be executed. There is a cost associated with these decisions, but there are also many examples of costly delays and disputes arising from rain and flooding of sites and difficult working conditions that resulted in inadequate-quality work. A sound construction platform upon which work can proceed is often a key to timely, high-quality construction in the field.

7.4 Reduction in Water Content

Water is one of the worst enemies of the geotechnical engineer and the construction process. Its very presence, not to mention its variation in quantity, can cause severe problems on the construction site. Excess interstitial water in soils leads to reduced shear strength, high compressibility, loss of mobility, reduced workability of soil, and major increases in construction time and costs, and may cause total abandonment of a construction site.

When the in situ water content of a soil is sufficiently high, construction cannot proceed directly on the site. In such cases, the soil will likely require a reduction in water content (which is normally accompanied by an increase in shear strength and/or a decrease in compressibility) prior to the initiation of construction. This can be accomplished by one of several site-improvement techniques such as surface trenching, precompression, or electro-osmosis, which are discussed in Secs. 7.4.1, 7.4.2, and 7.4.3, respectively.

A classic method of removing water from a site is through implementation of a comprehensive dewatering program. This often uses pumped wells, well points, and/or subsurface drains. Construction-site dewatering per se is a subject worthy of entire books and in fact has been the subject of several notable works (Cedegren 1989, Mansur and Kaufman 1962, Powers 1981). Classic dewatering as a method unto itself is beyond the scope of this book.

7.4.1 Surface trenching

When extremely wet soil deposits are present at a site, surface trenching may prove to be a very economical method of improving surface conditions. This technique has been used for hundreds of years to drain sites for agricultural cultivation as well as for construction purposes. Near the turn of the last century (well before the environmental movement in the United States), the U.S. government funded programs to trench and drain large tracts of land for development of highly productive farmland. The authors saw first hand the successful application of this technique over an extensive landmass southeast of Melbourne, Australia, where 10- to 12-ft-deep trenches constructed in the early 1900s still stand today, carrying water out of and away from the land that is used to raise cattle and crops. Trenching also has been used for the last several decades by the U.S. Army Corps of Engineers and the phosphate mining industry to dewater large hydraulic fills; these organizations have used progressively deeper trenches to develop thick desiccation crusts across the surfaces of their soft deposits. Thus surface trenching is an effective, economical technique that can be implemented successfully in various types of deposits with no (or minimal) specialized equipment.

Surface trenching assists in dewatering soil masses by providing rapid surface drainage of precipitation and by shortening the drainage path for water escaping from within the consolidating soil mass. When good surface drainage is established at a site, evaporative forces can work to dry the soil downward from the surface, even in regions with net negative evaporation (i.e., where precipitation exceeds evaporation). To have good drainage, trenches should be dug to an elevation below that of any desiccation cracks and should fall monotonically toward larger off-site drainage facilities. This allows precipitation to flow through the desiccation cracks, into the trenches, and off the site. As the site is dewatered, the trenches should be deepened progressively (thus the term *progressive trenching* was coined) to maintain good surface drainage; this often can be accomplished only in small stages as the underlying material consolidates and gains sufficient strength for the trenches to stand open.

When trenches are dug to a depth below the water table, gravity drainage may lower the water-table elevation. Drainage and lowering of the water table can occur fairly quickly in permeable materials. In fine-grained materials with relatively low permeabilities, this potential benefit of trenching may be minimal. However, if the trenches intersect pervious layers within the low-permeability material, significant drainage into the trenches may occur. As water is removed from the soil mass by drainage and/or evaporation, the effective stress of the drier (unsubmerged) soil mass increases, leading to consolidation of underlying layers. This causes an increase in excess pore water pressures, leading to additional consolidation and drainage. The successful use of progressive trenching in hydraulically dredged material is illustrated in Fig. 7.8, where distinct remnants of three separate disposal operations are visible along the sides of the 5-ft-deep trench.

The nature of the potential construction site will determine how a surface trenching program should be initiated and conducted. If the site is very soft and wet, such as a hydraulic fill, the initial dewatering may simply consist of a passive phase where ponded surface water is drained from the site through outlet weirs or other minimal activity. When the soft material nears the point of losing saturation (empirically this was found to be at a water content of about 1.8 times the liquid limit for dredged materials), a thin drying crust or skin will form with narrow desiccation cracks at 3- to 6-ft intervals (Department of the Army 1987). When hydraulic fills attain this consistency, an active dewatering (trenching) program can commence.

For extremely soft, wet sites, the process of trenching often must begin at the edge and work inward toward the center of the site. Thus an active construction-site dewatering program often starts with con-

Figure 7.8   Five-foot-deep trench cuts through three layers of dewatered dredged material.

struction of perimeter trenches. If the site to be trenched is a hydraulic fill contained within perimeter dikes, then initial perimeter trenches will be constructed by equipment working from the dikes. Draglines and backhoes are often used for this operation because they have relatively long boom lengths and can operate fairly easily on top of the dikes. For other sites without dikes and/or with overhead obstructions such as trees, utility lines, etc., low-ground-pressure trenching equipment might prove more useful.

For trenching to be an effective dewatering technique, it is necessary that trenches be constructed throughout the site at proper intervals to rapidly remove water from the entire site. This normally requires construction of a network of trenches inside the perimeter trenches. Interior trenches must intersect and drain into perimeter trenches, sumps at outlet structures (e.g., outlet weirs, etc.), or off-site drainage facilities. The pattern, spacing, and depth of interior trenches used will depend on the size and topography of the construction site, type of material to be dewatered, operating characteristics of trenching equipment, method of accessing the site while trenches are in place, and type of equipment that must navigate through or around trenches.

Although the trench pattern and depth used on a particular project must of necessity be site-specific, the following general statements can made regarding trenching. The physical layout of trenches at individual sites should facilitate expedient drainage from that site; i.e., the trenching pattern must be adapted to the topographic lay of the land. Parallel trenches have been used most commonly on relatively flat or gently sloping sites such as agricultural fields, large (future) construction sites, and hydraulically filled areas where the surface elevation is usually gently sloping from the inflow point toward the outlet structure (Fig. 7.9). When a low spot(s) occur at the outer edge of a site, trenches may be constructed to radiate out from this area(s). More closely spaced trenches generally will speed the dewatering process but will be more expensive and time-consuming to excavate. The optimal spacing of trenches can be determined by considering the rainfall/runoff quantities and maximum acceptable time for removal of runoff, as well as the permeability of the soil deposit; also factored into this process must be the cost of trenching. For most hydraulically dredged materials, an optimal spacing of 100 to 200 ft has been found to effectively remove water in a timely manner while minimizing the expense of trenching. The greater the depth of trenches, the greater will be the thickness of the dewatered deposit, but trenches cannot be dug effectively any deeper than the material's strength will bear. In very soft deposits where trenching is most likely to occur, trenches will likely collapse if initially constructed to any notable depth, e.g., more than 6 to 12 in. Therefore, periodic deepening (or *progressive trenching*) will likely be necessary.

Figure 7.9   Parallel trenches in dredged material containment area in Savannah, Georgia.

Several types of amphibious and low-ground-pressure equipment are produced that can be used for trenching in very soft soils. Both tracked and wheeled vehicles have been used successfully. The most important consideration in selecting this equipment is the ground contact pressure, because this factor, in combination with soil strength, controls vehicle mobility on the site. For production purposes, continuously operating rotary trenchers pulled by a low-ground-pressure vehicle may provide the best production rates. The U.S. Army Corps of Engineers recommends this equipment for routine dewatering in their dredged material containment areas (Department of the Army 1987). These mechanical excavation devices normally have a cutting wheel or wheels that can cut a trench as deep as 3 ft. The chassis may be either wheeled or tracked (Fig. 7.10). Some of the wheeled vehicles are amphibious and are capable of trenching in thick slurries. (This is sometimes useful in establishing a depression for early site drainage.) Care must be taken when wheeled equipment is used on a thin crust of material that overlies very soft, wet material because the drier material may fill the tire treads and cause a complete loss of traction if or when the tires punch through the crusted material (Willoughby 1977, 1978; Poindexter 1988).

More conventional construction equipment can be used for trench construction as drying proceeds at very soft sites and can be used ini-

Figure 7.10   Wheeled and tracked vehicles can be used for trenching in soft soil sites.

tially at sites with sufficient shear strength to support it. In many instances it may be necessary to provide some type of construction platform from which the equipment can operate. Several types of equipment that can be used at soft soil sites are listed in Table 7.3, along with some operational characteristics (Haliburton 1978). An important consideration in selection of equipment for use on low-shear-strength sites is the method of operation of the particular equipment. Some equipment, e.g., rotary trenchers and bulldozers, simply makes one continuously moving pass across the ground. Other types, such as backhoes and draglines, work from one location, swiveling and rotating as needed during excavation. Many soft sites can support one pass of continuously moving equipment but do not have adequate strength to support the dynamic and shear forces associated with sedentary equipment.

Prior to initiation of trenching, a site evaluation should be conducted to assess soil shear strength and to relate this to various vehicle support requirements. One such method is the U.S. Army Corps of Engineers system for relating the rating cone index RCI (soil strength) to vehicle cone index VCI (equipment support requirements); procedures are given by Willoughby (1977, 1978).

7.4.2   Precompression

Precompression is widely used to improve the characteristics of soils at construction sites where unacceptably large settlements are

TABLE 7.3 Operational Characteristics of Trenching Equipment

Equipment used for trenching	Crust thickness for effective operation, in		Maximum trench depth, in	Approximate trenching rate, lineal ft/h	Approximate rental cost,[a] $/h
	Minimum	Maximum			
Riverine utility craft[b]	0	12	18	2,000+	75–100
Low-ground-pressure tracked vehicle + plow	4	24	24	2,000+	35–45
Small dredge	4	10	30	25	50–75
Amphibious dragline	6	18[c]	Crust + 18	40	50–70
Small dragline on double mats	12	18[d]	Crust + 18	30	35–50
Medium dragline on double mats	12	18[d]	Crust + 18	40	40–50
Small dragline on single mats	18	24[e]	Crust + 18–24	50[f]	35–45
Medium dragline on single mats	18	30[e]	Crust + 18–24	60[f]	40–50
Large dragline on single mats	24	36[e]	Crust + 24	80[f]	45–55

Note: Vehicle or mat ground pressure also must satisfy critical layer RCI mobility criteria (Willoughby 1977). Low-ground-pressure tracked vehicle is assumed to pull drag plow with point set only 1 to 2 in below existing crust. Dragline sizes are assumed to be approximately: small dragline: $5/8$ yd^3 or smaller bucket and gross weight of 40,000 to 75,000 lb; medium dragline: $3/4$- to 1-yd^3 bucket and gross weight of 55,000 to 75,000 lb; large dragline: $1 1/4$-yd^3 or larger bucket and gross weight of 70,000 to 120,000 lb.

[a] Southeastern United States in 1977 dollars.
[b] Corps of Engineers' equipment described by Haliburton 1977.
[c] Above this crust thickness, conventional dragline is usually more efficient.
[d] Above this crust thickness, use single mats.
[e] Above this crust thickness, no mats required.
[f] Increase rates 10 linear ft/h if dragline is working from perimeter dike.
SOURCE: From Haliburton, 1978.

expected. The process involves placement of inexpensive, available fill material (often sand) on a site prior to construction for the purpose of preconsolidating and strengthening the foundation soils. This fill material must be of sufficient mass to produce stresses on the site's compressible strata to equal or exceed the stresses that will be induced by the permanent structure. When consolidation is complete, the fill is removed either partially or in its entirety. The site is then ready for construction to proceed.

Precompression is particularly useful at sites where the in situ soils are susceptible to large volume decreases and strength increases under sustained static loads and where sufficient time is available to permit complete precompression (Mitchell 1982). This technique is especially applicable to soft, compressible soils such as saturated soft clays, compressible silts, organic clays, and peats. It also has been used on loose sands, cinders, and "rubbish fills" (Johnson 1970).

Although it seems to be a historical site-improvement technique, precompression actually has been used routinely only since the 1960s. It became popular in the 1940s for use in highway construction and came into common use by this industry in the early 1950s (Johnson 1970). In 1949, the U.S. Army Corps of Engineers used precompression at the Morganza Floodway, a part of the Mississippi River flood-control project in Louisiana, to eliminate approximately 2.5 ft of settlement under the abutments; this was possibly the first use of precompression at a major hydraulic structure. The technique still had very limited use at structural construction sites until the early 1960s (Johnson 1970). More recently, precompression has been used extensively to improve poor subgrade soils prior to construction on many types of projects, including embankments, structures, tanks, highways, airfields, and abutments.

Precompression requires placement of a load that is often referred to interchangeably as a *preload* or a *surcharge load.* Technically speaking, when the fill load is equal to the future structural load, it is a *preload,* and when it exceeds the structural load, it is a *surcharge load* (Fig. 7.11). Although these loads are applied most often by earth fill (such as sand, gravel, or locally available soil), they may be applied to the compressible foundation soils in other ways. Water-filled tanks have been used to preconsolidate underlying soils, particularly when petroleum storage tanks must be pretested before filling with petroleum (which is lighter than water). Lined ponds can be filled with water to surcharge larger areas. More complex and site-specific methods of preloading include lowering of the water table, applying a vacuum to a soil that has been covered with an impermeable membrane, electro-osmosis (see Sec. 7.4.3), and anchor and jack systems (Mitchell 1982).

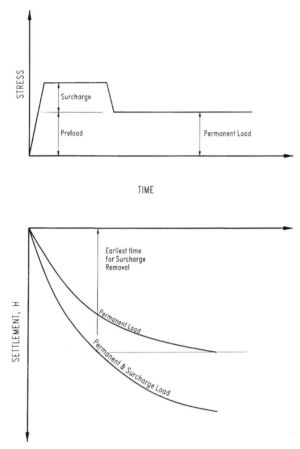

Figure 7.11 Temporary surcharge fill causes precompression of foundation soils before project construction.

Precompression normally should be used to eliminate all primary consolidation settlement ΔH_f expected to occur under the permanent loading. As shown in Fig. 7.11, this magnitude of consolidation settlement will occur while the soil is only partially consolidated under the surcharge load, i.e., before 100 percent consolidation occurs under the permanent plus surcharge load ΔH_{f+s}. The surcharge can be removed when settlement equals ΔH_f, but it should be recognized that the compressible stratum is not equally consolidated throughout its thickness at this time. In fact, for a layer drained on both top and bottom (the situation leading to most rapid consolidation), the center part of the compressible layer will be less consolidated and the upper and lower edges will be more consolidated than the average for the entire thickness. This is illustrated in Fig. 7.12 for an average degree of con-

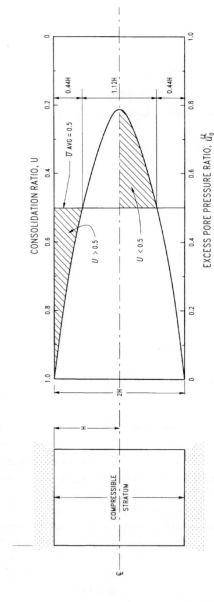

Figure 7.12 Soil layer conditions when average degree of consolidation equals 50 percent (*Johnson 1970*). ["Precomposition for Improving Foundation Soils," S. J. Johnson, *Journal of Soil Mechanics and Foundation Engineering*, 1970, reproduced by permission of American Society of Civil Engineers]

solidation U of 50 percent. The center of the layer would continue to consolidate under the permanent load, so it would be advisable to use the degree of consolidation at the center of the compressible stratum to determine when to remove the surcharge load. Procedures for these calculations are presented by Johnson (1970) and Mitchell (1982). Johnson (1970) states that "partial compensation for primary consolidation is discouraged as being generally unsound."

When the time required for primary consolidation is considered excessive, vertical strip drains or sand drains can be used to shorten the drainage path and speed consolidation (see Sec. 9.2.3). Drains normally will be needed for homogeneous deposits in excess of 15 ft thick. When drains are installed in the compressible foundation soils, some type of drainage layer is needed between the existing soil surface and the surcharge fill; this drainage layer may be either a free-draining sand, a thick geotextile, or a geonet (see Secs. 9.2.3, 9.5, and 9.6).

Multiple benefits can be gained from precompression. Surcharging not only increases the rate and magnitude of consolidation of the foundation soils but also may significantly reduce secondary compression (Lambe and Whitman 1969). Although the common goal of precompression is reduced postconstruction consolidation (namely, a reduced magnitude of settlement and decreased differential settlement), the resulting increase in shear strength may be the most beneficial effect at many sites (Johnson 1970, Mitchell 1982).

Illustrative Example 7.1: Precompression A 213-acre site located adjacent to a coastal river in the eastern United States was to be developed as a freight aircraft parking ramp with ancillary support buildings and facilities. The site was composed of interbedded alluvial deposits of sands and organic silts. Variable thicknesses of 2 to 16 ft of similar dredged material had been deposited on the site during development of adjacent waterways during the early 1900s. A basin for a shipyard was dredged in a portion of the site during World War I. By about 1935 the shipyard was no longer in use and was silting up with natural river deposits. The abandoned shipyard subsequently was used as a disposal area for dredged material. The dredged fill materials in the abandoned basin consisted of a variety of organic silts, clays, and sands. Throughout the site, individual layer thicknesses of specific materials varied from a few inches to a number of feet. Generally, the thickest cohesive deposits were the predominately organic silt fill in the abandoned shipyard basin, which could exceed 30 ft in thickness.

Analysis of laboratory consolidation test data and site stratigraphy produced the following approach to developing the site. Buildings were to be supported on piles, and building ground floors were to be slabs on grade that were separate from the building structure. Variable subsurface conditions under these floor slabs suggested that settlements would range from 1 to 4 in. A preload of 5 to 6 ft of fill was used over these areas for 4 months to induce the settlements and to minimize differential settlement under the floor slabs. About one-third of the planned aircraft parking apron extended over the abandoned shipyard basin with its relatively thick dredged organic silt fill, and the remainder of the apron spanned the less settlement-prone but variable alluvial and dredged material deposits. Calculated settlements under the anticipated pavement structure were as high as 34 in and could take as long as a year for completion under preloading. To accommodate these conditions, the ancillary support buildings and facilities were constructed first to allow time for the longer preloading needed for the pavement apron sections.

Without preloading, the magnitude of settlements and the differential settlement between points due to variable site stratigraphy would have caused major damage to floors and pavements. While it may have been feasible to find alternative ways to support the building ground floors, there was no reasonable alternative for the large aircraft pavement ramp.

Therefore, preloading was a key tool in site improvement by speeding consolidation of cohesive materials at this site. By phasing the construction sequence, time was made available to carry out the longer-duration preloading needed under the pavement sections on the site.

The rate at which primary consolidation occurs in fine-grained materials is a function of the soil permeability and the length of the drainage path. The fine-grained deposits in the preceding example had relatively high permeability for settlement-prone materials (e.g., silts compared with highly plastic clays) and throughout much of the deposit had relatively short drainage paths due to interlayered deposits of sand. Consequently, settlement could be induced relatively quickly by preloading and could be accommodated within the construction plan. However, thick deposits of relatively pure clay may require tens of years to achieve primary consolidation even under a surcharge load. This normally would not be acceptable within any reasonable construction schedule. In such cases, use of vertical drains to shorten the drainage path within the deposit can make construction feasible. For example, during development of a marine terminal on a silty clay dredged material, geosynthetic vertical drains reduced the time of primary consolidation under preloading from 20 years to 9 months.

Although preloading may reduce settlements under the structure to be built and will improve the soil strength, this does not imply that strengths at the end of preloading will necessarily be good. After preloading, the modulus of subgrade reaction from plate load tests for the aircraft ramp in the example ranged from 20 to 45 psi/in, and pavement design CBR values for the marine terminal were 1 for over half the site and 4 for the bulk of the remainder of the site.

Although precompression is a "tried and true" concept for construction-site improvement and in concept is a simple exercise involving placement/dumping of fill on a poor site, it is a complicated process that requires proper application of knowledge to reach a successful conclusion. The loadings required for precompression (both the magnitude and the rate) and the mechanisms involved in this type of site improvement (e.g., consolidation, excess pore water pressures, effective stresses, and shear strength considerations) necessitate thorough investigation of the site and a comprehensive knowledge of soil mechanics. Johnson (1970) summarized the situation nicely when discussing precompression for improving foundation soils in 1970:

A significant requirement when considering precompression techniques is the need to make especially detailed subsoil investigations, laboratory tests and design analyses, and to provide high-quality field inspection. Precompression is generally used where subsoils are weak and highly compressible. This introduces major questions concerning foundation stability under preloads and the need to predict relatively accurately both magnitude and time rate of consolidation. Precompression design is complex and requires a relatively high degree of competence and investment in time and money. However, since construction cost savings are often extremely large, additional design costs are of negligible importance, provided that the engineer and the owner recognize this. Hardly any other activity of the soils and foundation engineer earns such a large return on invested effort. While an almost incredibly large number of distressing failures have plagued precompression applications, these have resulted mainly from insufficient attention to the need for detailed investigation, design, and supervision.

7.4.3 Electro-osmosis

When an electric current is applied to a soil mass containing two electrodes, water within the soil will travel from the anode to the cathode. This results in a reduction in the water content near the anode. It also will result in dewatering and consolidation of the soil mass if water is removed at the cathode without any reintroduction of water at the anode. Use of a well point as the cathode can expedite removal of collected water. Electro-osmosis also has been used to increase the skin friction of piles shortly after driving and can be used to speed the migration of various fluids, including contaminants, stabilizing agents, and grouts, through (either into or out of) fine-grained soils.

The mechanism of water movement by electro-osmosis was studied as early as 1807 by Reuss (Casagrande 1952) and 1879 by Helmholtz (Mitchell 1993), and a treatise on the application of the topic to soils was written by Casagrande (1952). More recent work by a number of investigators is summarized by Mitchell (1993). Reuss discovered that application of an electric potential could make water flow through the capillaries of a porous diaphragm toward the cathode. He also discovered that the water flow stopped as soon as the electric current was turned off. Helmholtz developed a mathematical explanation of the phenomenon; his theory is still the most widely used one to explain the phenomenon of electro-osmotic water movement. Helmholtz found that in a capillary filled with water, there is "free water" in the center of the capillary and a boundary film of water adjacent to the sides of the capillary. The boundary film is composed of two layers (thus it is termed the *double layer*); the very thin outer part of the double layer is assumed to have a negative charge and to be rigidly attached to the capillary wall. The remainder (the much thicker part) of the double water layer has a positive charge and can be moved toward a negative

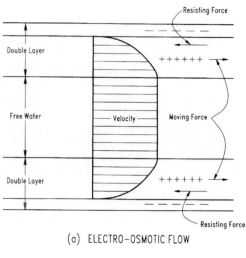

(a) ELECTRO–OSMOTIC FLOW

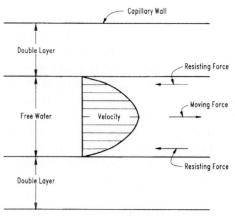

(b) HYDRAULIC FLOW

Figure 7.13 Comparison of electro-osmotic flow with hydraulic flow in a single capillary (*Casagrande 1953*).

pole (the cathode) when an electric potential is applied. As these positive charges move, they drag along the movable part of the double water layer, which in turn drags along the free water. Assuming that no other forces are acting on the free water, the distribution of velocity in the capillary will be as shown in Fig. 7.13. For comparison, the simplified velocity profile for hydraulic flow in a capillary is also shown. The thickness of the double water layer will vary with electrolyte concentration and type of clay mineral, but it is of the order of magnitude of 100 nm (1 nm = one-millionth mm = one-thousandth μm), which is about the size of the pores in fine-grained soils.

TABLE 7.4   Typical Coefficients of Electro-Osmotic Permeability

No.	Material	Water content, %	k_e in 10^{-5} cm²/s · V	Approximate k_h, cm/s
1	London clay	52.3	5.8	10^{-8}
2	Boston blue clay	50.8	5.1	10^{-8}
3	Kaolin	67.7	5.7	10^{-7}
4	Clayey silt	31.7	5.0	10^{-6}
5	Rock flour	27.2	4.5	10^{-7}
6	Na-montmorillonite	170	2.0	10^{-9}
7	Na-montmorillonite	2000	12.0	10^{-8}
8	Mica powder	49.7	6.9	10^{-5}
9	Fine sand	26.0	4.1	10^{-4}
10	Quartz powder	23.5	4.3	10^{-4}
11	Äs quick clay	31.0	2.0–2.5	2.0×10^{-8}
12	Bootlegger Cove clay	30.0	2.4–5.0	2.0×10^{-8}
13	Silty clay, West Branch Dam	32.0	3.0–6.0	1.2×10^{-8} to 6.5×10^{-8}
14	Clayey silt, Little Pic River, Ontario	26.0	1.5	2×10^{-5}

SOURCE: From Mitchell, 1993. [*Fundamentals of Soil Behavior*, J. K. Mitchell, Copyright © 1993. Reprinted by permission of John Wiley and Sons.]

The coefficient of electro-osmotic permeability k_e is a measure of the hydraulic flow velocity under a unit electrical gradient. This k_e is determined in a manner that is conceptually similar to determination of the coefficient of hydraulic permeability. Measurement is made of the rate of water flow through a soil sample of known length and cross-sectional area under a known electrical gradient. Past research has shown that the value of k_e is about 1×10^{-9} to 1×10^{-8} m²/s·V for most soils. Typical values are given in Table 7.4.

The most significant difference in flow characteristics resulting from hydraulic and electrical gradients relates to the coefficient of permeability. Whereas the hydraulic permeability varies with the size of a soil's pore spaces (as the square of the effective pore size), the electro-osmotic permeability does not. Investigations have shown that for most soils, the electro-osmotic coefficient of permeability can be assumed to be 5×10^{-5} cm/s for a constant potential gradient of 1 V/cm or 5×10^{-5} cm²/s · V (Casagrande 1952). Thus, when a soil has a hydraulic coefficient of permeability greater than 5×10^{-5} cm/s, dewatering is more easily accomplished by gravity drainage, while in fine-grained soils electro-osmosis can be a more efficient mechanism for water movement than a hydraulic gradient. Mitchell (1993) illustrates the point by calculating the hydraulic gradient needed to equal a given electrical gradient for both a fine sand and a clay. To equal an electric potential gradient of 20 V/m, the sand (with $k_h = 1 \times 10^{-5}$ m/s and $k_e = 5 \times 10^{-9}$ m²/s · V) would require a hydraulic gradient of 0.01. For the clay (with $k_h = 1 \times 10^{-10}$ m/s and $k_e = 5 \times 10^{-9}$ m²/s · V), a hydraulic gradient of 1000 would be needed to cause flow equal to that resulting from an electric potential of 20 V/m. However, many practical considerations such as economics, power requirements, and energy losses to the system often prevent electro-osmosis from being a viable dewatering technique.

If electro-osmosis is to be considered, an assessment of its potential efficiency in removing water and causing consolidation must be evaluated. The water-removal efficiency is quantified by the amount of water moved per unit electric charge; it is affected by soil type, water content, and electrolyte concentration in the pore water. The electro-osmotic flow rate for several soils is shown in Fig. 7.14 as a function of electrolyte concentration and water content. Viewed simplistically, consolidation occurs at the anode as water is moved away from that location, while the continued presence of water at the cathode inhibits consolidation. As the electro-osmotic process proceeds, a hydraulic gradient develops that tends to cause flow from cathode to anode. Consolidation will continue until the hydraulic gradient is sufficient to cause an equal water flow as the electro-osmosis causes but in the opposite direction. When the driving force from the hydraulic gradient equals the electro-osmotic driving force, consolidation will stop. The total amount of consolidation that will occur will equal the amount of water removed. Both the amount and rate of consolidation resulting from electro-osmosis can be predicted (Mitchell 1993). It should be noted that the actual amount of consolidation at different locations will vary; it will be largest in the region of anodes and smallest near cathodes.

Electro-osmosis is most effective in normally consolidated, saturated silts and silty clays with low electrolyte concentrations in the pore water (Mitchell 1982, Bjerrum et al. 1967). One of the most common applications of electro-osmosis is for stabilization of cut slopes in saturated silty soils (Casagrande 1952). Water flow begins as soon as the electric potential is applied, so seepage pressures are induced immediately. If these forces act into the soil (away from the cut slope), the stability of the slope is increased. This technique was used by Casagrande at Salzgitter, Germany, in 1939 to stabilize a long cut section in a clayey silt. When the excavation had progressed to a depth of 6 ft, an unstable condition had developed, as evidenced by "flowslides...and an oozy state [on] the bottom of the excavation" that halted construction. Casagrande installed a 300-ft-long trial section for electro-osmosis. He used 4-in slotted steel pipes for the cathodes, which were installed at the tops of both sides of the excavation; they were driven 22.5 ft into the ground at 30-ft intervals. Then 0.5-in-diameter gas pipe was installed for anodes at the centerpoint between the rows of cathodes. Casagrande (1952) reported that "within a few hours after application of 180 volts, the condition of the slopes

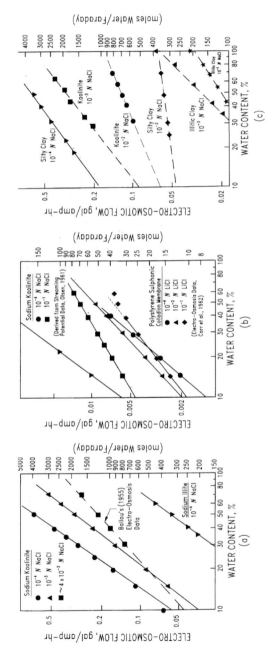

Figure 7.14 Electro-osmotic flow rate as a function of water content, soil type, and electrolyte concentration (*Mitchell 1993*). [*Fundamentals of Soil Behavior*, J. K. Mitchell, Copyright © 1993. Reprinted by permission of John Wiley and Sons.]

improved to such an extent that excavation work could be resumed; and one to two days later even the power shovels were able to work from inside the cut....In this trial section slopes could be cut to any degree of steepness without the slightest indication of instability."

The electrodes used in electro-osmosis are often pipes, rods, or railroad rails (O'Bannon 1977, Mitchell 1982). Well points are often used for cathodes to facilitate removal of water. The electrodes are typically arranged in some regular pattern at spacings of 6 to 15 ft. Mitchell (1982) recommends using a hexagonal pattern (with the cathode at the center of the hexagon) because this arrangement is most efficient in terms of power consumption, average voltage, and anode-to-cathode ratio.

Various problems have plagued electro-osmotic dewatering and have prevented this technique from being used extensively. Specific difficulties have included

- Gas generation from hydrolysis
- Corrosion of electrodes (especially iron and aluminum anodes)
- Extreme desiccation of soil, including extensive cracking, particularly around the anodes
- Electrochemical hardening of the soil resulting from corrosion/decomposition of the anodes and subsequent transport and deposition of metallic ions, causing permanent changes in the soil
- Severe changes in pH that dissolve soil constituents
- High power demands
- Concentrated contaminated effluent
- Decrease in efficiency over time

New techniques and/or major modifications to the direct-current method of electro-osmosis are under development. If these changes prove successful in field applications, they may eliminate many if not all of these problems (Joe Lynde, Technology Business Group, Middleton, Wisconsin, personal communication, 1994).

7.5   Site Strengthening

If a site does not provide soils of adequate strength and suitable consolidation characteristics, the soils themselves may be strengthened through physical densification or reinforcing. In the following subsections, several methods of strengthening soils over large volumes will be discussed briefly. These methods are generally very specialized, and obtaining the assistance of engineers and contractors experienced with these techniques generally will be prudent.

The methods of physically improving soil strength usually require some form of compaction. Conventional compaction, such as discussed in Chap. 5, generally is unable to achieve densification over a sufficiently great depth to be effective in site improvement. To overcome this, large amounts of energy can be used to densify the soil as discussed in Sec. 7.5.1. An alternative approach is to use one of several techniques discussed in Sec. 7.5.2 that employ vibratory techniques to achieve columns of densified material to strengthen the overall soil mass behavior. Section 7.5.3 discusses several methods of adding cementing materials to soil at depth. Section 7.5.4 examines methods of heating and cooling soils to obtain improved performance. Finally, Sec. 7.5.5 covers reinforcement of soils with geosynthetics. This latter technique has proven highly effective and is increasingly popular today. In each of these approaches, the soil mass will be strengthened physically by some form of densification, reinforcement, or a combination of these to improve its load-carrying capacity.

7.5.1 Dynamic compaction

Dynamic compaction can be used successfully to improve weak soils and is applicable to both noncohesive and cohesive materials, although the greater application has been to cohesionless soils. It is also effective in compacting large deposits of other materials such as construction debris landfills. Variations of this technique have been practiced around the world for hundreds of years, as summarized by Slocombe (1993). Notable users of dynamic compaction for site improvement include the ancient Chinese and the Romans. The method of delivering the dynamic load has varied to the extreme of using "an old war cannon" to cause compaction in 1871 (Slocombe 1993). Modern use of the technique was facilitated by development of large mobile cranes that are used to move and lift the large mass that causes the dynamic compaction. This method of site improvement has come into favor and use in Europe and in the United States since about 1970.

Dynamic compaction (also referred to as *high-energy tamping, heavy tamping, deep dynamic compaction, dynamic consolidation,* and *impact densification*) involves systematically dropping (in free fall) a large, heavy weight onto the surface of a soil deposit to cause compaction of underlying layers (Fig. 7.15). Normally, a standard large-capacity crawler crane is used to lift, move, and drop the compacting weight. The boom of the crane must be sufficiently long to allow an adequate lift height. For reasons of safety, the crane should not work at maximum capacity; in Great Britain, cranes are required to work at no more than 80 percent of their safe working load. Other specifics of crane equipage and operation "such as crane counterbalance

Figure 7.15 Dynamic compaction of construction debris landfill showing craters from compaction of deep layers.

weights, jib flexure, torque convertors, line pull, drum size, type and diameter of ropes, clutch, brakes, as well as many other factors and methods of working have been subjected to rigorous analysis by the major specialist organizations to improve reliability and productivity" (Slocombe 1993). Obviously, design and specification of crane capabilities are very involved and complicated and are critical to successful implementation of dynamic compaction.

The weights used for dynamic compaction normally are within the range of 5 to 20 tons, although weights as great as 35 tons have been used. Standard cranes are capable of lifting these weights. When much larger weights were needed at the Nice Airport, Menard used especially built equipment to handle the 187-ton load that was dropped from a 22-m height (Slocombe 1993). The weights are usually made of steel, concrete, or some combination of these two materials. The size and shape of the weight have a great impact on the dynamic compaction process; these effects have been investigated extensively (Slocombe 1993), but many of the details are proprietary.

Compactive energy is a major controlling factor in the success or failure of dynamic compaction, so requirements for a particular weight are normally tied to a corresponding drop height to indicate the compactive energy, e.g., 15-ton drop from a height of 20 m or 35 tons from 30 m. For most dynamic compaction, the drop heights range up to about 20 m. Slocombe (1993) reports that "full-scale dynamic compaction" is usually considered to be anything in excess of about a 15-ton drop from a height of 15 m. Weights of 6 to 10 tons usually comprise "mini" dynamic compaction. The total cumulative energy levels applied during dynamic compaction typically range from 100 to 400 $(t \cdot m)/m^2$, or 30 to 150 ft · ton/ft^2. It is interesting to note that the high end of the range was extended at the Nice Airport, where Menard's *giga tripod machine* was able to deliver 3900 t · m or 28,000,000 ft · lb of energy per blow (Mayne et al. 1984).

In addition to the weight and height of drop, the pattern of the drops is important. Both the "spacial distribution...and chronological sequence" (Mayne et al. 1984) of compactive energy application are important, especially during compaction of deep layers. To compact deep layers, a widely spaced grid is used, combined with an *appropriate* number of drops (up to 50 drops) from the maximum crane height at each location. The spacing of these drops is dictated by material type, depth to the deep layer, and location of the groundwater table; it is usually at least equal to the thickness of the compressible layer. To compact layers at intermediate depths, a smaller grid spacing is used (approximately one-half the initial spacing), and a lesser drop height and lesser number of drops are used. For the surface treatment, the number of drops and the drop height are further reduced, and the pattern of application is continuous across the site. The dynamic compaction process should be designed to proceed progressively from the deepest layers of interest to the most shallow layers.

Each time the weight is dropped, a crater is formed. As a practical matter, the craters are typically filled by bulldozing surficial material into them after completion of each phase or pass of dynamic compaction. When the site is leveled in this manner, the amount of subsidence resulting from each pass of compaction can be measured easily by surveying. The reported subsidence for a number of dynamic compaction sites is shown by soil type in Fig. 7.16.

The applicability of dynamic compaction to cohesive and noncohesive soils is summarized in Table 7.5. As shown in the table, the mechanism for site improvement varies not only by material type but also by location relative to the water table.

By proper design of the improvement program, significant site improvement can be attained at substantial depths. Using the standard range of weights and drop heights, compressible soils can be compacted to depths of about 50 ft. When special equipment is utilized (as Menard did at the Nice Airport), the depth of influence can be increased to approximately 100 ft (Mayne et al. 1984). The depth of influence from dynamic compaction can be estimated with calculations. Menard first suggested that the depth of influence was equal to the square root of the energy delivered per blow (Menard and Broise 1975). This has been modified several times by numerous investiga-

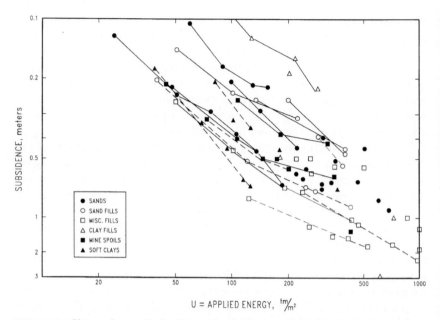

Figure 7.16 Observed magnitude of ground subsidence with level of applied energy per unit area (*Mayne et al. 1984*). ["Ground Response to Dynamic Compaction," P. Mayne, J. Jones, and J. Dumas, 1984, *Journal of Geotechnical Engineering Division*, reproduced by permission of American Society of Civil Engineers.]

tors; the most recent conservative estimate of the depth of influence from dynamic compaction (adapted from Leonards et al. 1980) is given by Mayne et al. (1984) as

$$d_{max} = \frac{1}{2} \sqrt{\frac{WH}{n}}$$

where d_{max} = depth of influence
W = weight
H = height of drop
n = units factor = 1 t · m = 672 lb/ft

Environmental impacts of dynamic compaction must be considered prior to implementation. Impacts include noise and vibrations. The noise is usually simply a short term nuisance. However, proximity to humans and sensitive sites such as hospitals, electronics installations, etc., should be evaluated before dynamic compaction is begun. Ground vibrations are the most serious potential impact, and both the magnitude and frequency are important. The impact of vibrations

TABLE 7.5 Mechanisms of Site Improvement with Dynamic Compaction

Soil type	Location relative to water table	Mechanism of improvement	Comments	Summary
Cohesionless	Above	Physical displacement of particles; low-frequency vibration.	A localized high-density zone may develop under impact site. This may dissipate energy of subsequent drops by bridging but also will help reduce stress on underlying soils for isolated footings.	Excellent results obtainable. Exercise care with high-silt-content soils.
Cohesionless	Below	Liquefaction caused by increased pore water pressure; low-frequency vibrations, then densify further.	Dissipation of excess pore water pressure may take 1 or more days in coarse soils or 1 or more weeks in fine soils. Time-dependent response of "free-draining" materials is crucial.	Excellent results obtainable. Exercise care with high-silt-content soils, especially below water table.
		Displacement without dilation or high excess pore water pressure.	Uses smaller number of drops from a lower height. Less costly than above.	Excellent results obtainable. Exercise care with high-silt-content soils, especially below water table.
Cohesive	Above	Instantaneous surcharge to pore water causing zones of positive pore water pressure and shear and hydraulic fracture of soil mass. Speeds drainage and consolidation.	Generally low in situ w_c so a small reduction in w_c results in large improvement.	Respond well.
Cohesive	Below	Same as above	High in situ w_c; requires large reduction in w_c to get significant improvement.	Technique is least successful in weak natural clays or clay fills below water table.

347

on structures in poor condition, schools, hospitals, utilities, and computer installations must be assessed (Slocombe 1993).

Other construction considerations that must be addressed include the possible necessity of a working platform for the crane (to provide a level, strong, free-draining surface), separation of cranes (by at least 100 ft) if more than one is used on a site, use of safety screens to control flying debris, and effect of various climates on the working surface (wet versus dry, warm versus cold)(Slocombe 1993).

A postcompaction testing program usually is required to determine the effectiveness of the dynamic compaction. This may be needed to determine conformance with project specifications, achievement of desired improvement in the deposit's engineering characteristics, or economical limits of site improvement. Table 7.6 shows several methods of carrying out such evaluations. Monitoring surface settlements is the easiest and fastest method. However, if some measure of the soil properties is needed or the achievement of densification at specific depths is important, the standard penetrometer tests, cone penetrometers, or pressure meters are more effective.

Dynamic compaction was used to improve a loose, silty sand deposit that was to provide foundation support for a large coal shipment facil-

TABLE 7.6 Methods for Determining Effectiveness of Dynamic Compaction

Method	Effectiveness	Comments
Conventional sampling and lab testing	Poor	Generally too slow to provide information for field decisions
Standard penetrometer test	Good	Efficiency of test important; recovers samples; gives results in terms commonly used by geotechnical engineers
Surface settlement	Fair	No evaluation of soil properties or of variation with depth
Plate load tests	Poor	Size of plate crucial to determine depth of effective testing
Geophysical tests	?	Limited experience in this application
Cone penetrometers	Good	Fast, provides a record of changes with depth; gravelly soils may be troublesome
Pressuremeter	Good	Provides complete stress-strain-strength relationship with depth
Full-scale loading	Excellent	Expensive; gives best measure of actual behavior

ity in Newport News, Virginia. The Massey Coal Terminal is a 33-acre facility that receives coal by rail from the northeast and central United States and transfers the coal to ocean-going vessels. Three cranes were used to dynamically compact the silty sand with an energy per blow of 400 to 480 t · m. The process took 6 months to complete. Verification testing was conducted using the standard penetration test (SPT), the cone penetration test (CPT), and the pressuremeter test (PMT). (Over 300 CPTs were conducted at the site because the test was rapid and easy to conduct.) Results from these tests are given in Fig. 7.17 and show the improvement that occurred at this site from use of dynamic compaction.

As with most other site-improvement techniques, the use of a specialty contractor with experience in the application of dynamic compaction is essential to a successful project. Although the process of applying a dynamic load to a soil mass is conceptually simple, the proper selection of the weight, drop height, spacing of impact loads, and overall design of the dynamic compaction program to accomplish the required site improvement is far from simple.

7.5.2 Vibratory methods

Various forms of vibration have been used for about 100 years to densify soil deposits. These include methods as diverse as vibratory rollers, vibroflotation, blasting, stone columns, and pile drivers. Most applications have been to densify loose granular deposits, although different forms have been applied with varying degrees of success to other types of deposits. This subsection will address construction-site improvement through the use of vibrocompaction, vibroreplacement, and vibrodisplacement. (Vibroreplacement and vibrodisplacement are often considered together as vibroreplacement but will be discussed separately here.) Use of and results to be expected from vibratory rollers are discussed in Chap. 5. Use of pile (casing) drivers to perform site improvement is discussed by Mitchell (1982) and Greenwood and Kirsch (1983). Blasting is not used widely for site improvement (Mitchell 1982, Department of the Army 1983) and consequently will not be addressed here.

The most common types of deep vibratory soil improvement involve use of a cylindrical (poker) vibrator. The vibrator is normally suspended from a crane and is inserted into the ground by a combination of vibration, water jetting/flushing, and/or air jetting (Fig. 7.18). Even with guidelines attached to the vibrator, it will tend to hang as a pendulum (Greenwood and Kirsch 1983); thus the hole will remain vertical, but it may vary in diameter as obstructions and lenses of varying

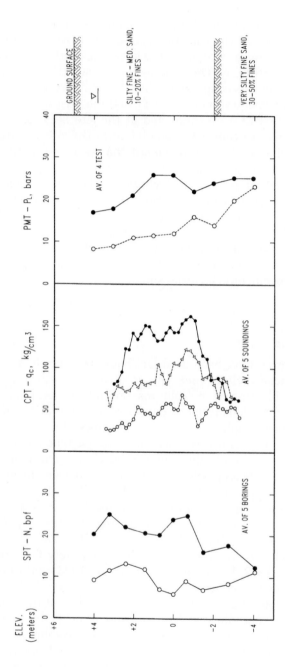

Figure 7.17 Massey coal terminal site improvement using dynamic compaction (*Mayne et al. 1984*). ["Ground Response to Dynamic Compaction," P. Mayne, J. Jones and J. Dumas, 1984, *Journal of Geotechnical Engineering Division*, reproduced by permission of American Society of Civil Engineers.]

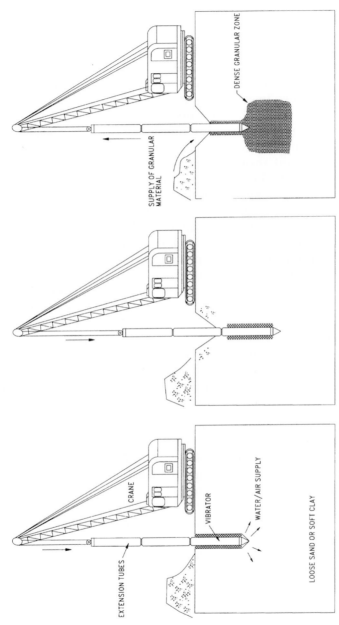

Figure 7.18 Equipment used in deep vibratory techniques.

EXTENSION TUBES

CRANE

VIBRATOR

WATER/AIR SUPPLY

LOOSE SAND OR SOFT CLAY

SUPPLY OF GRANULAR MATERIAL

DENSE GRANULAR ZONE

density are encountered. This technique is most applicable to soils in which rapid penetration can be attained, such as loose sandy soils and soft cohesive soils. Depths of soil improvement as great as 115 ft have been achieved with this vibratory technique. The version of this technique that is used most widely is often termed *vibroflotation.*

The modern version of vibroflotation, as developed from the original German version of the 1930s, uses a poker vibrator, called a *vibroflot,* that weighs 2 to 5 tons and has dimensions of 12 to 18 in in diameter and 10 to 16 ft in length. The vibroflot is a unique poker vibrator that contains an eccentric weight that causes a horizontal vibration. Because of the particular design of the vibroflot, the vibratory output remains constant throughout the depth of penetration. Currently used vibrators develop centrifugal forces of up to 20 tons at frequencies of 1200 to 3000 rpm; more powerful vibrators are under development (Moseley and Priebe 1993). The vibroflot device usually incorporates jets aimed downward at the tip and outward at various depths along the side to allow water and/or air jetting, which are used in advancing the hole and in removing material from the hole. Many vibroflots are now equipped for automatic recording of power output and depth of penetration; when stone columns are formed, the rate of stone consumption is also recorded (Moseley and Priebe 1993).

The vibroflot is used in all major types of deep vibratory treatment: vibrocompaction, vibrodisplacement, and vibroreplacement. What makes each of these techniques unique are the particulars regarding use of the water/air jets, removal or not of the loosened material, and whether or not a granular material different from site material is introduced into the hole as backfill.

Vibrocompaction.   Vibrocompaction is used principally in loose sands to densify the deposits. It is effective in deposits where the sand is looser than its maximum dry density (see Sec. 4.2.6 for a discussion of maximum, minimum, and in situ dry densities of granular materials). Vibrocompaction causes a reduction in the void ratio and compressibility and an increase in the shear strength of the treated deposit. It is often used to reduce the potential for liquefaction of sands under earthquake loads.

The sands on which vibrocompaction is effective are shown in Fig. 7.19. A general rule of thumb is that this technique is not effective on sands containing more than about 20 percent fine material. Saito (1977) reported data clearly indicating the influence of fines on degree of site improvement by vibrocompaction (Fig. 7.20).

To perform vibrocompaction, the vibrator is lowered into the ground through a combination of self-weight of the vibrator, vibration, and water jetting. The hole can be advanced at a rate of about 3.6 ft/min (Department of the Army 1983). The lower (downward-aiming) jets

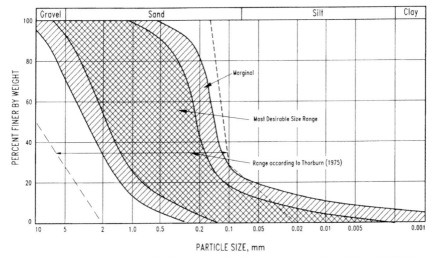

Figure 7.19   Particle size distributions best suited for vibrocompaction (*Mitchell 1982*).

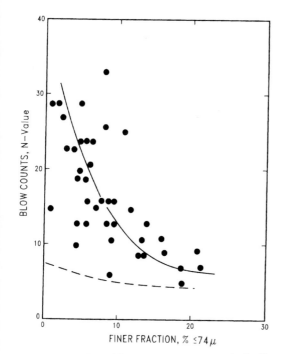

Figure 7.20   Effect of fines content on amount of soil improvement obtained by vibrocompaction (*Saito 1977*).

are normally used at low pressure but with sufficient volume of water to move soil particles up past the vibrator, thus allowing the vibrator to sink into the ground. If deep penetration (20 to 30 m) is to be attained, additional jets along the sides may need to be used to maintain the upward flow of water and sand. In unsaturated granular deposits, there will be outward flow through the sides of the borehole, and it may be necessary to increase the water input to maintain overflow at the surface. If water is not used during penetration to remove material and to reduce effective stresses, the vibration will cause compaction around the vibratory device, and penetration will stop; the point at which this occurs depends on the characteristics of the particular vibrator (Greenwood and Kirsch 1983).

When the required depth is reached, water flow is redirected from the lower jets to the side jets, and the flow is changed to low volume and high pressure. This undercuts the sand along the sides of the borehole and allows it to fall to the bottom of the hole, where it is compacted by vibration. The vibrator is slowly withdrawn, stopping at about 1-ft intervals for 30 to 60 seconds to complete compaction. The specific distances and times used will depend on the material type and properties. The effect of compaction is seen at the ground surface in the form of a depression centered on the borehole. After completion of vibration, this depression is filled with granular material from the site (if it is suitably clean), or it may be filled with imported granular material (see following paragraphs on vibrodisplacement). The process is monitored by observing power consumption to determine when compaction is complete.

The process of vibrocompaction is repeated on a grid pattern across the site; both square and triangular patterns often are used (Mitchell 1982). The area treated per borehole ranges from 3 to 10 m^2. By proper placement of the boreholes, the entire site can be uniformly compacted to the required degree, resulting in a relatively homogeneous soil mass. Figure 7.21 shows the effect of spacing of the vibrocompaction boreholes on probable relative density attained midway between boreholes; the effects of spacing on allowable bearing capacity to limit settlement to 1 in are shown in Fig. 7.22. These figures should be used to provide generalized guidance only. The actual results for specific sites will depend on material properties, water table depth, and vibrator characteristics (Mitchell 1982).

Problems that can cause less than completely successful compaction by vibrocompaction include not allowing enough time for compaction at each depth, allowing the vibrator to sink in the borehole (causing increased power consumption and falsely indicating completion of compaction), excessive overflow of water that prevents sand particles from settling in the borehole, poorly located side jets that

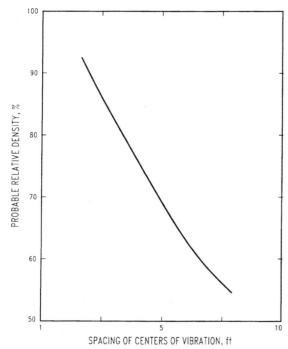

Figure 7.21   Relative density as a function of vibrocompaction hole spacing (*Department of the Army 1983*).

slow the process, and poor control of the compaction process (Greenwood and Kirsch 1983).

Vibroreplacement.   Vibroreplacement is used in soft, impervious cohesive soils (c_u = 15 to 50 kN/m^2) that are easily penetrated using low-pressure, high-volume bottom water jets. The loosened soil particles are moved out of the borehole by the upward flow of water. Direct impact of the vibratory machine can be used to loosen any stiffer layers.

When the required depth is reached, gravel backfill is introduced into the borehole, and it will move downward against the upward water flow. If the gravel tends to arch in the borehole, vibration combined with slight vertical movement of the vibrator should alleviate the problem. The combination of motions will compact the gravel at the bottom and will push it into the side walls of the borehole. If the strength of the soil is so low that these shear forces cause collapse of the adjacent soil, the continued water flow and gravel movement will remove the disturbed soil and replace it with gravel. This process will continue until an equilibrium condition is reached, indicated by

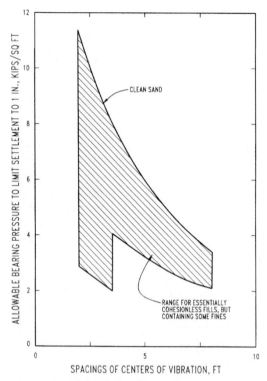

Figure 7.22 Allowable bearing pressure on cohesion-less soils stabilized by vibrocompaction (*Department of the Army 1983*).

increased power consumption. At this point, additional gravel should be added to the hole and the compaction process repeated. This technique will result in a compacted stone column with a diameter dependent on the in situ soil strength and the vibratory forces utilized. Column diameter is normally about 0.8 to 1 m.

The material used for backfill in vibroreplacement may be fairly uniformly graded subangular or rounded gravel. Since the coarsergrained fraction of the natural soil is not washed from the borehole, this material normally falls with the backfill and fills the voids in the gravel. Therefore, it is not necessary to use a well-graded gravel to obtain a rigid column.

Because of the mechanics involved in this process (i.e., sheared or disturbed soil is washed out and replaced with gravel), the soil remaining around the gravel should be in an undisturbed state. Greenwood and Kirsch (1983) reported that field investigations indicate that free drainage without smear zones occurs around these columns.

Vibrodisplacement. Vibrodisplacement is similar to vibroreplacement; each results in formation of a stone or sand column. Besides details of advancement of the borehole, the major difference is that vibrore-placement removes soil and replaces it with gravel, whereas vibrodis-placement displaces soil laterally to form a hole that is then filled with gravel. Vibrodisplacement is used in somewhat stronger cohesive soils. These soils have been termed *stable* and have unconfined compressive strengths of 30 to 60 kN/m^2 (Greenwood and Kirsch 1983).

In vibrodisplacement the borehole is advanced by vibration and self-weight of the vibrator. Soil is simply displaced laterally by this method; no material is removed from the hole. The hole thus formed is small in diameter, about the size of the vibratory device. Compressed air is introduced through the bottom jets during penetration, but the main purpose of the air is to assist in removal of the vibrator by relieving suction pressures that form below the vibrator upon its upward movement.

When the proper depth is reached, gravel is placed in the hole through a bottom-feed system that continuously feeds gravel directly to the tip of the vibrator (Moseley and Priebe 1993), or the vibrator must be removed from the hole while gravel is dumped in, after which the vibrator must be reinserted for compaction. The backfill is compacted and displaced radially to reach equilibrium. Compaction is continued at approximately 1- to 2-ft intervals until the entire depth of soil has been treated.

Backfill used in vibrodisplacement should be well graded, ranging in size from 10 to 100 mm; for use in many bottom-feed systems, the maximum particle size is 40 mm. Angular particles should be used to maximize particle interlock. Sufficient fines should be included in the backfill to prevent intrusion of the surrounding soil into the column.

Use of the bottom-feed system for bringing gravel into the borehole eliminates the need to remove the vibrator from the hole, decreases the occurrence of instability along the sides of the hole, speeds the process of compaction, and facilitates construction of a high-quality stone column. It also minimizes the chance of "damag[ing] the fabric of soft normally consolidated soils" (Greenwood and Kirsch 1983) by use of too much air or air at too high a pressure.

When weak soils are reinforced with stone columns, the deposit behaves as a composite material. This composite material will have increasing stiffness as the diameter of the columns increases, as the spacing between columns decreases, and as the angle of internal friction of the stone increases. These factors have been interrelated for design of raft and footing foundations (Moseley and Priebe 1993). According to these design analyses, the stiffness of a composite weak soil with stone columns comprising 20 percent of the cross-sectional area will be approximately 100 to 150 percent greater than the weak soil alone. This range of improvement is for angles of internal friction varying from 37.5 to 45 degrees.

Assessment of results.   The degree of improvement that can be expected from vibrocompaction, vibroreplacement, or vibrodisplacement will depend principally on the soil's grain size distribution, but it also will depend on other soil conditions, the type of vibratory equipment used, procedures adopted, and the skills of the site staff. Consequently, optimal methods and equipment for the task and final results cannot be determined explicitly at a project's design stage. Input from experienced contractors and field trials are valuable and generally necessary to specify and achieve successful results.

The importance of the preceding advice is illustrated by the following example. Initial testing and evaluation of the stability of an existing embankment at the Thermolito Afterbay Dam in California discovered that a foundation layer of silty sands would potentially liquefy during a magnitude 6.5 earthquake. A test program described by Harder et al. (1984) investigated whether these silty sands could be improved by vibrocompaction and if the improvement would be sufficient to prevent liquefaction. There were two sites under investigation: one composed of silty sand with more than 20 percent fines and one composed of silty sand with a median of 15 percent fines and 70 percent of the samples with less than 20 percent fines. The silty material at the first site had too many fines to be a good candidate for vibrocompaction, whereas the other site was borderline. In fact, the program of vibrocompaction trials, followed by standard penetration tests (SPT) and cone penetration tests (CPT), found that neither site could be improved effectively using this particular technique. As this example illustrates, it is prudent to verify site-improvement technique effectiveness with a planned program of field measurements.

To determine the effectiveness of vibrocompaction, it is essential that both pre- and postimprovement testing of the ground be conducted at the site. It is equally important that the same techniques and procedures be used for both sets of tests. The most useful and expedient method of testing the densification achieved in vibrocompaction is with standard penetration tests (SPTs) and/or cone penetration tests (CPTs).

Assessment of the site improvement resulting from construction of stone columns is best determined by full-scale load tests. However, these tests are usually prohibitively expensive even on small sites, and other methods must be used to determine the amount of improvement. Table 7.7 provides a general assessment of various testing approaches for stone columns.

TABLE 7.7   Methods for Determining Effectiveness of Stone Columns

Method	Effectiveness	Comments
Standard penetrometer test	Good	Efficiency of test important; recovers samples; gives results in terms commonly used by geotechnical engineers
Amount of stone used and power consumption of vibrator	Fair	No direct evaluation of soil properties
Plate load tests	Fair	Size of plate crucial to determine depth of effective testing
Cone penetrometers	Good	Fast; provides a record of changes with depth; gravelly soils may be troublesome
Pressuremeter	Good	Provides complete stress-strain-strength relationship with depth
Full-scale loading	Excellent	Best method; expensive

7.5.3 Stabilization and grouting

The basic techniques of soil stabilization that were described in Chap. 6 allow soil characteristics to be improved by strengthening, plasticity reduction, control of swell characteristics, and drying. These improvements can be accomplished easily and economically using the techniques described in Chap. 6 if the soils to be treated are at the surface. However, if the soils are below the surface or water table, if large depths and volumes must be stabilized over the site, or if the soils are inaccessible (e.g., adjacent to or under a building), the techniques of Chap. 6 are unusable, and new approaches are needed.

There are two basic approaches to stabilizing materials at depth: injection and deep mixing. The nature of the soil and the improvement needed will determine what stabilizer and technique will be used. Injection processes are particularly sensitive to the nature of the soil that is being reinforced because it is much harder to obtain penetration in a fine-grained soil than in a coarse-grained soil. Deep mixing, on the other hand, allows physical mixing of the stabilizer and soil and has proven particularly effective with soft clays that are essentially impossible to stabilize by injection of stabilizers in the clay soil's fine pores.

Injection methods.   Injection of soils with chemicals is normally used to improve their strength, to lower their permeability for water control, or to accomplish both. The nature of the site soils and the objectives of the injection will determine what chemicals are appropriate for the task. Table 7.8 summarizes characteristics of the basic com-

TABLE 7.8   Materials Commonly Used to Improve Soils by Injection Processes

	Suspensions	Gels	Liquids
Material	Cement, mixtures of cement and clay, lime, mixtures with fly ash	Sodium silicate with various additives, lignochromate	Polymers
Effect	Strengthen, with addition of clay lowers permeability	Strengthen, water control	Strengthen, water control
Suitable soils	Coarse sands and gravels	Medium to fine sands	Silty or clayey soils
Approximate lowest practical k (cm/s)	10^{-5}	10^{-6}	10^{-8}
Relative cost*	1–3.4	1.3–6.5	10–40

*Specific costs vary with actual composition of grout and additives used. Relative costs based on data reported by Littlejohn, 1993.

mon injectable materials used today. Chapter 5 discussed the impact of most of these materials in some detail, and injection simply is a change in delivery system and not in fundamental chemical impact.

There are several approaches to injection stabilization or grouting. The simplest concept is to inject the grouting material under pressure into holes drilled to the desired stabilization depth. The viscosity of the grouting material and the pore structure of the soil will determine how closely spaced the injection holes must be to achieve overlapping grout penetration. Obviously, any fissures and cracks will aid distribution of the material in the soil mass.

This technique is not new. Injection of silicates was used to control water inflow in the early 1900s for control of mining excavations (Littlejohn 1993). The specific techniques and equipment vary somewhat depending on the materials being injected (e.g., some are two-stage processes, some require special chemical mixing, etc.), but the construction process is well established. The most troublesome soils in construction tend to be fine-grained, where injection of suspension grouts such as cement tends to be of limited effectiveness. The need to extend the range of soils that could be injected led to development of a number of proprietary low-viscosity gels and then later polymeric materials for use in injection grouting. More extensive discussions of grouting and materials can be found in Karol (1983) and in two American Society of Civil Engineering conferences (1982, 1992).

Another approach to achieving grout penetration into a soil is by fracturing the material, and forcing the grout into the resulting fissures. To accomplish this, the specific soil layer to be stabilized is isolated along the drill hole using packers both above and below this layer. The grout is injected through a sleeve containing uniformly spaced holes. The pressurized grout compacts and fractures the soil, developing areas of compacted material with lenses of grout. The process can be repeated to gain more fracturing as needed. This technique also can effectively raise the surface to relevel settled areas.

The idea of compacting the soil by grouting developed in the United States in the 1950s (Warner 1982). With this approach, a stiff cement grout, typically having a slump of an inch or less, is forced into the borehole. This material does not mix with the soil but compacts it. It has proven effective in releveling settled areas, settlement control over soft soil tunneling, densifying loose granular material prone to liquefaction, and for repairs associated with sinkholes in karst topography (see Sec. 8.7) (Rubright and Welsh 1993).

There are several approaches to grouting: pressure-injection grouting, soil fracturing, and compaction grouting. This work is often carried out by specialists in the field, and obtaining their advice and assistance is prudent.

Deep mixing. If it is not feasible to achieve the desired results by injecting a stabilizing material into a soil, there are several techniques for physically mixing stabilizers with the soil at depth. The first technique, known as *jet grouting,* is an imaginative extension of the pressure-injection concept. After the drill hole is carried to the desired depth, one or more jets of high pressure grout are forced out horizontally and are rotated as the system is slowly drawn upward. This jet erodes the surrounding soil, and eroded material is carried to the surface. As the system is drawn to the surface, it leaves behind a mixture of grout and eroded soil. This approach is most effective in sands, though it has been used on a wide variety of soils, including gravelly soils, clays, peats, and organic soils (Bell 1993). Table 7.9 shows some of the limitations of this approach. Several systems have been developed to carry out this work and can use combinations of grout, air, and water to increase effectiveness of the erosive jet.

Use of augers and special mixing tools to make columns of lime- or cement-stabilized clays developed independently in Sweden (Broms and Boman 1979) and Japan (Kawaske et al 1981). In the Swedish method, a special auger is inserted to the desired depth of the column, and as the auger is withdrawn, lime is injected pneumatically through a hole in the auger. The auger mixes and compacts the lime with the soil as it is slowly withdrawn. This technique relies on the soil reinforced with lime columns behaving as a block, and it has proven effective for foundations for light buildings, to reduce settlement and increase bearing capacity, to provide protection against soil creep, and to improve excavation support and slope stability. The

TABLE 7.9   Soil Limitations on Jet Grouting

Soil	Jet grouting effectiveness	Comments
Clays	Effective	Waste slurry at surface much thicker and harder to handle; potential clogging problems
Fissured	Effective	Lumps exacerbate clogging; flushing with water may be needed
High cohesion	Limited	Standard techniques leave columns too small to be useful (when shear strength >8 psi); may need to use two-stage process with initial air/water jet followed by grout jet
Sensitive	Effective	Can cut large-diameter column, but varying sensitivity within deposit and large waste volume can be troublesome
Sands, silty sands, and silts	Highly effective	Best situation; easily eroded; lack of cohesion makes waste slurry easier to handle
Gravelly soils	Variable	May allow loss of grout and jet effectiveness due to high permeability; leaves gravel particles in the treated soil, which enhances strength

SOURCE: Based on information from Bell, 1993.

technique was developed for use in soft clays, and approximately 2 million linear feet have been placed in Sweden since 1975 (Broms 1993). It also has been used with other stabilizers such as cement, slag, and gypsum. It is a common technique in Scandinavia, but it is little known in the United States.

The Japanese deep mixing techniques for cement and lime were developed originally for offshore construction on soft marine sediments. The initial systems were barge-mounted and were quite massive. These have been adapted for land-based work also. The larger barge-based systems have been used to depths of 165 ft, while the land-based systems are limited to maximum depths of about 100 ft (Toth 1993). The Japanese approach is similar in concept to the Swedish lime column with two exceptions: (1) the layer to be stabilized is commonly mixed twice, and (2) cement is commonly added as a slurry rather than as a dry powder (Toth 1993).

7.5.4   Thermal methods

The load-carrying capacity of soil can be increased by either heating or freezing. As evidenced by our ancient heritage with both sun-dried and fired bricks, increases in strength with a decrease in water susceptibility can be achieved by heating fine-grained soils. However, applications of this concept to achieve large-scale, deep soil improvement generally have been limited to eastern Europe and Russia. There, thermal techniques have been used primarily to treat collapsible loess deposits (see Secs. 3.5.4 and 8.4) but also have been used to stabilize landslides, construct in situ vitrified piles, and provide foundation support. Several approaches to achieving thermal stabilization at depth have been used: Bored holes are heated by injection and burning of pressurized fuel and air (closed system), two intersecting holes are used with the burner in one and egress of combustion gases from the other (open system), or electric heaters may be used by either lowering the heater into the hole or blowing compressed air through the electric heater at the surface (Ingles and Metcalf 1973, Mitchell 1982). Commonly, several holes will be heated simultaneously. Table 7.10 shows some suggested minimum temperatures required for several applications.

The major limitation for heating to improve soil properties is the high energy cost involved. As energy costs have risen, the economic practicality of this approach has decreased. Ingles and Metcalf (1973) have suggested that the fuel consumption for stabilizing slips on slopes can be calculated as

$$F = \frac{100(6.4\ \omega\gamma_d + 0.25T\gamma_d)}{EC_f}$$

where F = fuel usage per unit volume stabilized
ω = soil moisture content
γ_d = soil dry density
T = burning temperature at the soil

TABLE 7.10   Suggested Minimum Temperatures for Several Thermal Applications

Application	Minimum treatment temperature	
	°C	°F
Treatment of collapsible loess	300–400	572–752
Reduction of lateral pressure	300–500	572–932
Control frost heave	500	932
Massive column construction below frost depth	600	1112
Manufacture of building materials	900–1000	1652–1832

SOURCE: Adapted from Jurdanov, 1978.

E = thermal efficiency of the system

≈ 35 percent for open burning

≈ 70 percent for closed burning

C_f = unit heat capacity of the fuel

The most economical hole spacing can then be calculated as

$$d = 2\sqrt{\frac{D_c}{\pi F H_c}}$$

where d = hole spacing for minimum cost

D_c = cost for drilling unit length of hole

H_c = cost for unit weight of fuel

These calculations, in conjunction with the suggested temperatures in Table 7.10, may allow a preliminary assessment of the economic viability of heating to achieve the desired site improvement. However, pilot holes will be required on site to verify effectiveness, economics, and requisite firing times.

Soil freezing dates back to the nineteenth century, when it was used to stabilize mine shaft excavations. Today it has developed into an effective method of temporary ground support. Two basic processes are in use: the Poetsch, or closed, processes that use circulating cooled brine in pipes to carry out the freezing and the open processes using direct application of liquid nitrogen or solid carbon dioxide (dry ice) to the zone to be frozen and allowing heat to be carried directly to the atmosphere. The closed processes allow the best control of the freezing and are the most economical. However, setup times are lengthy, and freezing typically requires weeks. The open processes are appreciably faster but are more costly and difficult to control. Consequently, open systems tend to be used for emergency or short-term soil improvement for a limited time, and closed systems tend to be used for more comprehensive and extensive programs.

Ground freezing essentially converts the in situ soil moisture into a cementing medium for the soil particles. This freezing can easily double the strength of a soil and may give it strengths approaching that of a weak rock. This will generally make it self-supporting for excavations. However, the behavior of the frozen soil is distinctly viscoplastic. Its strength under sustained loading is only a fraction of that under rapid loading. This behavior must be recognized in the design stage.

The success of a ground-freezing effort is affected by several factors:

1. *Soil type and moisture content.* Freezing will be most effective in soils below the water table and fine-grained soils with adequate capillary moisture above the water table. Strength is approximately proportional to the degree of saturation, and unsaturated soils generally

are relatively pervious, even when frozen, compared with saturated soils (Braun et al. 1979). The radius of the frozen zone around a freezing pipe generally will be larger for sands and gravels than for silts, clays, and organic soils (Mitchell 1982).

2. *Groundwater.* Flowing groundwater continuously adds heat to the thermal regime that the freezing system must overcome. Special methods (e.g., grouting to reduce permeability, reduced freezing pipe spacing to handle the added heat, direct freezing with liquid nitrogen) are needed to accomplish ground freezing when flows exceed about 4.9 ft/day (Braun et al. 1979). Water containing salts and other contaminants may require freezing at lower than normal temperatures, and the effects of the contaminants on the engineering characteristics of the frozen soil should be determined by testing.

3. *Stratigraphy.* Ground freezing is most effective when the frozen earth wall can be carried directly to an underlying impervious layer. In other conditions, particular care is needed to minimize water flow under the frozen soil (Braun et al. 1979).

4. *Position and alignment of freezing pipes.* The freezing system must build a solid wall of frozen soil to achieve structural strength and stop water flows. Frozen cylinders of soil that do not coalesce into a solid curtain offer openings for water flow and reduce the system's structural strength.

5. *Ground volume change.* When water freezes, it increases 9 percent in volume. In fine sands and silts, water flows to the site of freezing and forms lenses that continue to grow as ice accumulates (see Sec. 8.8). These can cause lateral pressure on structures and cause heaving of the ground. The direction of freezing also causes the strength of the soil to be weaker perpendicular to the heat flow (Knutsson 1982).

6. *Presence of utilities.* These generally should be insulated to prevent freezing of the utility and thawing of the surrounding frozen soil, or the line should be relocated (Braun et al. 1979).

7. *Structural considerations.* The frozen structure must be designed for the anticipated loads and conditions to include strength that can be achieved with the soil and saturation levels that are available, viscoplastic behavior of the frozen material, impact of ground volume change on loads and strength, structural systems compatible with the high compressive strength and low tensile strength of frozen soils, and size of frozen area.

8. *Thermal regime.* Analysis of the thermal regime in a ground-freezing project is complex, and analytical solutions are crude initial approximations at best. Information on thermal solutions published by Braun et al. (1979), Lunardini (1991), and Xanthakos (1994) may prove helpful. Braun et al. (1979) summarize the problem as follows:

Preconstruction determination of refrigeration requirements is largely an art tinged with science, much the same as the selection and sizing of dewatering systems or the determination of grouting requirements.

Ground freezing has proven particularly effective for tunneling and excavations in difficult ground conditions. More detailed information and specific examples of applications can be found in Jessberger (1979) and Xanthakos (1994). Ground freezing is a specialized technique, and its use requires the assistance of engineers and contractors experienced with the process.

7.5.5 Geosynthetics

Geosynthetics are specific man-made products that can be used in a number of ways to provide various types of site improvement (Fluet 1988). The major types of geosynthetics are geotextiles, geomembranes, geogrids, geonets, and geocomposites. The functions typically performed by geosynthetics include reinforcement, separation, drainage, filtration, cushioning, and isolation/barrier. Each product has certain primary functions that it is intended to perform (e.g., geomembranes act primarily as a barrier to moisture flow). Many geosynthetics also perform secondary and tertiary functions; e.g., while functioning primarily in a filtration role, geotextiles also may provide separation and reinforcement in a layered embankment. A more detailed discussion of the uses and functions of geosynthetics in geotechnical construction is contained in Chap. 9.

7.6 Selection of Methods

Table 7.11 summarizes the site-improvement techniques discussed in this chapter. Each one offers certain advantages and limitations that must be matched to the specific site requirements. The major factors in selecting a site-improvement technique or some combination of techniques include

1. Specific improvements needed (e.g., strengthening, reduced water flow, or reduced differential settlement)

2. Site conditions, including soils, groundwater conditions, environmental limitations, weather, surface mobility, and size and depth of area to be treated

3. Available time for site improvement

4. Cost

5. Availability of necessary equipment, materials, and contractors with the requisite specialized experience.

TABLE 7.11 Site-Improvement Techniques

Technique	Principles	Suitable soils	Remarks
Soil replacement techniques			
Remove and replace	Excavate weak or undesirable materials and replace with better soils	Any	Limited depth and area where cost-effective; generally ≤30 ft
Displacement	Overload weak soils so that they shear and are displaced by stronger fill	Very soft	Problems with mudwaves and trapped compressibles under the embankment; highly dependent on specific site
Water-removal techniques			
Trenching	Allows water drainage	Soft, fine-grained soils and hydraulic fills	Effective depth up to 10 ft; speed dependent on soil and trench spacing; resulting desiccated crust can improve site mobility
Precompression	Loads applied prior to construction to allow soil consolidation	Normally consolidated fine-grained soil, organic soil, fills	Generally economical; long time may be needed to obtain consolidation; effective depth only limited by ability to achieve needed stresses
with vertical drains	Shortens drainage path to speed consolidation	Same as above	More costly; effective depth usually limited to ≤100 ft
Electro-osmosis	Electric current causes water flow to cathode	Normally consolidated silts and silty clays	Expensive; relatively fast; usable in confined area; not usable in conductive soils; best for small area

TABLE 7.11 Site-Improvement Techniques (Continued)

Technique	Principles	Suitable soils	Remarks
Site-strengthening techniques			
Dynamic compaction	Large impact loads applied by repeated dropping of a 5- to 35-ton weight; larger weights have been used	Cohesionless best; possible use for soils with fines; cohesive soils below water table give poorest results	Simple and rapid; usable above and below water table; effective depths up to 60 ft; moderate cost; potential vibration damage to adjacent structures
Vibratory methods			
Vibrocompaction	Vibrating insert densifies soils	Cohesionless soils with <20 percent fines	Can be effective up to 100 ft depth; can achieve good density and uniformity; grid spacing of insertions critical; relatively expensive
Vibroreplacement	Jetting and vibration used to penetrate and remove soil; compacted granular fill then placed in hole to form support columns surrounded by undisturbed soil	Soft cohesive soils ($C_u = 15$–50 kN/m^2)	Relatively expensive
Vibrodisplacement	Similar to above except soil is displaced laterally rather than removed from hole	Stronger cohesive soils ($C_u = 30$–60 kN/m^2)	Relatively expensive

TABLE 7.11 Site-Improvement Techniques (Continued)

Technique	Principles	Suitable soils	Remarks
Stabilization and grouting			
Injection	Fill soil voids with cementing agents to strengthen and reduce permeability	Wide spectrum of coarse- and fine-grained soils	Expensive; more expensive grouts needed for finer-grained soils; may use pressure injection, soil fracturing, or compaction techniques
Deep mixing	Jetting or augers used to physically mix stabilizer and soil	Wide spectrum of coarse- and fine-grained soils	Jetting poor for high-cohesion clays and some gravelly soils; deep mixing best for soft soils up to 165 ft depth
Thermal			
Heat	Heat used to achieve irreversible strength gain and reduced water susceptibility	Cohesive	High energy requirements; cost limits practicality
Freezing	Moisture in soil frozen to cement particles together to increase strength and reduce permeability	All soils below water table; cohesive soils above water table	Expensive; highly effective for excavations and tunneling; high groundwater flows troublesome; slow
Geosynthetics	Use geosynthetic materials for filters, erosion control, water barriers, drains, or soil reinforcing	Effective filters for all soils; can be reinforcing on soft soils	Widely used to accomplish variety of tasks; commonly used in conjunction with other methods (e.g., strip drain with surcharge or to build a construction platform for site access)

Costs of different improvement techniques often will be of decisive importance for many projects. This is a function of not just the improvement technique itself but also of the depth and areal extent of the necessary improvement. This is illustrated in Fig. 7.23, where road construction excavation and replacement is cost-effective to depths of about 15 ft and displacement methods are more cost-effective for deeper soft soil deposits. Surcharging with vertical drains is the most cost-effective alternative shown for depths below 15 ft, and bridging is the most expensive option. This information is illustrative rather than absolute, and each site will require its own economic evaluation of practical alternatives.

Site soil conditions generally will determine what site-improvement alternatives are feasible. Figure 7.24 provides some conceptual guidelines on which site-improvement techniques are effective for different soil conditions.

Nature seldom provides ideal conditions for construction. We can either avoid building on undesirable sites, design facilities to accommodate poor site conditions, or improve the site conditions. Today we have a number of techniques available that will allow us to significantly improve site conditions and effectively build on sites that were unusable in the past.

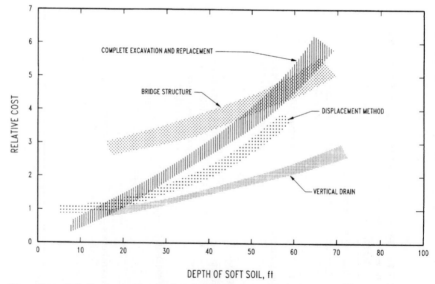

DEPTH OF SOFT SOIL, ft

Figure 7.23 Relative cost of several site-improvement techniques for different depths of soft soil (*Organization for Economic Cooperation and Development 1979*). [© OECD, 1980, Construction of Roads on Compressible Soils. Reproduced by permission of OECD.]

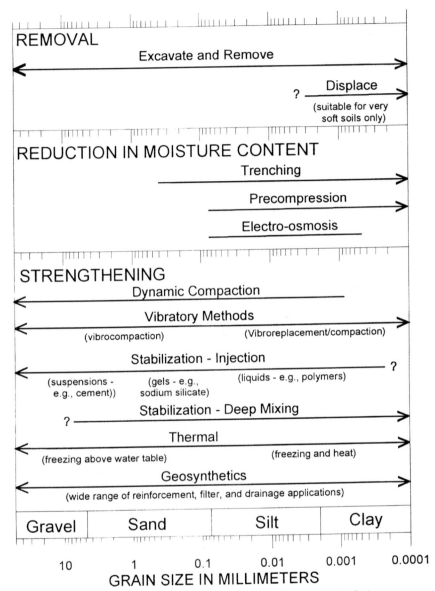

Figure 7.24 Conceptual ranges of soils for different site-improvement techniques.

7.7 References

American Society of Civil Engineers. 1992. *Proceedings of Specialty Conference on Grouting in Geotechnical Engineering,* ASCE, New York.

American Society of Civil Engineers. 1982. *Proceedings of Specialty Conference on Grouting, Soil Improvement, and Geosynthetics,* ASCE, New York.

American Society of Civil Engineers. 1978. "Soil Improvement: History, Capabilities, and Outlook," report by the Committee on Placement and Improvement of Soils,

Geotechnical Engineering Division, American Society of Civil Engineers, New York.

Bell, A. L. 1993. "Jet Grouting," in *Ground Improvement,* M. P. Moseley, ed., Chapman and Hall, London, pp. 149–174.

Bjerrum, L., J. Moum, and O. Eide. 1967. "Application of Electro-Osmosis on a Foundation Problem in Norwegian Quick Clay," *Geotechnique,* 17: 214–235.

Braun, B., J. Shuster, and E. Burnham. 1979. "Ground Freezing for Support of Open Excavations," *Engineering Geology,* 13: 429–453.

Broms, B. B. 1993. "Lime Stabilization," in *Ground Improvement,* M. P. Moseley, ed., Chapman and Hall, London, pp. 65–99.

Broms, B. B. 1987. "Stabilization of Very Soft Clay Using Geofabric," in *Geotextiles and Geomembranes,* vol. 5, Elsevier Applied Science Publishers, London, pp. 17–28.

Broms, B. B., and P. Boman. 1979. "Lime Columns: A New Foundation Method," *Journal of the Geotechnical Engineering Division, ASCE,* 105 (GT4): 539–556.

Casagrande, L. 1952. "Electro-Osmotic Stabilization of Soils," *Journal of the Boston Society of Civil Engineers,* 34 (1); reprinted in *Contributions to Soil Mechanics, 1941–1953,* Boston Society of Civil Engineers, Boston, Mass., pp. 285–317.

Cedegren, H. R., 1989. *Seepage, Drainage, and Flow Nets,* Wiley, New York.

Department of the Army. 1987. "Confined Disposal of Dredged Material," EM1110-2-5027, U.S. Army Corps of Engineers, Washington.

Department of the Army. 1983. "Soils and Geology, Procedures for Foundation Design of Buildings and Structures (Except Hydraulic Structures)," TM 5-818-1, Washington.

Fluet, J. E., Jr. 1988. "Geosynthetics for Soil Improvement: A General Report and Keynote Address," in *Geosynthetics for Soil Improvement,* R. D. Holtz, ed., Geotechnical Special Publication no. 18, American Society of Civil Engineers, New York.

Greenwood, D. A., and K. Kirsch. 1983. "State of the Art Report on Specialist Ground Treatment and Vibratory and Dynamic Methods," in *Conference on Piling and Ground Treatment,* Institution of Civil Engineers, London.

Haliburton, T. A. 1978. "Guidelines for Dewatering/Densifying Confined Dredged Material," Technical Report DS-78-11, USAE Waterways Experiment Station, Vicksburg, Miss.

Hammer, D. P., and E. D. Blackburn. 1977. "Design and Construction of Retaining Dikes for Containment of Dredged Material," Technical Report D-77-9, U.S. Army Engineer Waterways Experiment Station, Vicksburg, Miss.

Harder, L. F., W. D. Hammond, and P. S. Ross. 1984. "Vibroflotation Compaction at Thermolito Afterbay," *Journal of Geotechnical Engineering Division, ASCE,* 110 (1): 57–70.

Hausmann, M. R. 1987. "Geotextiles for Unpaved Roads: A Review of Design Procedures," in *Geotextiles and Geomembranes,* vol. 5, Elsevier Applied Science Publishers, London, pp. 201–233.

Ingles, O. G., and J. B. Metcalf. 1973. *Soil Stabilization, Principles and Practice,* Wiley, New York.

Jessberger, H. L., ed. 1979. *Engineering Geology, Special Issue: Ground Freezing,* vol. 13, nos. 1–4, Elsevier Scientific Publishing, Amsterdam.

Johnson, S. J. 1970. "Precompression for Improving Foundation Soils," *Journal of Soil Mechanics and Foundation Engineering, ASCE* 96 (SM1): 111–144.

Jurdanov, A. 1978. "Special Function of Deep Thermal Treatment of Soils and Its Development," *Osnovaniya, Fundamenty I Mekhanika Gruntov,* 20 (6): 14–16.

Karol, R. H. 1983. *Chemical Grouting,* Marcel Dekker, New York.

Kawaske, T., A. Niina, S. Saitoh, Y. Suuki, and Y. Honjko. 1981. "Deep Mixing Method Using Cement Hardening Agent," *Proceedings 10th International Conference on Soil Mechanics and Foundation Engineering,* International Society for Soil Mechanics and Foundation Engineering, Stockholm, Sweden, pp. 721–724.

Koerner, R. M. 1990. *Designing with Geosynthetics,* 3d ed., Prentice-Hall, Englewood Cliffs, N.J.

Knutsson, S. 1982. "The Shear Strength of Frozen Soil," *Proceedings of the Tenth International Conference on Soil Mechanics and Foundation Engineering,* vol. 3, International Society for Soil Mechanics and Foundation Engineering, Stockholm, Sweden, pp. 731–734.

Lambe, T. W., and R. V. Whitman. 1969. *Soil Mechanics,* Wiley, New York.

Leonards, G. A., W. F. Cutter, and R. D. Holtz. 1980. "Dynamic Compaction of Granular Soils," *Journal of Geotechnical Engineering Division, ASCE,* 106 (GT1): 35–44.

Littlejohn, G. S. 1993. "Chemical Grouting," in *Ground Improvement,* M. P. Moseley, ed., Chapman and Hall, London, pp. 100–130.

Lunardini, V. J. 1991. *Heat Transfer with Freezing and Thawing,* Elsevier, Amsterdam.

Mansur, C. I., and R. I. Kaufman. 1962. "Dewatering," in *Foundation Engineering,* G. A. Leonards, ed., McGraw-Hill, New York.

Mayne, P. W., J. S. Jones, and J. C. Dumas. 1984. "Ground Response to Dynamic Compaction," *Journal of Geotechnical Engineering Division, ASCE,* 110 (6): 757–774.

Menard, L., and Y. Broise. 1975. "Theoretical and Practical Aspects of Dynamic Consolidation," *Geotechnique,* 25 (1): 3–18.

Mitchell, J. K. 1993. *Fundamentals of Soil Behavior,* Wiley, New York.

Mitchell, J. K. 1982. "Soil Improvement: State-of-the-Art," *Proceedings of the Tenth International Conference on Soil Mechanics and Foundation Engineering,* International Society for Soil Mechanics and Foundation Engineering, Stockholm, Sweden, vol. 4, pp. 509–565.

Moseley, M. P., and H. J. Priebe. 1993. "Vibro Techniques," in *Ground Improvement,* M. P. Moseley, ed., Blackie Academic and Professional, Boca Raton, Fla.

O'Bannon, C. E. 1977. "Field Study to Determine the Feasibility of Electro-osmotic Dewatering of Dredged Material," Miscellaneous Paper D-77-2 USAE Waterways Experiment Station, Vicksburg, Miss.

Organization for Economic Co-Operation and Development (OECD). 1979. *Construction of Roads on Compressible Soils,* Paris, France.

Poindexter, M. E. 1988. "Equipment Mobility in Confined Dredged Material Disposal Areas: Field Evaluations," EEDP Technical Note EEDP-09-4, U.S. Army Engineer Waterways Experiment Station, Vicksburg, Miss.

Powers, J. P. 1981. *Construction Dewatering: A Guide to Theory and Practice,* Wiley, New York.

Raabe, A. W., and K. Esters. 1993. "Soilfracturing Techniques for Terminating Settlements and Restoring Levels of Buildings and Structures," in *Ground Improvement,* M. P. Moseley, ed., Chapman and Hall, London, pp. 175–192.

Richardson, G. N., and Koerner, R. M. 1990. *A Design Primer: Geotextiles and Related Materials,* Industrial Fabric Association International, Geotextile Division, St. Paul, Minn.

Rollings, M. P. 1994. "Geotechnical Considerations in Dredged Material Management," *Proceedings, First Environmental Geotechnics Congress,* International Society for Soil Mechanics and Foundation Engineering, Edmonton, Canada, pp. 21–32.

Rollings, R. S., M. E. Poindexter, and K. G. Sharp. 1988. "Heavy Load Pavements on Soft Soil," *Proceedings 14th Australian Road Research Board Conference,* vol. 14, part 5, Canberra, Australia, pp. 219–231.

Rubright, R., and J. Welsh. 1993. "Compaction Grouting," in *Ground Improvement,* M. P. Moseley, ed., Chapman and Hall, London, pp. 131–147.

Saito, A. 1977. "Characteristics of Penetration Resistance of a Reclaimed Sandy Deposit and Their Change Through Vibratory Compaction," *Soils and Foundation,* 17(4), The Japanese Society of Soil Mechanics and Foundation Engineering, Tokyo, Japan, pp. 32–43.

Slocombe, B. C. 1993. "Dynamic Compaction," in *Ground Improvement,* M. P. Moseley, ed., Chapman and Hall, London, pp. 20–39.

Thornburn, S. 1975. "Building Structures Supported by Stabilized Ground," *Geotechnique,* 25 (1): 83–94.

Toth, P. S. 1993. "In Situ Mixing," in *Ground Improvement,* M. P. Moseley, ed., Chapman and Hall, London, pp. 193–204.

USACE/USEPA. 1992. "Evaluating Environmental Effects of Dredged Material Management Alternatives: A Technical Framework," EPA/842-B-92-008, USEPA, Washington.

Warner, J. 1982. "Compaction Grouting: The First Thirty Years," *Proceedings of Specialty Conference on Grouting, Soil Improvement, and Geosynthetics,* American Society of Civil Engineers, New York, pp. 694–707.

Willoughby, W. E. 1977. "Low-Ground-Pressure Construction Equipment for Use in Dredged Material Containment Area Operation and Maintenance: Performance

Predictions," Technical Report D-77-7, U.S. Army Engineer Waterways Experiment Station, Vicksburg, Miss.

Willoughby, W. E. 1978. "Assessment of Low-Ground-Pressure Equipment for Use in Containment Area Operations and Maintenance," Technical Report DS-78-9, U.S. Army Engineer Waterways Experiment Station, Vicksburg, Miss

Xanthakos, P. P., 1994. "Artificial Ground Freezing," in *Ground Control and Improvement,* P. P. Xanthakos, L. W. Abramson, and D. A. Bruce, eds., Wiley, New York.

Specialized Construction Techniques

*"Where men of judgement creep and feel their
way, the positive pronounce without dismay."*
COWPER

8

Special Construction Topics

"He who accounts all things easy will have
many difficulties."   LAO-TSZE

8.1 Introduction

Because of the complexity and variability of geotechnical materials, there are a number of circumstances that can pose unusually difficult problems. If conventional methods of handling geotechnical materials are employed without incorporating procedures to address the special characteristics of the materials involved, the results can be quite distressing. This chapter will examine problems of construction in arid regions, with expansive soils, on collapsible soils, on soft and organic deposits, with sensitive clays, in karst areas, and with geotechnical materials subjected to freezing.

8.2 Arid Regions

Arid regions have an annual evaporation rate that exceeds the annual precipitation rates. They occur on every populated continent except Europe, with the largest desert areas lying between 20 and 30 degrees north and south of the equator. Their lack of water poses logistical problems in supporting construction crews and obtaining water for construction activities such as compaction and dust control or for portland cement concrete. Although rainfall amounts are low, they tend to be concentrated in a short period. These intense rains, coupled with high runoff rates, cause severe localized flooding in many arid regions. The low humidity combined with frequent windy conditions allows rapid evaporation that will make control of moisture loss in materials to be compacted or premature drying and cracking of concrete a recurring problem. The harsh environment is hard on personnel and equipment, which will affect the quality of work achieved.

Soils in arid regions are exposed primarily to mechanical weathering and may be further sorted by wind or water. Consequently, the reduced level of chemical weathering tends to produce predominantly coarse-grained materials with relatively little vertical soil profile development. Also, materials such as salts, sulfates, and carbonates that dissolve in humid temperate climates may exist unaltered in the desert and are often near the ground surface. A number of processes are at work, and the desert produces a variety of features, including alluvial fans, alluvial stream and river deposits, eolian deposits including both sand dunes and loess, residual soils, and lacustrine deposits. Although arid regions consist of predominantly coarse-grained material, there are areas of fine-grained soils and also high water tables that can cause problems. Cooke and Warren (1973) provide a detailed description of desert features, and Fookes (1976) and Fookes and Higginbottom (1980 a,b) provide in-depth engineering analysis of desert construction materials. Table 8.1 shows a general distribution of desert features for several major desert areas, and Table 8.2 shows some typical soil properties.

The dominant features of arid regions are their low rate of chemical weathering and high evaporation rate relative to precipitation. Many of the aggregates of the desert may contain weak, unstable minerals that normally would decompose in more humid climates. Consequently, aggregates should be evaluated carefully for stability and strength. The high evaporation rates also result in evaporite salts being deposited in the soils and intact rock. These evaporites can form cemented deposits known as *duricrusts* (see Sec. 3.2). Sometimes these duricrusts may be the best available aggregate, but

TABLE 8.1   Sample Areal Extent for Selected Desert Features

Feature	Sahara	Libyan desert	Arabia	Southwestern United States
Desert mountains	43%	39%	47%	38%
Volcanic cones	3	1	2	<1
Badlands (intricately eroded areas)	2	8	1	3
Wadis (dry stream bed, arroyo)	1	1	1	4
Alluvial fans	1	1	4	31
Bedrock pavements	10	6	1	1
Adjoining continuous rivers	1	3	1	1
Desert flats	10	18	16	20
Playas	1	1	1	1
Sand dunes	28	22	26	1

SOURCE: Modified from Fookes, 1976.

TABLE 8.2 Classification of Sample Desert Soils

Zone*	Formation	Particle size (in mm), % passing				Atterberg limits		CBR		Percent water soluble
		20	2	0.06	0.002	LL	PI	Soaked	Unsoaked	
II	Fan	87	38	—	—	—	—	53	39	10.8
II	Fan	63	27	—	—	27	6	46	39	0.1
III	Sandy desert	93	54	21	8	36	18	—	67	2.2
III	Sandy desert	83	30	—	—	—	—	59	—	1.6
III	Silty desert	97	77	31	—	112	50	16	3	41.9
III	Silty desert	—	99	76	35	39	18	22	5	1.0
IV	Silty desert	—	96	63	20	29	17	24	14	0.7
IV	Silty desert	—	100	57	15	39	24	30	14	0.5
IV	Loess	—	—	83	21	64	38	11	6	0.5
IV	Loess	—	—	89	18	33	6	10	4	0.6
IV	Loess	—	—	90	20	36	17	5	7	0.4
IV	Sandy desert	—	96	14	—	—	—	15	16	0.6
IV	Dune sand	—	99	83	9	39	17	4	3	1.8

*Zone I: mountain slopes—no samples in table; zone II: alluvial fans adjacent to mountains; zone III: alluvial plain; zone IV: base plain occupying central parts of basins and piedmont, probably largest of the desert zones, may have high water table, eolian deposits common.
SOURCE: Modified from Fookes, 1976.

they are often highly variable, weak, and contaminated. Materials contaminated with salts and evaporites will tend to be weak, may be soluble, and may cause durability problems such as salt blisters in asphalt or sulfate or alkali-aggregate attack in portland cement concrete. Selective quarrying and borrowing of materials combined with careful chemical and durability assessments are critical in arid desert environments.

8.3 Expansive Soils

8.3.1 Behavior

As clay particles in a soil absorb moisture onto the surface of the particle, the soil mass increases in volume, and if water is removed from the clay particles, the soil volume decreases. These volume changes on some clays and their effects can be very dramatic and cause damage to many engineered structures (Fig. 8.1). Some clay minerals, notably smectites, have a very strong ability to absorb water. This ability to absorb water is directly related to the clay's plasticity, and the variation in the clay mineral's plasticity index in Table 2.2 illustrates the wide variation found in clay minerals. Any soil that contains a significant quantity of highly plastic clay minerals will change

Figure 8.1 Residential building damage due to swell in the foundation soil.

in volume when its moisture content changes. These expansive soils that change volume with moisture fluctuations cause more damage in the United States than do the combined effects of other natural hazards, including earthquakes, floods, tornadoes, and hurricanes (Jones and Holtz 1973, FEMA 1982).

Dealing with expansive soils is a problem of control of moisture changes and not one of strength. Rollings and Rollings (1991) provided the illustrative example of repeated, ongoing failures of residential streets built on an expansive clay in a suburban area. As each enlargement of the residential area occurred, the design strength for the pavement was progressively reduced, resulting in thicker pavements. The pavement design method was based on providing sufficient thickness of pavement above the subgrade of a given strength so that the design traffic would not cause shearing and rutting in the subgrade. Since the problem causing distortion of the pavement surface was volume change from seasonal moisture fluctuations, the increasingly thick pavements designed to avoid traffic-induced shear failure were completely ineffective in preventing continuing failure of the pavement.

8.3.2 Identification

Expansive clays will develop wherever geologic processes allow accumulation of predominately silt- and clay-sized particles that contain large quantities of expansive minerals such as smectite. Consequently, if the proper parent materials are available to provide the minerals, expansive clays may be found in residual soils developed on fine-grained rocks such as shale or in transported soils that are essentially fine grained (e.g., backswamp alluvial deposits, lacustrine deposits, or deltaic deposits). Expansive soils are a potential problem wherever there are argillaceous rocks or fine-grained sediments (Department of the Army 1983a, Snethen et al. 1975).

Volume change in an expansive soil will be limited to the active zone where there are moisture changes occurring in the soil. These moisture changes may reflect environmental fluctuations, the thermodynamic effects of placing a floor slab or pavement over the soil surface, impact of vegetation or of changing vegetative cover, surface drainage, or transient effects such as leaking utilities or watering residential lawns. The normal active zone can be deep, e.g., 15 to 20 ft along the Front Range in Colorado, and analytical estimates of the depth of this zone are often too low (Nelson and Miller 1992).

Potentially expansive soils are usually identified by soil classification, mineralogy, swelling tests, their Atterberg limits, or a combination of these factors. CH clays under the Unified Soil Classification

System (USCS) and A-6 and A-7 soils under the AASHTO system are commonly expansive. The swell of either remolded or undisturbed samples can be measured in the laboratory using an odometer or other device after water is added to the sample. For instance, the Corps of Engineers measures the swell of the compacted samples as part of the development of soil compaction curves and subgrade strength determination (see Chaps. 4 and 5). The soil in the compaction molds is soaked in water for 4 days with a surcharge load equal to the anticipated overburden pressure of the pavement as part of the standard soaked CBR test. If the swell during the soaking exceeds 3 percent, the soil is considered expansive.

The soil's Atterberg limits reflect the activity of the clay minerals present and are therefore widely used to help identify expansive soils. Several methods using the Atterberg limits for estimating the swelling potential are presented in Table 8.3. These classifications use readily obtained soil parameters and will allow an estimate of the potential problems with swelling. These schemes are based on limited numbers of tests within specific geographic regions, so their accuracy when applied outside these regions is uncertain. Consequently, they should be treated as index tests only and will not be adequate for detailed design without more extensive investigation. References such as Nelson and Miller (1992) or Department of the Army (1983a) provide detailed recommendations on more in-depth testing. Nelson and Miller (1992) and Snethen et al. (1975) also provide summaries of several different classification schemes besides the ones shown in Table 8.3.

8.3.3   Remedies

There are basically six methods to control volume change in expansive soils:

1. Remove the expansive material.
2. Surcharge the material so that it cannot swell.
3. Control the moisture.
4. Prewet the soil.
5. Stabilize the soil with chemicals.
6. Control the construction process.

If the expansive soil deposit is thin, excavating and replacing it with nonexpansive material is a simple, effective solution. Often, however the material is too thick to remove economically. Also, if sufficient weight is placed on top of a swelling soil, it can counteract the

TABLE 8.3 Sample Atterberg Limit Identifiers for Expansive Soils

Potential Swell	Holtz and Gibbs (1956)			Ranganatham and Satyanarayana (1965)	Raman (1967)		Chen (1988)	Department of the Army (1983a)	
	Percent smaller than 1 μm	PI*	SL†	SI‡	PI*	SI‡	PI*	LL§	PI*¶
Low	<15	<18	>15	<20	<12	<15	<15	<50	<25
Medium	13–23	15–28	10–16	20–30	12–23	15–30	10–35	50–60	25–35
High	20–31	25–41	7–12	30–60	23–32	30–40	20–55	>60	>35
Very high	>28	>35	<11	>60	>32	>40	35	—	—

*Plasticity index, %
†Shrinkage limit, %
‡Shrinkage index = $PL - SL$, %
§Liquid limit, %
¶Suggests including soil suction in classification: <0.5 ton/ft² for low, 0.5 to 1.5 ton/ft² for medium, >1.5 ton/ft² for high; see Department of the Army (1983a) for more information on use of soil suction.

383

attempted swell, and there will be no change in the surface elevation. This is hard to accomplish reliably and will probably only be effective for soils showing relatively low expansive properties. Nelson and Miller (1992) note that swell pressures up to 600 lb/ft^2 have been controlled with 4 ft of fill and a concrete foundation, but soils can have swell pressures as high as 8000 lb/ft^2. Good high-quality testing is needed to allow reliable estimates of swelling pressures under the anticipated field conditions if this approach is used.

The volume change in an expansive soil occurs only if there is a moisture change. Therefore, if the moisture content is kept constant, there is no volume change. Horizontal and vertical moisture barriers using full-depth asphalt, asphalt membranes, or geomembranes to prevent access of water to the subgrade have proven effective in some installations. The moisture barriers can be effective in regions where the source of the water is only from the surface. In regions with high or fluctuating water tables or with high capillary rise, the source of the water is under the pavement or structure, and surface moisture barriers will not be effective. Low-volume roads on highly expansive black cotton soils have been built successfully in Australia simply by increasing the width of the sealed pavement surface and keeping traffic in the center. Moisture variation is concentrated along the outside edges of the pavement, where noticeable distortions occur. However, in the center zone where the traffic is operating, moisture conditions remain relatively uniform, and volume changes are minimal. The low permeability of the black cotton soil greatly slows water penetration to the center.

Ponding and prewetting have been used successfully to raise a clay's moisture content and achieve swelling before construction. However, the low permeability of expansive clays requires extended periods of soaking to achieve the needed moisture increases. In addition, the expansive clays will have very low strength in this wet state. Teng et al. (1972) used 20-ft-deep vertical sand drains successfully to soak a highly expansive montmorillonitic clay over a period of 140 days. This prewetting reduced the amount of swelling, but it was only effective in a fissured clay. A compacted clay fill area that also was flooded did not achieve the same success due to the lower permeability of the compacted clay compared with the fissured clay. There have been some successes with this approach, but the length of time needed to achieve the desired moisture changes and the low strength of the resulting wet clay are major limitations.

The expansive nature of soils can be neutralized or at least reduced by adding chemical admixtures to the soil. Although a number of chemicals have been tried, lime is the most economical and effective admixture for this purpose. The lime and soil may be mixed in place to a depth of 8 to 12 in, plant mixed, or placed up to 24 in deep by deep plowing. Lime slurry pressure injected into drill holes has given mixed results. This approach is most effective if the soil has cracks and fissures that allow easier penetration of the slurry into the soil. Lime stabilization was discussed more extensively in Chap. 6, and deep stabilization with lime was covered in Sec. 7.5.3.

If an expansive soil is allowed to dry out during construction, it will gain moisture after construction and will swell, causing disruption of the surface. Consequently, the swelling potential of a soil can be reduced if it is placed wet of optimum and maintained at that moisture content. Also, if it is placed at a lower density than normal and with compaction equipment such as a sheepsfoot roller that produces a dispersed particle structure, swelling may be reduced. It is often difficult to maintain this high moisture content in the field, but at least the upper 5 ft of fills made with expansive soils should be placed with tight moisture and density controls. The high moisture content and lower density recommended for expansive soils will result in a relatively low strength that must be addressed in design.

Additional information on expansive soils may be found in the proceedings of specialty conferences on expansive soils sponsored by the American Society of Civil Engineers' Expansive Soils Research Council and in references such as Holtz and Gibbs (1956), Gromko (1974), Snethen et al. (1975), McKeen (1976), Department of the Army (1983a), Chen (1988), and Nelson and Miller (1992).

8.4   Collapsible Soils

Collapsible soils are susceptible to sudden decreases in volume when they become saturated, and this tendency is magnified when saturation is combined with loading. Collapsible behavior typically will be possible when the available soil pore structure will hold a quantity of water greater than the soil's liquid limit. Figure 8.2 estimates when collapsible behavior may be encountered based on this concept.

Collapsible soils usually contain silt and fine sand-sized particles. Loess, which was discussed in Chap. 3, is probably the most widely distributed collapsible soil, but collapsible soils may be found deposited in other environments. Torrential floods may deposit materials quickly, leaving them in an unstable structure. The existing unstable structure may then be cemented by clay binders or evaporite deposits that will dissolve on saturation, allowing a dramatic decrease in volume. Knodel (1981) describes the difficulties of constructing the San Luis Canal in the arid southwestern San Joaquin Valley in California across alluvial fans adjoining the valley. These unconsolidated fan deposits had high porosity, low density, and montmorillonitic clay par-

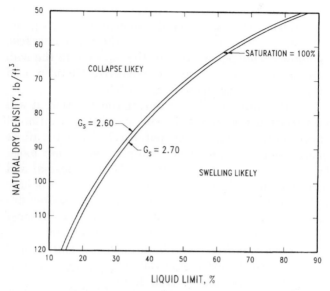

Figure 8.2 Guide to collapsible behavior of soils (*adapted from Gibbs and Bara 1962 and Mitchell and Gardner 1975*).

ticles binding the unstable structure together. Test ponds found settlements in these deposits could exceed 10 ft upon wetting. An extensive program of boring and testing, along with density criteria similar to Fig. 8.2, was used to differentiate between settlement-prone and safer areas. Ponds were built along approximately 20 miles of the canal alignment to collapse the unstable structures prior to the canal construction.

Residual soils also may be left with an unstable structure because of leaching of colloidal and soluble materials. Brink and Kantey (1961) described a collapsible structure that developed in a residual soil weathered from granite. The collapsible structure only developed on well-drained slopes where leaching of soluble and colloidal material was possible. This collapsible structure also decreased with depth.

Compacted soils are not immune to collapse problems. Soils containing between 10 and 40 percent clay seem to be the most susceptible to the problem, and low density, high stress, and compaction moisture contents 2 to 3 percent below Proctor density curve optimum moisture content seem to be major factors in determining if collapse will occur upon wetting (Lawton et al. 1992). Collapse of compacted soils has occurred in dams, embankments, fills, and foundations, and in Southern California such failures have resulted in approximately $100 million in damages (Lawton et al. 1992). Maintaining a sufficiently high moisture content to achieve good compaction in predominately sandy fills can be a particularly difficult task in arid regions due to high evaporation rates. The result of poor compaction on the dry side of optimum can be a soil fabric susceptible to collapse. Ponding or irrigation of borrow areas prior to excavating the material is helpful in overcoming these moisture-control problems in arid regions.

Collapsible soil structures can be found in a variety of depositional environments and may be found in compacted fills. Construction on these deposits with a collapsible soil structure requires that the collapse be induced prior to construction by ponding, infiltration wells, vibration, compaction (conventional, dynamic, or vibratory), or excavation and replacement.

8.5 Soft and Organic Soils

Clays generally cause construction problems because of their low strength, compressibility, low permeability, and stickiness in construction. Some clays rich in montmorillonite are particularly difficult to handle. Even simple construction tasks such as dumping the material out of a dump truck can prove frustrating when the clay's plasticity and moisture condition cause it to hang up in the bed of the truck. Other processed materials such as dredged material or phosphatic clays exist at high moisture contents and may be very plastic and cause serious engineering problems if used in construction. Often, though, if space is limited, the value of developing the land exceeds the cost of dealing with these problems. A common example is found in development of marine terminals that often must be located on reclaimed dredged materials or very soft coastal deposits.

Figure 8.3 shows the relative occurrence of clayey deposits that were considered to provide poor highway subgrade support based on a soil distribution analysis by Witczak (1972). This map was based on a physiographic unit approach (see Chap. 3). There can be significant localized clay deposits such as coastal lagoonal or estuary clay deposits that may be important locally but would not be reflected as areally significant at the larger physiographic unit level. Clays pose problems varying from minor to severe and, if they are present, are usually the controlling factor on most geotechnical design and construction projects.

Organic soils are notable for their high moisture contents, which may exceed 1000 percent for peat. They have very low strength and are best avoided, if possible. This is not always feasible, however. Contrary to popular impression, the occurrence of organic soils decreases toward the tropics because vegetation rapidly deteriorates in the intense tropical climate. Figure 8.4 shows the distribution of organic soils in the

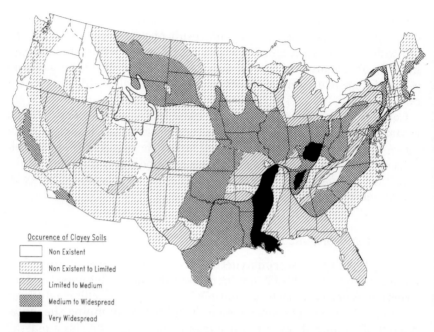

Figure 8.3 Frequency of occurrence of clayey subgrade soils providing poor subgrade support (*Witczak 1972*).

Figure 8.4 Frequency of occurrence of organic deposits (*Witczak 1972*).

United States based on an analysis by Witczak (1972). Organic deposits are present to a significant degree in about 13 percent of the United States, notably in coastal areas, the younger glaciated regions, and the Mississippi River alluvial plain. In Canada, there are over 500,000 mi^2 of organic deposits (MacFarlane 1969), and the Organization of Economic Cooperation and Development (1980) estimates that in some areas of Norway and Sweden between 8 and 11 percent of the roads are founded on marsh, bog, or soft marine clays. More detailed information on organic materials can be found in MacFarlane (1969) and American Society for Testing and Materials (1982).

Removal of the soft material, site improvement (see Chap. 7), building a construction platform, or some combination of these methods often will be needed to allow construction on these difficult soft soils (Rollings et al. 1988). MacFarlane (1969) estimates that at least 3 ft of granular fill is needed over peat to support construction traffic. Use of lightweight fill such as sawdust or foamed materials also may prove helpful in reducing loads on soft materials. "Corduroy roads" made of transversely laid logs are a very old but reliable method of crossing swampy terrain. Today, geotextiles provide a convenient method of bridging over soft soils and are highly effective in providing a reinforcing medium and preventing soft material from extruding into the overlying fill. Stone columns covered with a geotextile and granular fill proved effective for crossing mangrove swamps in the Northern Territory of Australia. Other potential techniques of building on soft organic deposits include complete bridging, excavation and replacement, displacement of the soft material with better materials, and surcharging the site, possibly with vertical drains added to speed the consolidation. However, fibrous organic deposits such as peat can have poor permeability that may make surcharging somewhat slow. Chapter 7 examined some of the problems of construction on poor sites in more detail.

Fill that contains organic materials such as garbage, trash, vegetation, etc., is a particular problem. These organic materials decompose with time and often have been dumped into place with minimal compaction. Unfortunately, these old dump sites are sometimes used for new construction, and there is always the problem of settlement of the old organic fill, and methane production from the decaying organic materials can pose a construction hazard. Old fill sites that contain organic materials are a major construction headache and require careful engineering.

8.6 Sensitive Clays

A clay's loss of strength upon remolding is the *sensitivity* of the clay, and the clay's sensitivity may be classified in accordance with Table

8.4. Most clays exhibit some degree of sensitivity, but the very large loss in strength characteristic of the quick clays in Table 8.4 poses special engineering problems. Also, the liquidity index of sensitive clays often will be well above 1 (see Sec. 2.3.2). This has serious engineering implications. For instance, if a slide occurs in such a material, the soil behaves essentially as discrete blocks of soil floating in a thick, viscous liquid. The slide is in the nature of a flow, is very rapid, and can extend over large areas. Consequently, excavations, loading of crests of slopes, and natural stream erosion along banks all have the potential to trigger catastrophic slides in such materials. Because of the difficulty of anticipating behavior of these materials, Tschebotarioff (1973) recommended adjusting the factor of safety for engineering design as a function of sensitivity, as shown in Table 8.4.

Sensitive clays have a particulate structure that carries load until remolding destroys the structure. The significance of a clay's sensitivity is its very low strength when remolded, and it is not due to any abnormally high strength in the in situ condition. It was a marine quick clay known as Leda clay that caused some of the problems encountered during construction of the St. Lawrence Seaway described in Sec. 3.5.3. These quick clays traditionally pose major problems due to their loss of strength on remolding, propensity for landslides, and difficulty in excavating with normal equipment.

TABLE 8.4   Classification of Clay Sensitivity

Terzaghi (1955)		Tschebotarioff (1973)			
				Factors of safety	
S_t^*	Description	S_t^*	Description	Permanent structures[†]	Temporary structures[‡]
>16	Quick				
8–16	Extra sensitive				
4–8	Sensitive	≥4	High	3.0	2.5
		2–4	Medium	2.7	2.0
1.5–4	Medium sensitive	1–2	Slight	2.5	1.8
		≤1	Not sensitive	2.2	1.6

*S_t = in situ unconfined compressive strength ÷ remolded unconfined compressive strength.
†Permanent structures would be foundations, permanent slopes, etc.
‡Temporary structures would include temporary slopes for an excavation and similar items.

Highly sensitive clays, particularly quick clays, are relatively rare and are associated with postglacial deposits. These are typically glacial lacustrine deposits and marine clays that were subsequentially uplifted (Terzaghi 1955, Mitchell 1993). Most of the quick clays that have been reported have been in Canada and Scandinavia. Six items have been identified that are associated with development of sensitivity in clays (Mitchell 1993):

1. *Fabric.* The clay must have an open, "cardhouse" structure of particles, although this alone is not the cause of sensitivity.

2. *Cementation.* Precipitates of carbonates, iron oxide, alumina, and organic matter may act as cement at particle contacts, and remolding destroys these bonds.

3. *Weathering.* Weathering may change the relative proportions of ions in the soil pore fluid, which may alter the soil's particle flocculation characteristics.

4. *Thixotropic hardening.* This is an isothermal, reversible stiffening that contributes to sensitivity in clays and may account for some of the lower levels of sensitivity observed.

5. *Leaching.* Removal of salt by leaching is a clearly recognized contributor to the sensitivity of uplifted marine clays. This leaching decreases the electrolyte concentration, increases the clay particle's double-layer thickness, and increases the individual particle repulsion forces. The net effect of the leaching is to reduce the strength of the undisturbed soil dramatically and to have an even more dramatic effect when the soil is remolded—possibly resulting in a quick clay. The selective removal of divalent cations (Ca^{2+} and Mg^{2+}) with an increase in the concentration of monovalent cations (Na^+ and K^+) also plays a significant role in development of the quick-clay characteristics.

6. *Formation or addition of dispersing agents.* Pore fluid chemistry determines the tendency of clay particles to disperse or deaggregate from one another.

More than one mechanism probably contributes to the development of sensitivity of clays. The quick clays that show the most dramatic loss of strength upon remolding appear to be soft glacial marine clays that have been uplifted and in which leaching of salts from freshwater occurs (Mitchell 1993). This leaching causes a further increase in particle repulsion due to an increase in monovalent cations and in pH (Mitchell 1993). Fortunately, these deposits are relatively rare. However, as recommended by Legget (1979), prudence dictates that every clay encountered in a physiographic region where geologic processes are known to develop quick clays should be treated as quick until proven otherwise.

8.7 Karst Topography

In regions with highly soluble rock, the surface topography may develop into a sequence of closed basins and dry valleys from cavity development in the underlying soluble rock. Such topography is known as *karst* after the Karst region inland from Trieste in the former Yugoslavia where such formations are developed to a striking degree. Approximately 15 percent of the United States is underlain by soluble rock formations (Beck and Sinclair 1986), and karst topography may be found extensively in Alabama, Georgia, Tennessee, Kentucky, Missouri, Indiana, West Virginia, Virginia, Pennsylvania, and Florida and to a lesser extent in other areas. Generally, such terrain is associated with limestones, but other soluble deposits such as salt or gypsum may develop similarly, and human activities such as pumping or extraction of minerals such as salt may induce solution-related problems. Underground mining also may lead to similar problems.

The development of sinkholes is a major problem in regions with soluble underlying rock. The sinkhole may take the form of a collapse sinkhole, which tends to occur rapidly and dramatically, or it may be a subsidence or raveling sinkhole, which develops inexorably with time. Figure 8.5 illustrates possible sinkhole development in a clean sand.

Sinkholes pose massive potential problems for structures such as buildings, highways, embankments, etc. While sinkhole formation is a natural process, human activities often precipitate the formation of these destructive features. Williams and Vineyard (1976) reported that 46 of 97 catastrophic surface failures in Missouri since the 1930s were triggered by some form of human activity.

Identification of potential sinkhole formation remains a major problem. Conventional boring programs are inadequate because the likelihood of placing a boring directly over a cavity is slim. Also, the loss of ground may be into a number of relatively small solution cavities rather than into a massive cavern. This makes location of cavities by physical sampling highly problematic. Considerable work has been done with geophysical and remote sensing equipment, but the problem of evaluating the likelihood of sinkhole formation remains difficult. An American Society of Civil Engineers publication (Sitar 1988) and an international conference (Beck 1984) provide additional details and case histories of problems in karst topography and exploration techniques, and remedial measures that have been used to deal with sinkholes.

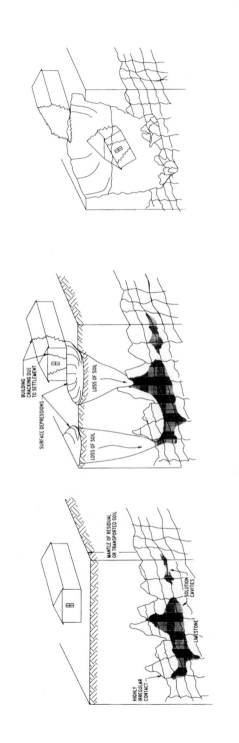

a. INITIAL CONDITION

b. SUBSIDENCE OR RAVELING SINKHOLE

c. COLLAPSE SINKHOLE

Figure 8.5 Development of sinkholes in sandy soils.

8.8 Freezing and Permafrost

8.8.1 Freezing effects

When water in a soil freezes, the ice occupies 9 percent greater volume than the unfrozen water in the soil, and this expansion pushes the soil particles apart. If no other water can enter the soil, the soil surface will rise by an amount of not more than about

$$h \leq 0.09 \, n \, H$$

where h = amount of surface heave
n = average porosity of the soil
H = thickness of the frozen material

If the freezing progresses slowly in a fine-grained soil, lenses of clear ice will form parallel to the freezing front, forming a sequence of frozen soil and ice. Water from the underlying soil will be drawn toward these lenses, gradually adding to their thickness until the available water is exhausted. If there is a source of water such as a water table, these lenses will continue to grow, resulting in substantial surface rise known as a *frost heave*. This is illustrated in Fig. 8.6.

The ability to transmit water is a function of the soil's capillary system. Clays have a strong ability to raise water from the water table by capillary forces, but the small size of capillaries and the resulting low soil permeability only allow upward migration toward the ice lenses to proceed slowly. Consequently, the ice-lens growth with resulting heave will be relatively slow. Coarse sands and gravels, on the other hand, are unable to generate capillary suction because of the large size of the pores, and growth of the ice lens will not be possible. Silts and fine sands can generate strong capillary forces and have appreciably better permeability than clays, so ice-lens formation and frost heaves will be worst among these soils. Frost heaves can result in surface distortion of pavements, displacement of piles and utility poles, heaving and rupture of utility lines, etc. In the spring when these ice lenses melt, the surficial soils will be very wet and soft, and if thawing proceeds slowly, there will be a viscous layer of soil and water trapped on top of still-frozen underlying material. Such conditions play havoc with roads and other pavements.

The potential frost susceptibility of soils is commonly rated using the U.S. Army Corps of Engineers classification shown in Table 8.5 (Department of the Army 1985). There is some overlap between the frost groups in Table 8.5. The system was developed for rating materials for pavements, and the NFS and PFS materials normally would be suitable for pavement base and subbase courses. The S1 gravelly soils would be suitable for subbase courses and have little heave and better strength during thaw periods than similar gravelly soils in the F1

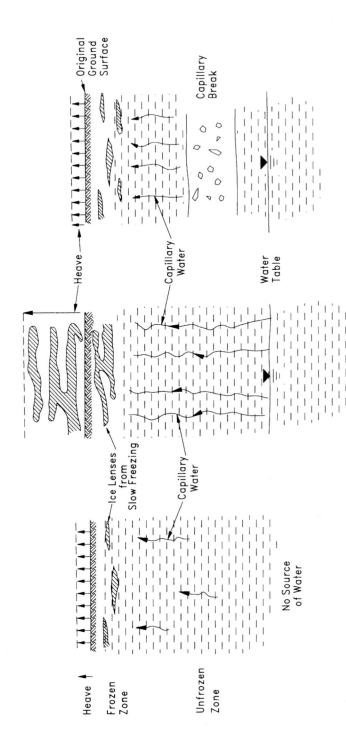

Figure 8.6 Effects of freezing in soils.

TABLE 8.5   Classification of Frost-Susceptible Soils

Frost group	Frost susceptibility[a]	Soil	Percent fines[b]	Typical USCS[c]
NFS[d]	Negligible to low	Gravels:		
		Crushed stone	0–1.5	GW, GP
		Sands	0–3	SW, SP
PFS[e]	Negligible to medium	Gravels:		
		Crushed stone	1.5–3	GW, GP
		Sands	3–10	SW, SP
S1	Very low to medium	Gravelly soils	3–6	GW, GP, G*-GM
S2	Very low to medium	Sandy soils	3–6	SW, SP S*-SM
F1	Very low to high	Gravelly soils	6–10	GM, GW/GP-GM
F2	Low to high	Gravelly soils	10–20	GM, G*-GM
		Sands	6–15	SM, S*-SM
F3	Medium to very high	Gravelly soils	>20	GM, GC
		Sands	>15	SM, SC
		Clays, $PI>12$	—	CL, CH
F4	Medium to very high	All silts	—	ML, MH
		Very fine, silty sands	>15	SM
		Clays, $PI<12$	—	CL, CL-ML
		Varved clays	—	f

[a]As rated in Fig. 8.7. Categories listed in increasing order of frost susceptibility so that likelihood of being in the upper range of given frost susceptibility increases as one proceeds down from NFS to F4.
[b]< 0.02 mm.
[c]Unified Soil Classification, G*-GM means GW-GM or GP-GM, similarly for sands.
[d]Non-frost-susceptible.
[e]Potentially frost-susceptible, requires laboratory determination to be certain of classification. Largest potential problem for heave is with soils approaching maximum density gradation curve (see Sec. 2.5.2) with 1.5 to 3 percent finer than 0.02 mm.
[f]Bands of clays, silts, and/or sands.
SOURCE: Modified from Department of the Army, 1985.

group. Similarly, sandy S2 materials might be used in the subbase and would be superior to F2 sandy materials. F1 soils generally will show superior strength over F2 soils during thaw periods.

Soil behavior under freezing and thawing conditions is a complex phenomenon, and classification systems such as Table 8.5 are guidelines only. Figure 8.7 shows results of laboratory-measured heave of remolded samples of soils, and as reflected in Table 8.5, any of the frost groups can show a range of heaving reactions. Any soil showing

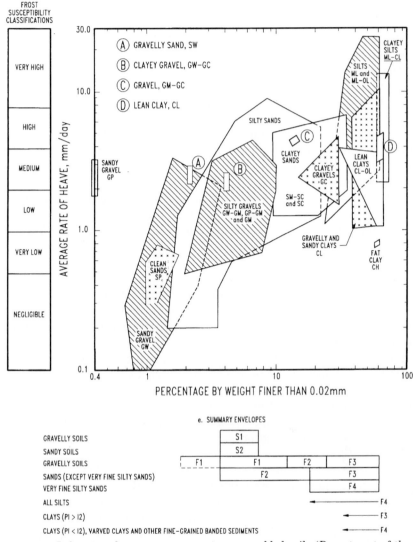

Figure 8.7 Laboratory heave measurements on remolded soils (*Department of the Army 1985*).

a heave rate of 1 mm/day in the laboratory probably will show measurable heave in the field (Department of the Army 1985). Therefore, use of this classification of frost-susceptible soils will help mitigate against problems, but it will not necessarily avoid all heaving.

Generally, any soil with more than 3 percent finer than 0.02 mm is susceptible to some degree of frost problems. To avoid problems with frost heave and springtime thaw weakening, construction within the expected frost zone can use materials that are not frost susceptible or are at least minimally susceptible (i.e., <3 percent finer than 0.02 mm). Alternatively, the source of water might be cut off, as shown in Fig. 8.6, or the design must allow for heave and thaw weakening. For structures such as building foundations and utilities that are very vulnerable to differential elevation change, design practice is to place them below the local zone of freezing.

Under extended intense cold conditions, soils will crack. These cracks may be spaced about 400 to 500 ft apart and play havoc with linear features such as roads. This type of cracking is observed with some frequency in Alaska and Canada and to a lesser extent in the northern United States.

Actual frost penetration into the ground is a complex phenomenon. It is obviously a function of temperature and duration of freezing—frost penetration will be deeper in colder climates and where duration of freezing is longer. Penetration also depends on the specific soil material being frozen and its moisture content. Similarly, insulation serves to decrease frost penetration—a grass-covered field has less frost penetration than does bare ground. Snow acts as an insulator, so ground that is covered with snow and maintained in that condition through the winter will have much less penetration than an adjacent road that is kept clear of snow by plowing all winter. Figure 8.8 shows a typical depth-of-freezing map for the United States, but there can be significant local variations depending on local climates and terrain. Local building codes often will dictate specific depths for foundations and utilities based on local experience to avoid problems with frost heave. Alternatively, there are methods of calculating the depth of frost penetration based on climatologic data (e.g., Department of the Army 1988).

Frost-susceptible materials cannot always be avoided, and often it will not be economical to excavate and replace them with better materials within the frost zone. The adverse effects of frost on the materials can be precluded by preventing the frost from penetrating to the frost-susceptible material. This may be accomplished by placing an adequate thickness of non-frost-susceptible materials to keep the freezing zone within the non-frost-susceptible materials and above the frost-susceptible material. This approach will be most practical in areas with relatively shallow frost penetration or for small areas. The required thickness can be reduced by using insulating materials that will be more efficient (and costly in most cases) than natural non-frost-susceptible materials such as clean gravel.

Nonetheless, sometimes it will be necessary to build on and with frost-susceptible materials, and construction techniques will need to

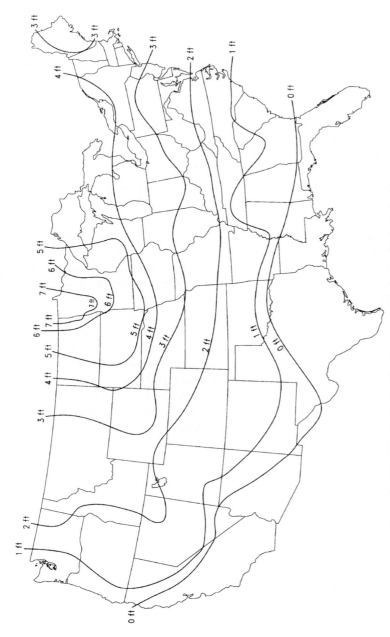

Figure 8.8 Approximate depth of frost penetration based on a survey of selected cities (*after Bowles 1977*). [J. E. Bowles, *Foundation Analysis and Design*, 1977. Reproduced with permission of McGraw-Hill, Inc.]

minimize the adverse effects of frost heave and particularly to minimize differential heave. To minimize differential heave, the frost-susceptible material should be as uniform as possible. Consequently, it would be prudent to process a nonuniform soil by windrowing, blading, rototilling, and similar means until a relatively uniform blend is achieved for materials that will be within at least the upper zones of frost penetration. All large cobbles, boulders, and roots should be removed because repeated frost action will tend to work these to the surface. The transition between cut and fill areas is a potential zone of differential frost heave, so a tapered wedge of fill should be placed into the cut area to transition from one material to the other. Utility cuts, culverts, drains, and similar necessary structures often cause differential heaving problems. If possible, utilities should be placed prior to the overlying fill materials to ensure that the overlying soil conditions are as uniform as possible. Utility trenches cut into existing pavements or filled areas often are backfilled and compacted poorly and may use inferior or different backfill material from the surrounding soils. The result is often severe differential frost heave at the cut. If culverts, utility ducts, drains, etc., cannot be placed below the frost line, they must be bedded on a non-frost-susceptible material that extends below the frost line. Failure to do so may result in progressive heaving of the solid inclusion much like the boulders mentioned earlier. To minimize the differential movement between the culvert or other structure bedded on non-frost-susceptible material and the surrounding frost-susceptible material, the non-frost-susceptible material should be extended beyond the culvert or other buried structure for 5 ft, and then the non-frost-susceptible materials should gradually transition into the frost-susceptible material.

8.8.2 Permafrost

When the mean annual air temperature is approximately $-4°C$ or below, the ground is permanently frozen, and a discontinuous zone of frozen ground extends to a mean annual air temperature of approximately $-1°C$ (Brown 1967). Such frozen ground is commonly referred to as *permafrost,* which describes the state of the ground and is not any specific material. While most permafrost is in the Arctic or Antarctic regions, there are isolated regions of permafrost due to local microclimates (e.g., Mt. Washington in New Hampshire). Approximately half of Canada and Russia and three-fourths of Alaska are within the permanent permafrost region, and major construction projects are undertaken in these regions (e.g., the Alaskan oil pipeline). At the boundaries between the discontinuous and continuous zones, permafrost in Canada is typically 200 to

330 ft thick compared with 820 to 985 ft in Siberia (Brown 1967). Figure 8.9 shows the approximate boundaries between permafrost, discontinuous permafrost, and the seasonal frost line in North America.

The soils of the permafrost region may be a variety of either transported or residual soils since, ironically, significant areas in the far north were never glaciated. Frozen rock and clean granular soils pose minimal problems, but they may contain lenses or masses of ice that leave voids if they melt. Water-bearing soils such as sands, silts, and clays will revert to a viscous liquid upon melting, and these pose a major threat. Much of the permafrost area is covered with a layer of organic material (peat or muskeg) that provides crucial insulation to the frozen underlying soils. If this layer is disturbed, then thawing of the underlying soils with dramatic loss in strength may occur. Consequently, much of the construction in permafrost regions is oriented at maintaining the soil in a frozen state. Buildings, pipelines, and similar structures are supported above the ground surface on piles placed in drilled holes. Horizontal construction such as roads and airfields must protect the insulating organic cover from construction traffic, and typically layers of brush followed by gravel or crushed stone are placed directly over the organic layer. Sheets of insulating material are also frequently used to help prevent unwanted thawing of the frozen subgrade.

Construction in permafrost is a specialized undertaking, and there are a number of useful references to the topic (Terzaghi 1952, MacFarlane 1969, Johnson 1981, Jordan and McDonald 1983, Ryan 1986, Michalowski 1989, Sodhi 1991, Department of the Army 1988, 1987a, 1987b, 1983b). Legget and Karrow (1983) offered some valuable advice on dealing with the world's intensely cold regions:

> The lesson is clear. In cold regions especially, one can take nothing for granted in subsurface geology....In all cold regions the presence of permafrost must be assumed until positive proof is available that the ground is not so perennially frozen. Unfortunately, all too often the approach to northern work has been the reverse of this—permafrost was not anticipated until it made itself inescapably obvious.[1]

---

[1] R. F. Legget and P. F. Karrow, *Handbook of Geology in Civil Engineering*, McGraw-Hill, New York, 1983. Reprinted with permission of McGraw-Hill, Inc.

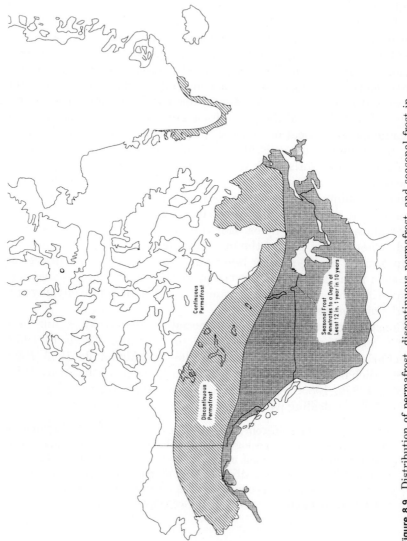

Figure 8.9 Distribution of permafrost, discontinuous permafrost, and seasonal frost in North America (*adapted from Department of the Army 1988*).

8.9 References

American Society for Testing and Materials. 1982. Testing of Peats and Organic Soils, STP820, ASTM, Philadelphia.

Beck, B. F., ed. 1984. *Proceedings of the First Multidisciplinary Conference on Sinkholes*, Rotterdam, The Netherlands.

Beck, B. F., and W. C. Sinclair. 1986. "Sinkholes in Florida: An Introduction," report no. 85-86-4, The Florida Sinkhole Research Institute and the U.S. Geological Survey, Tallahassee, Florida.

Bowles, J. E. 1977. *Foundation Analysis and Design*, McGraw-Hill, New York.

Brink, A. B., and B. A. Kantey. 1961. "Collapsible Grain Structure in Residual Granite Soils in Southern Africa," *Proceedings of the Fifth International Conference on Soil Mechanics and Foundation Engineering*, vol. 1, International Society for Soil Mechanics and Foundation Engineering, Paris, France, pp. 611–614.

Brown, R. J. 1967. "Comparison of Permafrost Conditions in Canada and the USSR," technical paper no. 255, Division of Building Research, Ottawa, Canada.

Chen, F. H. 1988. *Foundations on Expansive Soils*, American Elsevier Scientific Publications, New York.

Cooke, R. U., and A. Warren. 1973. *Geomorphology in Deserts*, Batsford, London.

Department of the Army. 1988. "Arctic and Subarctic Construction Calculation Methods for Determination of Depths of Freeze and Thaw in Soils," TM 5-852-6, Washington.

Department of the Army. 1987a. "Arctic and Subarctic Construction, General Provisions," TM 5-852-1, Washington.

Department of the Army. 1987b. "Arctic and Subarctic Construction, Utilities," TM 5-852-5, Washington.

Department of the Army. 1985. "Pavement Design for Seasonal Frost Conditions," TM 5-818-2, Washington.

Department of the Army. 1983a. "Foundations in Expansive Soils," TM 5-818-7, Washington.

Department of the Army. 1983b. "Arctic and Subarctic Construction, Building Foundations," TM 5-852-4, Washington.

FEMA. 1982. "Special Statistical Summary: Data, Injuries and Property Loss by Type of Disaster 1970–1980," Federal Emergency Management Agency, Washington.

Fookes, P. G. 1980. "An Introduction to the Influence of Natural Aggregates on the Performance and Durability of Concrete," *The Quarterly Journal of Engineering Geology*, 13 (4): 207–229.

Fookes, P. G. 1976. "Road Geotechnics in Hot Deserts," *The Highway Engineer*, 23 (10): 11–23.

Fookes, P. G., and I. E. Higginbottom. 1980a. "Some Problems of Construction Aggregates in Desert Areas, with Particular Reference to the Arabian Peninsula: 1. Occurrence and Special Characteristics," *Proceedings*, part 1, vol. 68, Institution of Civil Engineers, London, pp. 39–68.

Fookes, P. G., and I. E. Higginbottom. 1980b. "Some Problems of Construction Aggregates in Desert Areas, with Particular Reference to the Arabian Peninsula: 2. Investigation, Production, and Quality Control," *Proceedings*, part 1, vol. 68, Institution of Civil Engineers, London, pp. 69–90.

Gibbs, H. J., and J. P. Bara. 1962. Predicting Soil Subsidence from Basic Soil Tests, STP322, American Society for Testing and Materials, Philadelphia.

Gromko, G. J. 1974. "Review of Expansive Soils," *Journal of Geotechnical Engineering*, 100 (GT6): 667–687.

Holtz, W. G., and H. J. Gibbs. 1956. "Engineering Properties of Expansive Clays, " *Transactions, ASCE* 121: 641–677.

Johnson, G. H. 1981. *Permafrost Engineering—Design and Construction*, Wiley, New York.

Jones, D. E., and W. G. Holtz. 1973. "Expansive Soils: The Hidden Disaster," *Civil Engineering*, 43 (8): 49–51.

Jordan, D. F., and G. N. McDonald. 1983. "Cold Regions Construction," a state of the practice report prepared by the Technical Council on Cold Regions, American Society of Civil Engineers, New York.

Knodel, P. C. 1981. "Construction of Large Canal on Collapsing Soils," *Journal of the Geotechnical Engineering Division, ASCE,* 107 (GT1): 79–94.

Lawton, E. C., R. J. Fragaszy, and M. D. Hetherington. 1992. "Review of Wetting-Induced Collapse in Compacted Soil," *Journal of the Geotechnical Engineering Division, ASCE,* 118 (9): 1376–1394.

Legget, R. F. 1979. "Geology and Geotechnical Engineering," *Journal of the Geotechnical Engineering Division, ASCE,* 105 (GT3): 342–391.

Legget, R. F., and P. F. Karrow. 1983. *Handbook of Geology in Civil Engineering,* McGraw-Hill, New York.

MacFarlane, I. C., ed. 1969. *Muskeg Engineering Handbook,* National Research Council of Canada, University of Toronto Press, Canada.

McKeen, G. 1976. "Design and Construction of Airport Pavements on Expansive Soils," FAA-RD-76-66, Federal Aviation Administration, Washington.

Michalowski, R. L. 1989. *Proceedings of the Sixth International Conference Sponsored by the Technical Council on Cold Regions Engineering,* American Society of Civil Engineers, New York.

Mielenz, R. C. 1978. "Petrographic Examination," in *Significance of Tests and Properties of Concrete and Concrete-Making Materials,* STP69B, American Society for Testing and Materials, Philadelphia.

Mitchell, J. K. 1993. *Fundamentals of Soil Behavior,* 2d ed., Wiley, New York.

Mitchell, J. K., and W. S. Gardner, 1975. "In Situ Measurement of Volume Change Characteristics," *Proceedings of the Specialty Conference on in Situ Measurement of Soil Properties,* vol. 2, American Society of Civil Engineers, Raleigh, N.C., pp. 278–345.

Nelson, J. D., and D. J. Miller. 1992. *Expansive Soils: Problems and Practices in Foundation and Pavement Engineering,* Wiley, New York.

Organization for Economic Cooperation and Development. 1980. *Construction of Roads on Compressible Soils,* OECD, Paris, France.

Ranganatham, B. V., and B. Satyanarayana. 1965. "A Rational Method of Predicting Swelling Potential for Compacted Expansive Clays," *Proceedings of the Sixth International Conference on Soil Mechanics and Foundation Engineering,* vol. 1, Montreal, Canada, pp. 92–96.

Raman, V. 1967. "Identification of Expansive Soils from the Plasticity and the Shrinkage Index Data," *The Indian Engineer,* 11 (1): 17–22.

Rollings, R. S., M. E. Poindexter, and K. G. Sharp. 1988. "Heavy Load Pavements on Soft Soil," *Proceedings of the 14th Australian Road Research Board Conference,* part 5, Australian Road Research Board, Canberra, Australia, pp. 219–231.

Rollings, R. S., and M. P. Rollings. 1991. "Pavement Failures: Oversights, Omissions and Wishful Thinking," *Journal of the Performance of Constructed Facilities,* 5 (4): 271–286.

Ryan, W. L., ed. 1986. *Proceedings of the Fourth International Conference Sponsored by the Technical Council on Cold Regions Engineering,* American Society of Civil Engineers, New York.

Sitar, N. 1988. "Geotechnical Aspects of Karst Terrains: Exploration, Foundation Design and Performance, and Remedial Measures," geotechnical special publication no. 14, American Society of Civil Engineers, New York.

Snethen, D. R., F. C. Townsend, D. M. Patrick, and P. J. Vedros. 1975. "A Review of Engineering Experiences with Expansive Soils in Highway Subgrades," FHWA-RD-75-48, Federal Highway Administration, Washington.

Sodhi, D. S., ed. 1991. *Proceedings of the Sixth International Conference Sponsored by the Technical Council on Cold Regions Engineering,* American Society of Civil Engineers, New York.

Teng, T. C., R. M. Mattox, and M. B. Clisby. 1972. "A Study of Active Clays as Related to Highway Design," MSHD-RD-72-045, Mississippi State Highway Department, Jackson, Miss.

Terzaghi, K. 1955. "Influence of Geological Factors in the Engineering Properties of Sediments," in *Economic Geology,* Fiftieth Anniversary Volume, pp. 557–618; also reprinted as Harvard Soil Mechanics Series No. 50, Harvard University, Cambridge, Mass.

Terzaghi, K. 1952. "Permafrost," *Journal of the Boston Society of Civil Engineers,* 39 (1): 1–50.

Tschebotarioff, G. P. 1973. *Foundations, Retaining and Earth Structures,* 2d ed., McGraw-Hill, New York.

Williams, J. H., and J. D. Vineyard. 1976. "Geologic Indicators of Catastrophic Collapse in Karst Terrain in Missouri," Transportation Research Record 612, Transportation Research Board, Washington.

Witczak, M. W. 1972. "Relationships Between Physiographic Units and Highway Design Factors," NCHRP report 132, Highway Research Board, Washington.

"No hammers fell, no ponderous axes rung,
Like some tall palm the mystic fabric sprung,
Majestic silence!"   REGINALD HEBER

9.1   Introduction

Geosynthetics is a term that is applied to a myriad of manufactured products that are currently used extensively in geotechnical engineering. The term is broad-based and includes geotextiles (filter fabrics), geomembranes (liners), geowebs (confinement and strength), geogrids (reinforcement), geonets (drainage), and geocomposites. Several of these are shown in Fig. 9.1. Each type of geosynthetic has a generic function or functions, although the product may be used in numerous applications limited only by the designer's imagination. The principal functions performed by geosynthetics are reinforcement, separation, cushioning, filtration, transmission/drainage, and isolation/barrier. Figure 9.2 illustrates the functions performed by each geosynthetic product.

The applications for geosynthetics have multiplied profusely in recent years and will likely continue to grow. Some of the most common uses of geosynthetics are shown in Table 9.1, along with the basic functions they perform in each application. The following sections describe the currently available types of geosynthetics, their uses, and tips for construction.

Numerous books, articles, and symposium and conference proceedings have been written on geosynthetics and their properties, uses, and performance. Some of the notable contributions include Christopher and Holtz (1985), Department of the Army (1995), Giroud and Peggs (1990), Industrial Fabrics Association International (IFAI

Figure 9.1 Typical variety of geosynthetic products: nonwoven geotextile fabrics, no UV protection on left, UV protected on right.

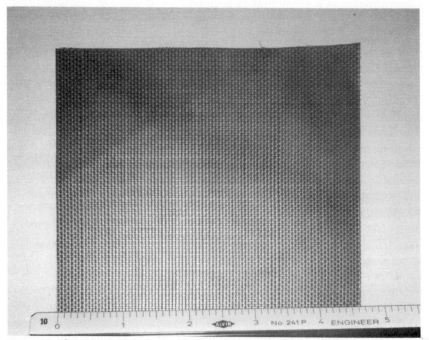

Figure 9.1 (*Continued*) Typical variety of geosynthetic products: woven geotextile fabric.

Figure 9.1 (*Continued*) Typical variety of geosynthetic products: geomembrane, textured for added friction.

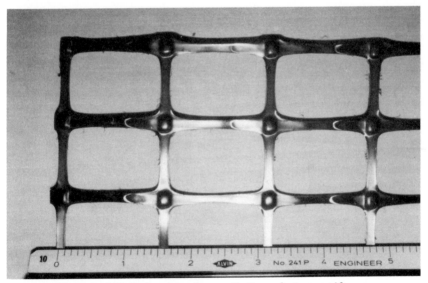

Figure 9.1 (*Continued*) Typical variety of geosynthetic products: geogrid.

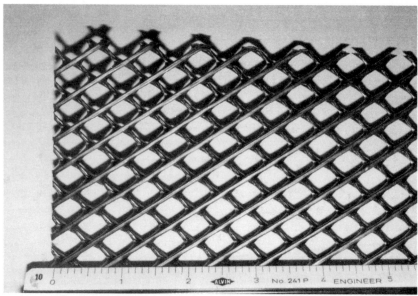

Figure 9.1 (*Continued*) Typical variety of geosynthetic products: geonet.

Figure 9.1 (*Continued*) Typical variety of geosynthetic products: geosynthetic composite structure: textured geomembrane barrier, geonet drainage layer, and geotextile fabric filter.

Geosynthetic Type   Function

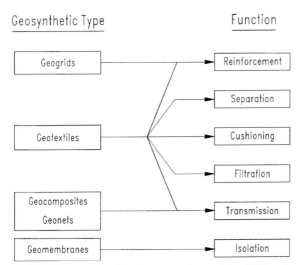

Figure 9.2   Primary functions performed by different geosynthetics (*Fluet, 1988*). ["Geosynthetics for Soil Improvement: A General Report and Keynote Address," J. Fluet, 1988, *Geosynthetics for Soil Improvement*, edited by R. D. Holtz, reproduced by permission of American Society of Civil Engineers.]

TABLE 9.1   Various Geosynthetic Applications and Functions

Application area	Geosynthetics involved*	Functions performed†
Subgrade stabilization	GT, GG	S, R, F
Embankments on soft soils	GT, GG	R, S
Slope reinforcement	GG, GT	R
Retaining walls	GG, GT	R
Drainage	GC, GN, GT	D, F, S, B
Drainage and filtration	GT	F, S
Erosion control—reinforcement	GC	R, S
Erosion control—rip rap	GT	F, S
Erosion control—mats	GT	S, F
Sediment control—silt fence	GT	B, R, S
Asphalt overlay	GT, GC	B, R, S
Geomembrane protection	GT	S, R
Moisture barrier	GM	B

*GC = geocomposite; GG = geogrid; GM = geomembrane; GN = geonet; GT = geotextile.
†B = barrier; D = drainage; F = filtration; R = reinforcement; S = separation.
SOURCE: Modified from Richardson and Koerner 1990.

1990), Ingold and Miller (1988), Koerner and Welsh (1980), Koerner (1984, 1989, 1990), ASTM STP952, ASTM STP1081, *ASCE Geotechnical Special Publication Nos. 12, 18,* and *26, Journal of Geotextiles and Geomembranes,* International Conference on Geosynthetics from 1984 to present, *Geotechnical Fabrics Report* (bimonthly publication), and USEPA 1985-87-89.

9.2 Geotextiles

One of the earliest, most widely used geosynthetics was the *geotextile.* This product is known alternatively as *filter fabric, construction fabric, fabric, synthetic fabric,* and *road-reinforcing fabric.* It is a porous fabric/textile material that is manufactured from synthetic materials, including polyester, polypropylene, polyethylene, polyvinyl chloride, nylon, glass, and combinations of these. Geotextiles are usually categorized as either woven or nonwoven depending on their method of manufacture. Geotextiles are available from many manufacturers with a variety of physical, mechanical, hydraulic, and durability properties. Koerner (1984) gave the following information to illustrate the variation that is available; in the ensuing years this list has only diversified further.[1]

Physical properties

Mass: 3 to 30 oz/yd^2

Thickness: 10 to 300 mil

Specific gravity: 0.9 to 1.4

Percent open area (POA): nil to 36 percent

Equivalent opening size (EOS): U.S. sieve No. 30 to 300

Mechanical properties

Grab strength: 50 to 5000 lb/in

Grab elongation: 20 to 200 percent

Modulus (10 percent secant): 100 to 10,000 lb/in

Mullen burst: 50 to 200 psi

Trapezoidal tear: 20 to 300 lb

---

[1] R. M. Koerner, *Construction and Geotechnical Methods in Foundation Engineering,* McGraw-Hill, New York, 1984. Reproduced with permission of McGraw-Hill, Inc.

Hydraulic properties

Cross-plane permeability: 0.01 to 5 cm/s

In-plane permeability: nil to 2 cm/s

Durability properties

pH resistance: 3 to 11 (excellent)

Biologic resistance: generally excellent

Ultraviolet (UV) stability: poor to reasonable

Moisture absorbance: nil to 3 percent

These properties should be examined and employed to aid in selection of the most appropriate geotextile for use in each and every project. A fact to keep in mind during initial geotextile selection is that woven fabrics generally have high tensile strength, high modulus values, and low elongation, whereas nonwoven geotextiles have high permeability and elongation. Geotextiles are extremely versatile products and can be used in many different ways. They are used most widely for separation, filtration, reinforcing, and erosion control.

9.2.1 Separation

When used for separation, geotextiles typically are placed between two dissimilar materials such as soil and aggregate to prevent the intermixing or intrusion of one material into the other. This use may be necessary to prevent a reduction in permeability and/or shear strength of the granular material such as railroad ballast, granular base course under a pavement, or stone or sand used as a surcharge load on soft soils (see Fig. 7.7). Geotextiles also may be used to separate two frost-susceptible soils, thus providing a capillary break to moisture flow. Koerner (1990) summarizes the separation function succinctly: "...the introduction of a flexible synthetic barrier placed between dissimilar materials so that the integrity and functioning of both materials can remain intact or be improved." A detailed discussion of requirements for designing geotextile applications for separation is given by Koerner (1990).

9.2.2 Cushioning

In Fig. 9.2, Fluet (1988) showed geotextiles being used in a cushioning function. Others consider this to be another type of separation. However considered, geotextiles can be used to cushion or protect adjacent materials such as geomembranes. The protective effect that geotextiles can have on geomembranes has been evaluated by Koerner et al. (1986); one example of the beneficial effect is shown in Fig. 9.3. The protective cushioning provided by geotextiles can be

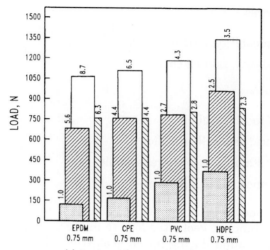

(a) Puncture test results of 600–g/m geotextile
with various geomembranes

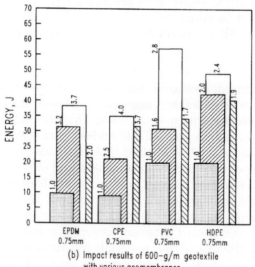

(b) Impact results of 600–g/m geotextile
with various geomembranes

Geotextile both sides

Geomembrane alone

Geotextile front

Geotextile back

Figure 9.3 Test results illustrating the cushioning benefit of
various geotextiles to different geomembranes (*Koerner et al.
1986*).

remarkable, and it varies with fabric thickness, compressibility, and strength.

9.2.3 Filtration and drainage

Because of the controlled permeability and equivalent opening (pore) size of geotextiles, they are used extensively in many drainage situations to perform filtration and/or drainage functions. Filtration is performed by the geotextile when flow occurs across (perpendicular to) the plane of the geotextile; drainage occurs when planar flow occurs within the geotextile.

When geotechnical materials of widely different grain sizes must be used adjacent to each other in drainage situations, a graded filter is normally required to prevent intrusion (or squeezing) of the finer material into the coarser material, migration of fines (carried by seepage water) into the coarser material, and subsequent clogging of the coarser, more permeable material (Cedegren 1989). Geotextiles can be used in many instances to replace a graded filter. This can be helpful and alleviate potential problems in two important ways: (1) construction of graded filters is difficult, time-consuming, and expensive because significant hand work is often required, and (2) the required thickness of graded filters can be quite large because two, three, or four filter layers are sometimes required to transition between the fine and coarse geotechnical materials.

Geotextile filters usually can be placed easily, their thickness is less than 300 mils, and since less material is required, they are often much less expensive than graded filters. Geotextile filters are often used around crushed stone or pipe underdrains, under erosion-control structures, behind retaining walls, and to replace some or all of the graded filters needed in specific earthwork construction projects, sometimes even in earth dams.

Proper placement of geotextile filters is sometimes a problem, particularly when used in an underdrain. An easy rule of thumb passed on to many geotechnical engineers by Joe Fluet states that "you never place a geotextile in a position where it will restrict flow." For example, a geotextile placed between the open-graded crushed rock underdrain and the perforated pipe that transmits the water out of the underdrain impedes water flow (see Fig. 9.4). However, if the geotextile is placed around the outside of the crushed rock, i.e., between the rock and the adjacent soil that is to be drained, the water is constantly flowing into a medium that is more permeable, and thus there is no restriction to flow.

The drainage function requires a geotextile to have a high in-plane ability to transmit water. Nonwoven geotextiles, particularly the

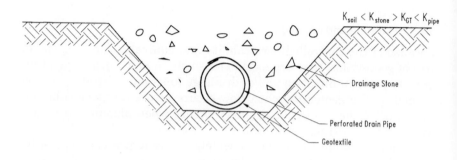

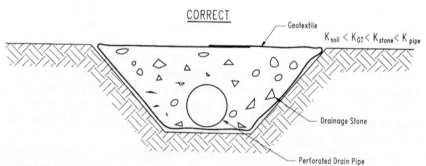

Figure 9.4 Geotextile performs as filter around crushed rock drain.

resin-bonded and needle-punched varieties, have this ability. The needle-punched nonwoven fabrics are the ones most often used in drainage applications. Some typical filter criteria used with geotextiles are shown in Table 9.2.

As with filtration, the important considerations in designing a drainage system that incorporates a geotextile include not only adequate flow capacity but also proper soil retention capability (see Table 9.2) and long-term soil-to-geotextile compatibility (Koerner 1990). Generally, nonwoven geotextiles should have at least five times the permeability of the adjacent soil, and woven geotextiles should have a percentage open area of 10 percent or more for clean soils (<5 percent passing No. 200 sieve) or 4 percent or more for other soils (Department of the Army 1995). Test procedures for determining percentage open area for woven geotextiles are given in Department of the Army (1995).

Geotextiles can be used for drainage under surcharge loads to move water from vertical drains laterally across the site to outlets, for

TABLE 9.2   Typical Filter Criteria Used with Geotextiles

Source	Required AOS*	Soil conditions†
Department of the Army (1995)	$\leq D85$	Fines $\leq 85\%$
	≥ 0.212 mm (No. 70 sieve)	Fines $> 85\%$
Giroud (1982)	$< 2 \cdot C_U \cdot D_{50}$	$1 < C_U < 3; D_r > 0.65$
	$< (18 \cdot D_{50})/C_U$	$C_U > 3; D_r > 0.65$
Christopher and Holtz (1985)	$< D_{85}$	$1 < C_U < 2; D_{50} > 74\ \mu m$
	$< 1.8 \cdot D_{85}$	$1 < C_U < 2; D_{50} > 74\ \mu m$
	< 1 to $2 \cdot D_{85}$	$2 < C_U < 8; D_{50} > 74\ \mu m$
	$< 1.8 \cdot D_{85}$	$2 < C_U < 8; D_{50} < 74\ \mu m$
	$> 3 \cdot D_{15}$	Optional requirement for gap-graded soils

*Apparent opening size in mm (ASTM D 4751). Equivalent opening size or EOS was a synonymous term that was commonly used in the past.
†Effective soil particle size in mm; see Sec. 2.3.1.

chimney drains in dams, as flow interceptors, for drainage behind retaining walls, and for gas or water collection systems in landfill covers (see Secs. 10.7.3 and 10.7.4). Geotextiles can be used effectively to perform a drainage function as long as the planar flow rates are "modest" (Koerner 1990). In situations where a drainage medium is needed to transmit large quantities of liquid and/or where it is critical to move the liquid rapidly (as in landfill leachate collection and leak detection systems), a material with a higher flow capacity than a geotextile should be used; geonets or geocomposites would be likely candidates to use in such situations.

9.2.4   Reinforcement

Since soil has compressive strength but not tensile strength and geotextiles have tensile strength but not compressive strength, the performance of a soil mass can be improved by using a geotextile to impart tensile strength to the mass, i.e., to reinforce the soil. Geotextiles can be used to provide reinforcement in several ways. They can be placed under an embankment constructed on soft soil or on permafrost, used to construct a fabric retaining wall, or used under a new asphalt concrete overlay to minimize reflective cracking. The use of a fabric for soil reinforcement dates back to 1926 when the South Carolina Highway Department used a heavy cotton fabric to reinforce eight unpaved roads in the state; by 1935, sufficient success had been achieved to warrant a paper in *Engineering News Record*

(Beckham and Mills 1935). Current technology, along with the manufacture of geotextiles with specific and known properties, allows present-day engineers to design geotextile reinforcement for particular project needs and imposed loads; this design process may sometimes be very sophisticated, as presented by Bonaparte, Holtz, and Giroud (1987). In many instances, geotextiles used for separation also may perform a reinforcement function.

When placed under an embankment (whether for a roadway, retaining dike, or railroad), the geotextile reduces the stresses transmitted to the underlying layers when it goes into tension upon loading, thus spreading the load over a larger area and reducing its intensity. Geotextiles have been used extensively in construction of embankments over soft soils. One of the softest soils to contend with is hydraulic fill. The U.S. Army Corps of Engineers has developed techniques for constructing embankments for interior separation dikes within its dredged material containment areas (Haliburton et al. 1980, Fowler and Koerner 1987). The technique is also applicable to construction over less soft foundation materials, and it proceeds as follows. The fabric is placed in transverse strips along the proposed alignment with about 20 ft of excess fabric left at each end. The material is overlapped slightly and sewn together. Fill is placed to form anchorage strips along the outer edges of the future embankment as the geotextile is positioned, and the excess geotextile is lapped back over the anchorage strips (Fig. 9.5). Small dikes are then constructed along the outside edges of the embankment to anchor the geotextile. As the dikes settle, the geotextile is stretched across the center section. Fill is then placed in the center section, and finally, the remaining center section is filled to the design elevation.

For about 25 years geotextiles have been used to construct "fabric walls" (Bell et al. 1975, Mohney 1977, Bell and Steward 1977). These walls (shown in Fig. 9.6) are built by placing a horizontal layer of geotextile that is then covered with 9 to 18 in of soil fill. The exposed end of the geotextile is then wrapped up around the edge of the soil layer to form the wall face, and the remainder of the geotextile is laid back on top of the soil layer. Subsequent layers are gently stepped back to form a slanted wall face, being sure to cover at least 12 to 36 in of the geotextile from the lower layer. Because of the potential for vandalism or degradation of the geotextile when exposed to UV light (e.g., sunlight), the completed face of fabric walls is usually coated with a spray such as an asphalt emulsion or gunite. Such walls have been used extensively in construction of low-volume roads, particularly for forest roads in the Pacific Northwest region of the United States. This type of construction is best suited to relatively small walls (usually less than 200 ft long and 20 ft high, more often 20 ft long and 10 ft

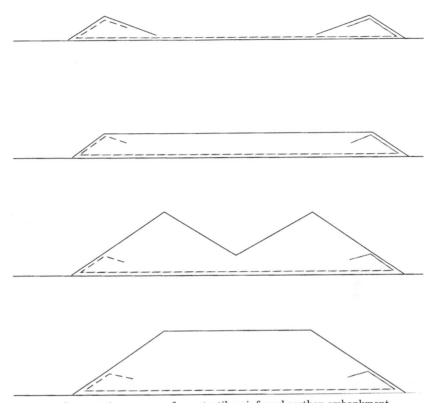

Figure 9.5 Construction sequence for geotextile-reinforced earthen embankment.

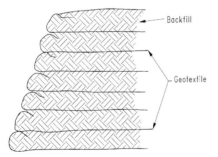

Figure 9.6 Fabric wall constructed using geotextile.

high) on low-volume roads where inexpensive construction is required.

Geotextiles can be used to minimize the occurrence and severity of reflective cracking of asphalt overlays. When an existing pavement (either portland cement concrete or asphalt concrete) requires rehabilitation, an asphalt overlay may be selected as the preferred alternative. After overlaying, the surface of the pavement will be in much better condition until existing cracks in the old pavement are reflected up through the new overlay, and this will inevitably happen. It is generally postulated that a geotextile used between an old pavement and an overlay bridges over the old cracks and acts as a stress-absorbing layer and reinforcement medium (Richardson and Koerner 1990, Koerner 1994), as well as a potential moisture barrier. Since the geotextile is normally placed on the tack coat, it will absorb some of the asphalt; therefore, a sufficient quantity of asphalt cement must be used to permeate the geotextile and to provide adequate bond between the existing pavement, the geotextile, and the overlay. While the geotextile potentially will lengthen the time until reflective cracking occurs and/or will minimize the number and severity of the cracks (Fig. 9.7), it is not a panacea. Also, the effectiveness of a geotextile in this application depends on climate; it is more effective in warm climates than in cold ones. There are a number of geotextiles manufactured specifically for use in pavement rehabilitation. All these geotextiles, as well as all other geosynthetics manufactured, are listed in the *Annual Specifier's Guide* published by the *Geotechnical Fabrics Report* (1993).

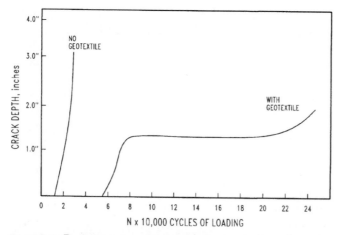

Figure 9.7 Resistance to reflective crack propagation of unreinforced and geotextile-reinforced overlays (*Richardson and Koerner 1990*).

9.2.5 Erosion control

Various types of geotextiles have been used for a quarter of a century for erosion control, and the applications are numerous. They may be used under rip rap or other materials installed for erosion control along channels of flowing water, for shore and beach protection, in marsh establishment in coastal areas, for silt fences, on steep banks subjected to erosive runoff, or formed into bags for use in stabilizing coastal sediments. A wide variety of geotextile types has been manufactured specifically for erosion control—from coarse nonwoven polymeric mesh netlike material $\frac{1}{2}$ to $\frac{3}{4}$ in thick to plastic netting interwoven with brown paper to the more "standard" geotextiles. In many cases, the geotextiles used in erosion control also perform as separators and drains.

When geotextiles are used for erosion control, they are often subjected to some of the harshest conditions possible. They may be subjected to point loads, impacts from construction, sunlight, and/or other potentially damaging factors. Geotextiles used in this environment often must have high permeability, high tensile strength, high elongation, high frictional resistance with the adjacent soil, good resistance to puncture and tear, resistance to degradation from UV light, dimensional stability, and long life (Koerner and Welsh 1980).

To design and construct with geotextiles under erosion-control materials such as rip rap, extreme care must be taken in selection of the geotextile for filtration, drainage, and other necessary properties; placement to ensure intimate contact of the fabric with the underlying soil; placement of the rip rap or other material to prevent damage to the fabric; and protection of the geotextile from UV degradation and vandalism (Richardson and Koerner 1990, Koerner 1984). When used in flowing water, the geotextile should be placed with the longitudinal (machine) dimension in the direction of the flowing water, and any overlaps should be made in the downslope or downstream direction to minimize water damage (Richardson and Koerner 1990). The geotextile must be anchored adequately along its entire underwater length to prevent ripping loose and possible tearing of the material (Koerner 1984). The U.S. Army Corps of Engineers recommends that geotextiles in erosion-control situations be pinned at intervals ranging from 2-ft spacings on 3:1 slopes to 6-ft spacings on slopes flatter than 4:1 (Calhoun 1972). If protective armor is used over the geotextile, it should be placed from the bottom of the slope and from the center of the geotextile-covered region. The armor must be placed very carefully to prevent puncturing or tearing the geotextile; the exact method of placement will depend on the type of armor used (Frobel et al. 1987).

When used alone for erosion control on steep slopes, geotextiles provide "turf reinforcement" (Richardson and Koerner 1990) to control sheet, gully, and rill erosion. In this application, the geotextile is used mainly to help anchor young vegetation by entangling its roots and stems and thus prevent its removal by flowing water. This form of erosion control is often used in ditches, storm drainage channels, on channel and aquifer banks, for shoreline protection, and on cut or fill slopes. Suggested combinations of water velocities and soil particle sizes for which "turf reinforcement" is applicable are shown in Fig. 9.8. Special geotextiles are usually manufactured for this particular application and may take many unusual forms: heavy woven geotextiles, three-dimensional nonwoven webbing, three-dimensional open cells, biodegradable paper woven into a loose-knit fabric, or various other filler materials included in an open synthetic netting. Guidance on use of these specialty products is normally provided by the manufacturers; generic guidance is available from Richardson and Koerner (1990).

More typical geotextiles are often used as a surface covering in intertidal areas to stabilize sediments for marsh creation. The geotextile is normally spread on the bottom of the water body in depths of less than 2 ft, and it is anchored with sand bags and/or pins. Small slits or crosses are then cut in the fabric, and young aquatic plants are planted through the holes. In this case, the geotextile is intended to remain intact only until the vegetation has become well established

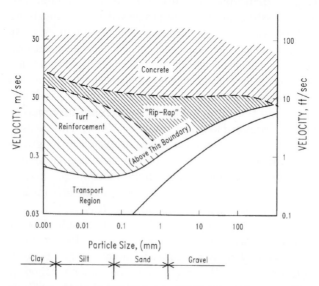

Figure 9.8 Range in which geotextile "turf reinforcement" is effective for soil erosion control (*Richardson and Koerner 1990*).

(a few months to 2 years), so long-term survivability is not a major consideration. Also, in many cases the geotextile will become covered with not only water but also with a thin layer of sediment that will retard UV degradation (Allen et al. 1984).

Geotextiles can be used effectively to contain soil particles eroded by wind or water. Such "silt fences" are often needed on new construction sites, freshly cut slopes, recently graded areas, and other sites where eroded soil must be contained, especially near water bodies and trafficked pavements. The fencing material must be capable of retaining soil particles while allowing free flow of the carrier water or wind. Thus geotextiles are a natural choice for this situation. In many states, even the smallest construction project must submit an erosion-control plan prior to project initiation showing the design and location of silt fences.

A silt fence is formed by placing a geotextile vertically, supported by fence posts at specified spacings and possibly by a metal fence on the downstream side. The bottom edge of a silt fence should be placed in a 6- to 8-in-deep anchor trench. When properly functioning, a silt fence will initially retain silt-and sand-sized particles. These accumulated particles will form a soil filter that will then remove finer particles and will retard flow through the geotextile. For water-borne sediments, a pond of water will accumulate behind the fence. Thus a silt fence must be designed for initial screening of larger particles, water storage, and sediment loading (Richardson and Koerner 1990). The requirements for water and sediment storage will depend on the size of the drainage basin, design rainstorm, and anticipated soil loss during the storm.

The geotextiles used in these applications are usually low- to medium-strength materials with high UV stability. For a specific installation, the geotextile selected must have an appropriate apparent opening size (AOS) for the soil that is to be retained (see Table 9.2), and it must have adequate tensile strength to resist bursting under the imposed sediments and ponded water. Geotextiles commonly used for silt fences are woven polypropylene fabrics with an apparent opening size of 0.2 to 0.7 mm (Richardson and Koerner 1990, Koerner 1990, Koerner and Welsh 1980).

9.2.6   Mattresses, forms, and bags

Geotextiles also can be used to create erosion control mattresses, flexible forms, and bags (Koerner and Welsh 1980, Richardson and Koerner 1990). The (woven) geotextile used in each application must be chosen to meet the requirements of that particular site. The geotextile must have proper strength, filtration, and drainage characteristics to resist rupture from the pressure of grout injection, to contain the fill material, and to be capable of discharging water from within the form or from beneath it.

In the early 1960s, the Dutch were placing geotextiles on slopes, pinning them in place, and pumping concrete into the space between the fabric and the soil. The system has since been refined substantially, and numerous patented processes are available. When used to construct erosion-control mattresses, two sheets of geotextile are commonly employed. The sheets are joined at discrete points on a grid pattern to create flexible forms that, after being positioned in the field, are filled with pumpable grout. The permeable geotextiles allow water to escape from the grout which is under pressure from the pumping process. This lowers the water–cement ratio of the grout with commensurate improvements in stiffness, strength, and durability. The final configuration of the erosion-control mattress (thickness—up to 20 in—and shape) is controlled by the spacing and length of the connecting threads (see Fig. 9.9). The connections serve as filter points or filter bands for escape of pore water (release of excess pore water pressure) from the underlying soil.

Geotextiles also are used to create forms for placing concrete for pile rehabilitation, underpinning bridge piers after undergoing scour, armoring coastal shores with mattresses or bags, and constructing in-ground concrete columns (Richardson and Koerner 1990, Koerner and Welsh 1980, Koerner 1990). The advantage of using these flexible geotextile forms is that little or no grading or shaping of the ground surface is required prior to form and concrete placement. This is particularly advantageous when working under existing structures, under water, or in tight quarters on rehabilitation projects.

Another application for geotextiles involves constructing bags into which sand, grout, concrete, hydraulically dredged material, or other geotechnical materials can be placed (Koerner 1990, Koerner and Welsh 1980). These bags are usually used for erosion control and are constructed of UV-stabilized, high-strength woven geotextiles. The bags may be discrete small bags similar to the traditional "sand bag," or they may be tubes of unlimited length, e.g., Longard tubes. When the bags or tubes are filled with geotechnical materials (sand, dredged material, etc.), they should have long-term UV stability, but when filled with grout or concrete, the bags and tubes are sacrificial and do not require UV stability.

9.2.7 Construction considerations

Successful construction with geotextiles requires a good design, selection of the proper geotextile for the application, careful construction, and vigilant inspection by qualified inspectors and/or engineers. In addition, the ability of the geotextile to survive the construction

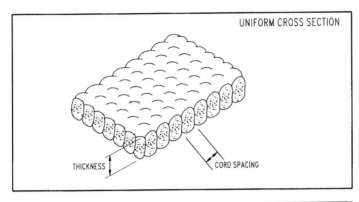

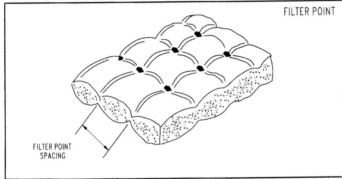

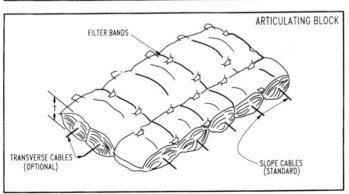

Figure 9.9 Typical cross sections of fabric-formed erosion-control mats (*Richardson and Koerner 1990*).

process must be considered during geotextile selection. The geotextile properties that are important for survivability in drainage, erosion, separation, and reinforcement functions are identified in Table 9.3.

Careful storage and handling are necessary to prevent damage to geotextiles. The fabric is usually delivered to the construction site in rolls with a protective covering of plastic on each roll. The rolls may be stored on the ground near the site of geotextile use if the protective covers are intact. The storage area should be smooth, free of sharp objects, and free-draining. If the geotextile must be stored for an extended period of time, it is best to store it in an enclosure.

It is essential that geotextiles be protected from UV light; this is even more important if the geotextile is to be used for some type of long-term reinforcement. The detrimental results of exposure to sunlight (particularly UV radiation) reported by Raumann (1982) indicate that very significant and rapid degradation of material strength and elongation can occur within a matter of a few weeks. For instance, the strengths of some polyester fabrics were reduced by 80 percent within 28 weeks of exposure to sunlight, while some polypropylene geotextiles lost 100 percent of their strength and elongation within only 8 weeks (Raumann 1982, Koerner 1990). The rate of degradation depends on material type, geographic location, time of year, and cloud cover, among other factors. Many of the current geotextiles (referred to as *UV-stabilized geotextiles*) are manufactured containing carbon black, which provides some protection from UV radiation, but they also will degrade under long exposure. Thus the user must know the level of UV protection and the degradation characteristics of the particular geotextile being considered. As a matter of good practice, any geotextile should be protected from unnecessary exposure by storage in its protective bag and/or in an enclosure. When the geotextile is unrolled and placed, it should be covered immediately with backfill.

Construction with geotextiles requires other specialized considerations. Not only are punctures, water absorption, and UV degradation to be guarded against, but the geotextile must be kept free of wind- or water-borne soil particles that may reduce or eliminate the fabric's filtering/drainage capabilities. If water has entered the geotextile, it should be protected from freezing. The site should be prepared correctly by removal of debris, careful grading, and removal of sharp objects prior to deployment of the geotextile. During installation, proper widths (minimum of 300 mm) and directions of overlaps must be maintained. When the geotextile will be subjected to tensile stresses, sewing of seams should be accomplished using approved methods. Careful placement of the overlying layer is also critical to successful performance of the geotextile, whether this layer is soil, rip rap, or

TABLE 9.3 Geotextile Properties Important for Constructability and Survivability

Physical property	Role: Drainage	Erosion	Separation	Reinforcement	Test procedure
Adsorption	Yes	Yes	Yes	Yes	None
Cutting resistance	Yes	Yes	Yes	Yes	None
Flammability	—	Yes	—	—	None
Flexibility	Yes	Yes	Yes	—	ASTM D 1388
Modulus	—	—	Yes	Yes	Proposed by ASTM Committee D-35
Puncture resistance	Yes	Yes	Yes	Yes	ASTM D 751
Roll dimensions	Yes	Yes	Yes	Yes	N/A
Seam strength	—	Yes	—	Yes	ASTM D 1682, method G
Specific gravity	—	Yes	Yes	—	ASTM D 854
Tear strength	Yes	Yes	Yes	Yes	ASTM D 1117
Tensile strength	Yes	Yes	Yes	Yes	ASTM D 1682, method G
Temperature stability	—	—	—	Yes	None
UV stability	Yes	Yes	Yes	Yes	ASTM D 4355
Weight	Yes	Yes	Yes	Yes	N/A
Wet and dry stability	—	—	—	Yes	None

SOURCE: From Richardson and Wyant, 1987. Copyright ASTM. Reprinted with permission.

some other material. More detained guidance on field installation of geotextiles is available from Koerner (1990), Ingold and Miller (1988), Richardson and Wyant (1987), Richardson and Koerner (1990), and Frobel et al. (1987), as well as the often excellent guidance available from the various geotextile manufacturers.

Problems related to construction with geotextiles are usually associated with these conditions (Richardson and Wyant 1987):

1. Fill placement or compaction techniques damage the geotextile.

2. Installation loads are greater than design loads, leading to failure during construction.

3. Construction environment leads to a significant reduction in assumed fabric properties, causing failure of the completed project.

4. Field seaming or overlap of the geotextile fails to fully develop desired fabric mechanical properties.

5. Instabilities during various construction phases may render a design inadequate even though the final profile would have been stable.

9.3   Geomembranes

Geomembranes, also known as *liners, membranes, plastic sheets, pond liners, impermeable sheets,* etc., are very low permeability synthetic sheets used as liners or barriers to control fluid migration in geotechnical materials. As a primary function, they are used exclusively as liquid and/or vapor barriers. Although not absolutely impermeable, geomembranes are usually considered to be impermeable because they are much less permeable than the geotextiles and soils with which they are normally used. (Geomembranes have a permeability of 10^{-10} to 10^{-13} cm/s as compared with clays, where the lowest normal permeability is about 10^{-8} cm/s.) Geomembranes are used extensively in civil, geotechnical, and environmental engineering. Probably the largest single use of geomembranes at present is for lining and capping systems in landfills (see Chap. 10).

Geomembranes usually are composed of flexible thermoplastic or thermoset polymeric materials (Table 9.4). They are produced in factories as thin sheets that are transported to the job site in rolls. During field installation, the geomembrane sheets are connected together by one of several seaming processes (see Sec. 10.3.5) to produce the required size and shape of barrier.

For any geomembrane to perform acceptably, it is necessary for the material to remain in good condition throughout manufacture, transport, handling, and installation. The properties of the geomembrane that enhance survivability during installation are thickness, tensile

TABLE 9.4   Major Types of Polymers Used for Geomembranes

Polymer* type	Characteristics	Examples
Thermoplastic	Softens when heated and can be re-formed	Polyethylene,† polyvinyl chloride (PVC), chlorinated polyethylene (CPE)
Thermoset	Once cured, cannot be softened without permanently breaking molecular bonds	Polyisobutylene and butyl rubber, polychloropene (neoprene), ethylene vinyl acetate (EVA)
Thermoplastic elastomer	Has elastomeric properties at ambient temperatures but will soften when heated to allow flow and processing	Block copolymers of styrene and butadiene such as SBS rubber, chlorosulfonated polyethylene (CSPE or Hypalon)

*Long-chain molecules composed of repeating units, from the Greek *poly* ("many") and *meros* ("parts").
†Includes very low density (VLDPE), low-density (LDPE), linear low-density (LLDPE), medium-density (MDPE), high-density (HDPE), high-molecular-weight (HMWPE), and ultra-high-molecular-weight (UHMWPE) polyethylene.

strength, tear resistance, puncture resistance, and impact resistance (Koerner 1990).

Geomembranes are used extensively as impermeable liners for containment of liquids in surface impoundments, as covers for reservoirs, and as an integral part of landfill liner and capping systems. Use in landfill lining and capping systems, as well as for lining of surface impoundments, is discussed in Chap. 10. Geomembranes often are used to cover reservoirs when it would be desirable to place a roofed structure over the reservoir but the cost would be prohibitive. Reservoir covers may be needed to reduce evaporation; to prevent migration of vapors, liquids, and/or solids either into or out of the reservoir; to prevent human intrusion into the reservoir; or to reduce the need for site drainage and cleaning. Geomembrane covers may be placed at a fixed elevation, allowed to float on the surface of the liquid, or suspended over the reservoir. For each application, anchorage of the cover edges is critical. Details of design and construction for reservoir covers are given by Koerner (1990).

Geomembranes can be used as liners in canals. Most fluid-carrying canals are used for water, although other fluids sometimes may be transported in canals. The design and construction of geomembrane liners for canals are essentially the same as for surface impoundments with only slight modifications. As with impoundments (discussed in Chap. 10), the type and thickness of geomembrane used

must be selected to be most compatible with the fluid being transported and with the surrounding environment. (Historically, 20-mil-thick
PVC has been the most widely used geosynthetic liner for water-bearing canals.) Design and construction considerations particular to
canal liners include lack of an anchor trench—the geomembrane is
simply "run out" for a distance of about 4 ft past the canal walls and
is covered with soil; the need for a nonerosive cover soil (usually a
well-graded sandy gravel, as shown in Fig. 9.10); use of longer
geomembrane overlaps (10 to 12 in), which should be placed in the
downstream direction; and use of a geotextile between the geomembrane and a concrete cover (if used) to cushion the geomembrane and
to collect any fluid escaping from the concrete liner. Use of a cover
(either soil or concrete) is usually required for erosion resistance, to
hold the geomembrane in place, to provide UV protection, and to protect the geomembrane from water action, plant growth, animals, vandalism, and canal maintenance equipment (Koerner 1990).

Geomembranes also are used to provide secondary containment
around underground storage tanks (USTs). Two methods are available: wrapping the geomembrane around an individual tank or lining
an excavation that will hold one or more tanks. When the geomembrane is wrapped around an individual tank, a geonet (sometimes
referred to as a *stand-off mesh*) normally is placed between the tank
and the membrane. A leak-monitoring and removal pipe would be
placed to intersect the geonet at its lowest point, thus providing a
leak-detection system. If the entire excavation is lined, a geotextile
often is used on each side of the geomembrane to protect and cushion
it. The excavation usually is filled with drainage stone, which acts as
a bedding layer for the tanks and as a leak-detection system when fit-
ted with a monitoring pipe. An advantage of the latter approach is
that the geomembrane can be extended under the piping network
that connects the underground tanks to the gas pumps, thus inter-
cepting any leaks that may occur.

Koerner (1990) reports that geomembranes have been used in earth
dams, along the upstream face of leaking concrete dams, to create
dams by filling tubes with water, to waterproof tunnels, and for seepage control in cutoff trenches. Obviously, geomembranes can be used
to prevent water movement in other novel ways, with the applications
limited only by the designer/constructor's imagination.

9.4 Geogrids

A *geogrid* is a relatively stiff netlike polymeric material composed of
relatively high-strength sets of longitudinal and transverse ribs (connected at their intersections) and having a large, open grid structure

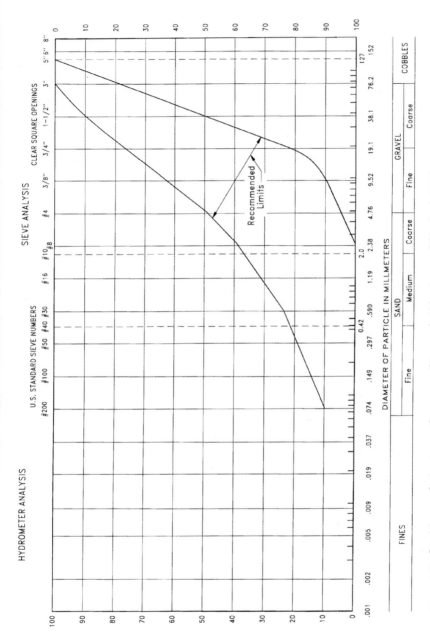

Figure 9.10 Limiting gradation curves for cover soil in canals (*Morrison and Starbuck 1984*).

(see Fig. 9.1). The ribs of a geonet may vary in polymer type, cross-sectional dimensions, and method of manufacture of the grid network. The openings are called *apertures* and are usually 0.5 to 4 in in length and/or width. Geogrids are used primarily for tensile reinforcement of soil or aggregate (Richardson and Koerner 1990).

Geogrids are excellent materials to use for reinforcement because of their relatively high strength, high modulus, and low creep character-istics. They have been used to reinforce soil masses (in roads, behind retaining walls, in earth dams, and on slopes), under unconfined geo-technical materials (surcharge fills, sand, railroad ballast, and unpaved roads), to reinforce disjointed rock and concrete sections, as inserts between other geosynthetics, and to reinforce landfills for ver-tical expansion. Design methods have been developed for using geogrids in these various applications and are summarized by Koerner (1990).

Construction with geogrids is similar to that for other geosynthet-ics. The geogrid is transported to the construction site in 3- to 10-ft-wide rolls. When taken from storage, geogrids should be installed and covered as quickly as possible. Seaming by sewing or bonding is not possible because of the large opening size in geogrids. Therefore, a mechanical system is usually used. Hog rings, pipe, staples, or rein-forcing bars can be used to connect the adjacent geogrids and permit load transfer to occur.

9.5   Geonets

Geonets, the newest of the geosynthetics, are three-dimensional net-like polymeric materials used exclusively for drainage. Almost all are made of medium- to high-density polyethylene; they are almost pure resin, having only 1.5 to 3 percent other ingredients (Koerner 1990). Geonets most often are made by extruding two sets of ribs over one another and joining the crossover points to form a diamond-shaped opening that is approximately 0.5 in long and 0.4 in wide; the ribs are typically 0.2 to 0.3 in thick (see Fig. 9.1). Geonets differ from geogrids "not in the material or its configuration, but in its function" (Koerner 1990).

Geonets are used for drainage in many applications where sand or gravel layers traditionally have been used. These include use in land-fill leachate collection and leak-detection systems, in landfill covers, behind retaining walls, under building foundations and sports fields, and as a drainage blanket under surcharge fills. When used in these applications, geonets are always used with a geotextile, geomem-brane, or other manufactured material on their upper and lower sides; they are not used directly adjacent to soil. Geonets also are used in some geocomposites to provide the required drainage capacity.

The most important characteristic of a geonet is its ability to transmit flow. This is referred to as the geonet's *transmissivity* or *planar flow rate*. The transmissivity is affected not only by geonet thickness (the main physical property used to characterize geonets) but also by cross-plane compressibility of the geonet and intrusion of adjacent materials into the geonet openings. The flow behavior of a geonet can vary significantly depending on the particular field installation conditions. For instance, the flow rate in a geonet will decrease as hydraulic gradient decreases, normal stress increases, and/or intrusion of adjacent material into the geonet increases. The type of material placed adjacent to the geonet significantly affects the amount of intrusion and thus the effective flow rate of the geonet. If a geonet is placed adjacent to a geomembrane, there is no significant intrusion, but if it is placed beside a geotextile, normal forces will cause intrusion of the geotextile into the geonet, thus reducing the planar flow capacity. In many cases when a geotextile must be used against a geonet, two or more layers of geonet may be placed together to ensure adequate long-term flow capacity. Because of the large effect on geonet flow rates, any proposed geonet design should be evaluated by laboratory testing prior to installation in the field. Current test methods are discussed by Koerner (1990). As the tests become more well-defined and routine, ASTM standard test procedures will be developed.

Construction with geonets requires a few special considerations. Geonets, like most other geosynthetics, are shipped in rolls inside a protective cover. When removed from the protective cover, they should be placed into position and covered immediately. This protects the geonet from UV degradation, accidental damage, or fouling by soil, debris, tools, etc. When used on slopes, geonets should be unrolled to run up and down the slopes, not across the slope, since they are strongest in the long (machine) direction, and any seams would not hinder flow. Conventional geosynthetic seaming is not used on geonets, and generally they are not overlapped because the rolls are narrow. Instead, they are often connected with staples, "hog rings," threaded loops, or wires. When a geonet is used adjacent to a geomembrane, metal hog rings should never be used (Koerner 1990).

9.6 Geocomposites

Geocomposites are manufactured products that combine geotextiles, geogrids, geonets, and/or geomembranes in laminated or composite form. Some of the more common geocomposites are geotextile-geonet, geotextile-geomembrane, geomembrane-geogrid, and geotextile–polymer core combinations. Each is created for a particular purpose or to enhance a particular function of geosynthetics—separation, reinforcement, drainage, filtration, or barrier. For instance, the geotextile-geonet has enhanced drainage capabilities, the geomembrane-geogrid composite gives the impermeable geomembrane increased strength and friction properties, and the geotextile-geomembrane combination provides drainage adjacent to the geomembrane while also providing increased tensile strength and increased resistance to puncture, tear propagation, and sliding. Many of these features can be attained to some smaller degree simply by placing the appropriate geosynthetics adjacent to one another in the field. There are two geocomposites, however, that are more unique and cannot be duplicated easily; they are the geotextile–polymer core composite and the geoweb.

The geotextile–polymer core composite is made of a semirigid polymer sheet that is deformed or extruded to have macrotexture that provides flow channels within its structure. This core is covered on one side or is wrapped completely with a geotextile that acts as a filter. These geocomposites come in a variety of types and sizes, but there are two basic systems. The first system incorporates a deformed sheet of polymeric material with a geotextile on one side. This system is used to provide drainage behind retaining walls, adjacent to basement walls, and under structures or athletic fields, and as capillary breaks or drainage interceptors. Some of these geocomposites now include a geomembrane on one side that can be placed adjacent to a structure to act as a vapor barrier. Installation of these products would be similar to that of other sheet geosynthetics.

The second geocomposite drainage system uses a polymer core that is about 4 in wide and is wrapped with a geotextile. This system is often referred to as a *strip, vertical, prefabricated vertical,* or *wick drain.* These geocomposite drains have virtually eliminated the use of sand drains in civil engineering construction.

Strip drains can be installed quickly and easily by machines (similar to sewing machines) using special mandrels. The drains, which are shipped in rolls, are placed on the installation rig (called a *sticker*) in dispensers similar to a spool of thread or a roll of toilet paper. The end of the drain is threaded into a hollow steel lance (which must be as long as the depth to which the drain will be installed) and is attached to a base plate. The base plate is used to keep the strip drain at the bottom of the lance and to prevent soil from entering the lance. The lance containing the drain and base plate is pushed into the ground; when the desired depth is reached, the lance is withdrawn, leaving the base plate and drain in place. The entire process is very fast, averaging about 1 minute per cycle. Drain spacings of 3 to 15 ft are common, and depths up to 150 ft can be achieved (Mitchell 1982). Vertical strip drains often are used in combination with surcharge loading to speed consolidation of soft, compressible materials prior to construction on a site. This technique was used very successfully in hydraulically dredged material at the Seagirt Marine Terminal in Baltimore, Maryland, to shorten the time for primary consolidation from 20 years to 9 months, at which time terminal construction could proceed (Thomas Shafer, STV Lyons, Baltimore, personal communication, 1987). They also have been used extensively during port expansions in the soft, compressible deposits around the Port of New Orleans.

As with other drainage geosynthetics, one of the major considerations in use of geocomposite drains is their ability to transmit flow under field stresses and conditions. Hansbro (1993) and Koerner (1994) provide guidance on the compressibility behavior of geocomposite drains and methods for designing with them. A concern associated with vertical strip drains is the potential for folding (or kinking) of the drain as consolidation of the surrounding soil occurs with a potential reduction in or elimination of flow. Reductions in flow of up to 100 percent have been reported for various kinked drains tested in the laboratory under differing conditions (Lawrence and Koerner 1988).

In many ways, strip drains are superior to sand drains. Strip drains impart tensile strength to the soft soil, do not restrict water flow once it enters the drain, and can be installed easily and quickly using relatively small, low-ground-pressure construction equipment. Conversely, installation of conventional sand drains requires use of larger construction equipment and large quantities (and weights) of sand; is slow, messy, and difficult to properly complete, and does not ensure free flow of water through the drain.

Besides the geotextile–polymer core geocomposite, the other unique geocomposite is the geoweb, which is sometimes called *geogrid* or *sand grid*. It is composed of three-dimensional cells that create a honeycomb structure and can be filled with granular material (Fig. 9.11). These cells physically contain and confine the material to greatly increase its shear strength rather than relying mainly on friction. These geowebs are made of 50 mil HDPE strips that are 8 in wide and are ultrasonically welded together at about 13-in intervals. The geoweb is shipped in a collapsed or flat state (11 ft×5 in×8 in); the sections are placed directly on the subgrade soil at the construction site, where they are opened up or expanded into a honeycomb-like

Figure 9.11   Example of a geoweb or sand grid.

structure (8 ft×20 ft×8 in) and filled with granular material. The granular material (usually sand) is compacted by a vibratory plate compactor, and an asphalt emulsion is often sprayed on the surface. This type of expedient roadway construction has been used very successfully by the U.S. military. In tests, these geoweb systems over soft subgrade soils have supported 10,000 passes of tandem axle trucks weighing 53,000 lb with only minor rutting. Had the geowebs not been used, the trucks would have been able to make no more than 10 passes before becoming immobilized (Webster 1981). Geowebs of this type also have been used to construct earthen embankments and to provide an inexpensive mat foundation.

9.7   References

Allen, H. H., J. W. Webb, and S. O. Shirley. 1984. "Wetlands Development in Moderate Wave-Energy Climates," in *Dredging and Dredged Material Disposal, Proceedings of the Conference on Dredging '84,* American Society of Civil Engineers, New York, pp. 943–955.

Bell, J. R., and J. E. Steward. 1977. "Construction and Observations of Fabric Retained Soil Walls," *C.R. Coll.Int. Sols Text,* vol. 1, pp.123–128.

Bell, J. R., N. Stilley, and B. Vandre. 1975. "Fabric Retained Walls," *Proceedings of the 13th Annual Engineering Geology and Soil Engineering Symposium,* Moscow, Idaho, April 1975, pp. 271–287.

Calhoun, C. C. 1972. "Development of Design Criteria and Acceptance Specifications for Plastic Filter Cloths," report no. AEWES-72-7, U.S. Army Engineer Waterways Experiment Station, Vicksburg, Miss.

Cedegren, H. R. 1989. *Seepage, Drainage, and Flow Nets,* Wiley, New York.

Christopher, B. R., and R. D. Holtz. 1985. *Geotextile Engineering Manual,* FHWA-TS-86/203, Federal Highway Administration, Washington.

Department of the Army. 1995. "Engineering Use of Geotextiles," TM 5-818-8, Washington.

Fluet, J. E., Jr. 1988. "Geosynthetics for Soil Improvement: A General Report and Keynote Address," in *Geosynthetics for Soil Improvement,* R. D. Holtz, ed., Geotechnical Special Publication no. 18, American Society of Civil Engineers, New York.

Fowler, J., and R. M. Koerner. 1987. "Stabilization of Very Soft Soils Using Geosynthetics," *Proceedings of Geosynthetics '87,* Industrial Fabrics Association International, St. Paul, Minn., pp. 289–300.

Frobel, R. K., G. Werner, and M. Wewerka. 1987. "Geotextiles as Filters in Erosion Control," in *Geotextile Testing and the Design Engineer,* ASTM STP 952, J. E. Fluet, Jr., ed., American Society for Testing and Materials, Philadelphia, pp.45–54.

Geotechnical Fabrics Report. 1993. *1993 Specifier's Guide,* December 1993, Industrial Fabrics Association International, St. Paul, Minn.

Geotechnical Fabrics Report, published bimonthly by Industrial Fabrics Association International, St. Paul, Minn.

Giroud, J. P. 1982. "Filter Criteria for Geotextiles," *Proceedings of the Second International Conference on Geotextiles,* Industrial Fabrics Association International, Las Vegas, Nevada, pp. 103–108.

Giroud, J. P., and I. D. Peggs. 1990. "Geomembrane Construction Quality Assurance," in *Waste Containment Systems: Construction, Regulation, and Performance,* R. Bonaparte, ed., ASCE Geotechnical Special Publication no. 26, American Society of Civil Engineers, New York.

Haliburton, T. A., J. Fowler, and J. P. Langan. 1980. "Design and Construction of a Fabric Reinforced Test Section at Pinto Pass, Mobile, Alabama," Transportation Research Record 79, Transportation Research Board, Washington.

Hansbro, S. 1993. "Band Drains," in *Ground Improvement,* M. P. Moseley, ed., Chapman and Hall, London, pp. 40–64.

Ingold, T. S., and K. S. Miller. 1988. *Geotextiles Handbook,* Thomas Telford, London.

Koerner, R. M. 1994. *Designing with Geosynthetics,* 3d ed., Prentice-Hall, Englewood Cliffs, N.J.

Koerner, R. M. 1990. *Designing with Geosynthetics,* 2d ed., Prentice-Hall, Englewood Cliffs, N.J.

Koerner, R. M. 1989. *Durability and Aging of Geosynthetics,* Elsevier Applied Science Publications, London.

Koerner, R. M. 1984. *Construction and Geotechnical Methods in Foundation Engineering,* McGraw-Hill, New York.

Koerner, R. M., M. J. Monteleone, J. R. Schmidt, and A. T. Roethe. 1986. "Puncture and Impact Resistance of Geosynthetics," *Proceedings of the 3rd International Conference on Geotextiles,*Vienna, Austria, IFAI, St. Paul, Minn., pp. 677–681.

Koerner, R. M., and J. P. Welsh. 1980. *Construction and Geotechnical Engineering Using Synthetic Fabrics,* Wiley, New York.

Lawrence, C. A., and R. M. Koerner. 1988. "Flow Behavior of Kinked Strip Drains," in *Geosynthetics for Soil Improvement,* R. D. Holtz, ed., ASCE Geotechnical Special Publication no. 18, American Society of Civil Engineers, New York, pp. 22–39.

Mitchell, J. K. 1982. "Soil Improvement: State-of-the-Art," *Proceedings of the Tenth International Conference on Soil Mechanics and Foundation Engineering,* vol. 4, International Society for Soil Mechanics and Foundation Engineering, Stockholm, Sweden, pp. 509–565.

Mohney, J. 1977. "Fabric Retaining Walls: Olympic National Forest," *Highway Focus,* 9 (1): 88–103.

Morrison, W. R., and Starbuck, J. G. 1984. "Performance of Plastic Canal Linings," REC-ERC-84-1, U.S. Department of the Interior, Bureau of Reclamation, Denver, Colo.

Raumann, G. 1982. "Outdoor Exposure Tests on Geotextiles," *Proceedings of the 2nd International Conference on Geotextiles,* Industrial Fabrics Association International, St. Paul, Minn.

Richardson, G. N. and R. M. Koerner. 1990. *A Design Primer: Geotextiles and Related Materials,* St. Paul, Minn.

Richardson, G. N., and D. C. Wyant. 1987. "Geotextiles Construction Criteria," *Geotextile Testing and the Design Engineer,* J. E. Fluet, Jr., ed., ASTM STP 952, American Society for Testing and Materials, Philadelphia, pp. 125–138.

Webster, S. L. 1981. "Investigation of Beach Sand Trafficability Enhancement Using Sand-Grid Confinement and Membrane Reinforcement Concepts," reports 1 (1979) and 2 (1981), Technical Report TR GL-79-20, USAE Waterways Experiment Station, Vicksburg, Miss.

10

Impoundments, Landfills, and Liners

"One generation passeth away, and another generation cometh; but the earth abideth for ever."

Old Testament: Ecclesiastes 1:4

10.1 Introduction

Containment facilities, in various forms, have been used for many years. During the period 1300 to 700 B.C., the prehistoric people who inhabited what is now northeast Louisiana built their villages on constructed concentric, semicircular mounds, allowing for ponding of village wastes in the depressions between mounds. During settlement of the American West, farmers sometimes impounded water during periods of abundant rainfall to provide a ready reservoir for use during drier times. As industrialization of nations occurred, many containment facilities were constructed to retain various types of raw materials and/or waste products. Most of these containment facilities were not designed and almost none were lined to prevent leakage of wastes into the surrounding environment. With the increasing environmental awareness and emphasis of the past 25 years, the situation has changed drastically. Nowadays not only must new waste containment facilities meet stringent government requirements, often involving elaborate double composite liner systems, but many existing facilities must either be remediated (cleaned up) and closed or retrofitted with pollution-reduction/prevention systems and monitored to ensure that current legal requirements for nonpollution are met.

A number of types of containment facilities are currently used by industry and government to hold both clean and waste materials. These include municipal drinking water reservoirs and sewage lagoons, petroleum storage tank berms, dredged material containment areas, mining waste ponds, and landfills; landfills may contain municipal solid waste, ash monofill, or hazardous waste. These containment facilities generally can be categorized as either surface impoundments (usually containing liquids) or landfills (containing solid materials). They may provide either temporary storage (as in the case of some surface impoundments) or permanent storage (i.e., surface impoundments and landfills).

Although impoundments may be used to retain clean drinking water, impoundments and landfills are used principally to contain wastes, including many hazardous wastes. According to a U.S. Environmental Protection Agency (U.S. EPA) survey of 14 selected manufacturing industries (Quarles 1982), almost 50 percent of hazardous wastes were disposed of in surface impoundments in the 1970s, while 33 percent were placed in landfills (see Table 10.1).

10.2   Types of Containment Facilities

The two types of containment facilities most often used are impoundments and landfills. Impoundments are typically used to contain liquids, and landfills are used for solid wastes and solidified liquid wastes. In the past, both impoundments and landfills most often were constructed below the ground surface or in low topographic areas (Fig. 10.1a, d). (*Landfills* were so named because the process of disposing of waste resulted in the "filling" of topographically low areas.) Excavated containment facilities usually are found in relatively flat areas where the native materials are easily excavated and the groundwater table is at considerable depth. Excavated containment facilities are still used in many cases, but because of concern for the possibility of contaminant transport by groundwater, containment facilities often are constructed above the existing ground elevation to prevent interception of and to distance them from the groundwater table (see Fig. 10.1b, c). Diked or above-ground containments often

TABLE 10.1   Disposal Practices of 14 Manufacturing Industries

Disposal method	Percent of disposed waste
Surface impoundment	48
Landfill/dump	33
Incineration	15
Deep-well injection	2
Land treatment	0.3
Road application	<0.1
Sewer	<0.1
Other (mainly resource recovery at site)	2

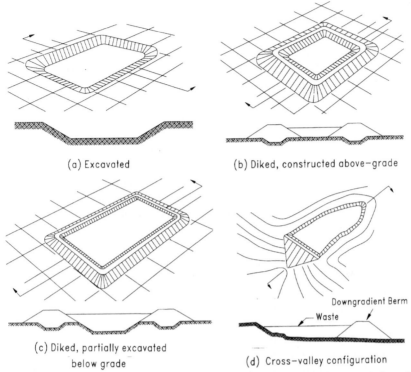

(a) Excavated

(b) Diked, constructed above-grade

(c) Diked, partially excavated
below grade

(d) Cross-valley configuration

Figure 10.1 Types of containment facilities as classified by construction type (*adapted from USEPA 1988b*).

are used at sites with near-surface bedrock, high water tables, and/or soils having a large capacity for capillary movement of soil moisture. For example, the only "mountains" visible in the flat, high-water-table South Florida landscape are m untains of waste. It should be noted that use of cross-valley impoundments (see Fig. 10.1*d*) requires special considerations for control of surface runoff and subsurface moisture movement through the valley.

Both surface impoundments and landfills usually must be lined. Regulatory requirements for liners are often stringent, causing the design and construction of many lining systems to be the single most expensive part of the containment facility. Surface impoundments and landfills are discussed in Secs. 10.2.1 and 10.2.2, respectively; various types of lining systems are discussed in Sec. 10.3.

10.2.1 Surface impoundments

Surface impoundments for containment of liquids are often called *ponds, lagoons,* or *pits*. They may be used to contain clean materials (e.g., drinking water), raw materials, or industrial or municipal wastes. These facilities normally are sited (located) in the general vicinity of the entity they are intended to serve. Municipal drinking water reservoirs and sewage lagoons will be located near the municipalities for which they were constructed. Industrial and mining wastes and dredged sediments usually will be placed in surface impoundments at their site of generation.

Because the original sources from which slurried wastes are generated are very different, the physical and chemical characteristics of impounded slurries will vary considerably from waste to waste (Carrier 1994). However, individual waste streams often have a fairly consistent composition over time (with the composition being dependent on the particular mine or industry), and all slurried wastes will have a high liquid content. For most wastes, the potential contamination hazard associated with the specific waste product usually can be established through a review of the literature and an appropriate laboratory testing program. Dredged materials are somewhat different because they consist principally of soil particles and water, and their chemical makeup can vary significantly between industrialized harbors and pristine streams; the greatest potential for contamination from dredged material is from saltwater contamination of freshwater aquifers (in coastal areas) and/or from contaminant migration if polluted sediments are dredged (Rollings 1994).

Design and operation of surface impoundments require several considerations specific to these facilities. Most of the special requirements result from impounding liquids (instead of solid waste) and are necessary to prevent problems associated with the behavior of liquids. These requirements include design and operation to prevent or control the following:

1. *Overtopping from rainfall, run-on, filling activities, wind-induced interior waves, tidal fluctuations (outside the dikes), and other causes.* Overtopping prevention is necessary for stability of the retaining structures and to prevent contamination of the surrounding environment. Thus it is important to maintain some amount of *freeboard* (the distance from the top of the waste to the top of the dikes). For example, the U.S. Army Corps of Engineers recommends that a minimum of 2 ft of freeboard be maintained in its dredged material containment facilities. (See Hodge et al. 1986 for detailed guidance of hazardous waste impoundments.)

2. *Seepage caused by hydraulic head differential.* Seepage from impoundments into the surrounding soil represents the greatest potential for contamination from surface impoundments and can lead to groundwater contamination. Thus some type of impermeable liner usually is required for these facilities.

3. *Air emissions of volatile materials.* The potential for air emissions is highly dependent on the chemical makeup of the impounded fluid. Either the type or the quantity of emissions may require control. This contaminant migration pathway should be assessed during the design phase.

4. *Wave action inside the containment facility.* Lagoons and ponds may contain bodies of liquids or semisolids with large surface areas. Consequently, the design of such systems should consider the potential for wave action and should allow adequate freeboard to prevent overtopping and provide necessary erosion protection on the inside of the retaining dikes.

Although past use of impoundments for waste containment normally did not involve application of constructed impermeable membranes or liners, surface impoundments constructed under present environmental laws and technical guidelines are almost always lined, especially if there is any possibility of contamination of the surrounding environment. The major exceptions to this are municipal drinking water reservoirs that impound naturally occurring waters and impoundments containing uncontaminated dredged material, which is not categorized as a waste in the United States.

If surface impoundments are to be used to store waste temporarily until it can be treated or recycled, the lining system may be simple, and penetration into the liner itself may be acceptable, e.g., a clay liner may be used. In this case, when the wastes and residues are removed or decontaminated at closure of the facility (the official legal end of active use and the decommissioning of the facility), the liner must be removed or decontaminated also. More likely, a synthetic liner will be employed to provide greater environmental protection and to minimize cleanup activities and costs. At permanent facilities, the waste pond may be the final repository for the material, or it may stay in continuous use, such as in a sewage holding pond. At such permanent disposal facilities, the liner should be of a type to prevent waste from passing into it. Thus a synthetic liner should be used; in fact, a double liner is often needed to provide greater reliability of containment. If hazardous wastes are to be contained, a double liner system is required, and it must have a leak detection system between the two liners, as shown in Fig. 10.2.

10.2.2 Landfills

Landfills, including those designed to contain municipal solid waste (MSW), ash monofill, and hazardous waste, constitute a special class of containment facility that is heavily regulated by federal and state laws. Requirements are spelled out regarding types and

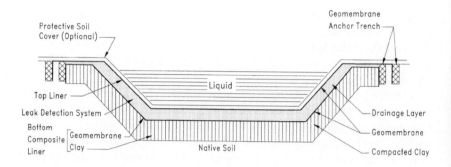

Note: Leachate collection system is not
used in surface impoundments

Figure 10.2 Basic components of a typical surface impoundment for liquid containment.

conditions of acceptable waste materials, methods for waste placement and compaction, lining system design and construction,
leachate collection systems, and monitoring both during and after
active operation. There are also specific requirements for docu-
menting design and construction plans, activities, and emergency
operations with regulatory agencies. A myriad of permits typically
must be obtained prior to initiation of construction of these facilities. The regulatory and permit requirements are steadily evolving
and are quite dynamic.

The basic components of a landfill typically consist of a lined
"basin" that may be either above or below the existing ground surface,
the waste, a gas collection system, and a final cover system. During
filling of the facility, daily and intermediate covers are placed on the
waste to minimize rainfall infiltration, control nuisance vector populations (e.g., rodents and insects), and provide a working platform for
future equipment operations. The basic components of a landfill are
shown in Fig. 10.3.

The types of waste and their characteristics vary significantly,
often within a single landfill. Such materials as municipal garbage,
vegetation, construction debris, tires, and old appliances may be
found in almost any municipal solid waste landfill. Hazardous waste
landfills contain many types of concentrated, toxic, flammable, cor-
rosive, or otherwise hazardous materials; other materials that are
not hazardous but have been contaminated by spills or other exposure are also placed in hazardous waste landfills, e.g., soils contaminated with PCBs and building materials contaminated with
asbestos. Materials placed in these landfills are normally required

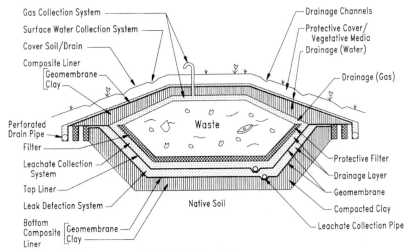

Figure 10.3   Basic components of a typical landfill.

to be in a solid form, thus requiring hazardous liquids to be solidified by some procedure before they are accepted for landfilling.
Because the contents of landfills vary, the leachate generated by
percolation of water through landfills can have extremely different
chemical compositions and may require different lining systems and
handling techniques.

As the consequences of improper waste disposal are better understood, there is a trend toward increasing governmental regulation
and stronger national minimum standards. However, regulations for
landfill lining systems can vary considerably from country to country
and from state to state within the United States. Sixteen of 50 states
have no design standards (down from 32 states in 1988), and the
states with requirements vary considerably (e.g., South Carolina
requires only a 300-mm-thick clay liner, while New York requires double composite liners and leachate collection and leak detection systems). Federal regulations in the United States now require that all
landfills be lined; this has had a major impact on the nation's approximately 6000 landfills, causing closure of many smaller municipal
landfills. The current minimum containment requirements normally
imposed by the United States Environmental Protection Agency (U.S.
EPA) require a double liner system with leachate collection and leak
detection systems for hazardous waste and a single composite liner
with a leachate collection system for nonhazardous waste; these
requirements are shown in Fig. 10.4. The more stringent liner, cover,
and monitoring requirements, coupled with strong public resistance

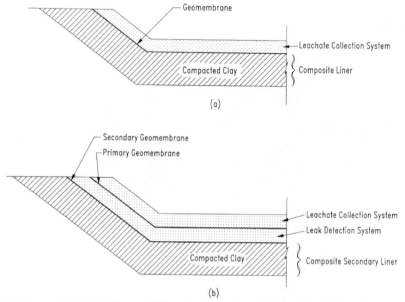

Figure 10.4 U.S. EPA minimum landfill liner requirements for (*a*) nonhazardous wastes and (*b*) hazardous wastes.

to new waste facilities and escalating costs, are leading to new emphasis on recycling, waste minimization, incineration, etc., in the future.

Because of land availability and costs, government regulations, and public opposition, it has become almost impossible in some geographic areas to get permits for new landfills (sometimes termed "green field" landfills because these sites are currently undeveloped). This has led to increased reliance on expansion of existing landfills and/or costly shipping of waste to distant landfills. Landfill expansions may take either of two forms: vertical expansion (building on top of an old landfill) or lateral expansion (building beside an existing landfill). Another result of the current regulatory and public atmosphere is that many of the companies involved in waste disposal commonly exceed minimum lining requirements in an effort to mollify public opposition and to remain in compliance with possible future regulatory requirements.

10.3 Types of Lining Materials

Many types of materials can be used for liners or as components in composite lining systems. These materials have particular properties, installation requirements, and compatibilities that affect the applications in which they may be used most effectively. Some of the more common types of lining materials used in landfills and surface impoundments are discussed below.

10.3.1 Asphaltic materials

In the past, asphalt has been used in several forms for lining impoundments and canals. Asphalt cement, used as a binder in dense graded asphalt concretes and asphalt macadams and sprayed without fillers, has been used as a membrane. The asphalt concrete mix used for hydraulic liners is typically high in asphalt cement and fine filler content to provide low permeability and high erosion resistance; this product is referred to as *hydraulic asphalt concrete.* Asphalt macadams have proven useful in the southwestern United States when some degree of permeability is desired to reduce hydraulic stresses and when high erosion resistance is needed, typically in canals. The asphaltic concretes and macadams are typically 3 to 4 in thick, whereas the sprayed asphalt cement membrane is less than $\frac{1}{2}$ in thick and usually has no aggregate or fillers. Presently, the major use of asphalt is as hydraulic asphalt concrete and sprayed-on flexible membrane liners.

When used in hydraulic asphalt concrete (HAC), asphalt cements are typically harder than paving-grade asphalts, e.g., penetration grades of 40–50 or 60–70 (see Chap. 6). Additionally, a higher percentage of asphalt cement is required, usually 6.5 to 9.5 percent by weight of aggregate. Table 10.2 gives some typical asphalt cement contents and other properties for HAC mixes. A higher content of mineral filler is also used in these mixes. The HAC mix requires use of high-quality dense-graded aggregate, which is necessary to produce a nearly void-less mix to ensure low permeability; the aggregate also must be compatible with the particular waste liquids to be contained. It should be noted that these HAC mixes are designed to minimize permeability; they are not designed for strength, as are paving mixes.

HAC liners have been used extensively in the western United States by the Department of Interior/Bureau of Reclamation (Hickey 1971). The major uses have been for water storage ponds and transport canals and for desalinization ponds. The existing installations have proven to be resistant to light vehicular traffic, wave action, and freeze-thaw cycles. The durability of HAC to wastes is not as well documented. The major concern in waste applications is the compatibility of the waste with both the asphalt and the aggregate. Many organic wastes pose a particular threat to the integrity of the asphalt, and some aggregates may be adversely affected by the chemical characteristics of liquid waste; e.g., carbonaceous aggregates may be dissolved by acidic liquids.

TABLE 10.2 Permeability of Hydraulic Asphalt Concrete to Water

Asphalt, %	Compaction, %	Voids, %	Specific gravity	Unit weight, lb/ft³	Maximum specific gravity	Coefficient of permeability	
						cm/s	ft/yr
7.5	99.2	2.8	2.248	140.3	2.313	8×10^{-7}	0.82
7.5	98.0	3.9	2.223	138.7	2.313	2×10^{-7}	0.18
7.5	93.8	8.0	2.128	132.8	2.313	1×10^{-4}	112
7.5	91.4	10.4	2.072	129.1	2.313	2×10^{-3}	1630
7.75	96.0	6.9	2.147	134.0	2.306	2×10^{-6}	2.1
7.75	99.0	2.9	2.240	139.8	2.306	1×10^{-6}	1.0
8.0	93.2	8.0	2.115	132.0	2.299	1×10^{-4}	136
8.0	93.0	8.4	2.107	131.5	2.299	1×10^{-3}	1340
8.0	98.7	2.6	2.240	139.8	2.299	$<2\times10^{-9}$	<0.002
8.5	90.6	9.5	2.067	129.0	2.285	3×10^{-7}	0.32
8.5	94.4	6.0	2.147	134.0	2.285	5×10^{-8}	0.056
8.5	94.0	6.2	2.144	133.8	2.285	4×10^{-5}	46
8.5	96.0	4.2	2.189	136.6	2.285	1×10^{-5}	14
8.5	96.0	4.2	2.189	136.6	2.285	8×10^{-6}	8.8
8.5	97.0	3.2	2.313	138.0	2.285	$<5\times10^{-9}$	<0.005
8.5	98.0	2.1	2.236	139.5	2.285	$<4\times10^{-9}$	<0.004
8.5	98.0	2.6	2.224	138.8	2.285	$<6\times10^{-10}$	<0.0005
8.75	99.0	2.3	2.226	138.9	2.279	$<2\times10^{-9}$	<0.0016
8.75	99.8	1.7	2.240	139.8	2.279	$<9\times10^{-10}$	<0.0009
8.75	99.5	2.0	2.232	139.3	2.279	$<8\times10^{-10}$	<0.0007
8.75	98.0	3.6	2.197	137.1	2.279	1×10^{-9}	<0.001

SOURCE: Modified from Hinkle, 1976, as cited in USEPA, 1988c.

HAC liners that are constructed in multiple lifts should be laid out such that construction joints in overlying lifts are not directly above the joints in the previously placed lift; i.e., HAC construction joints should be staggered like the vertical mortar joints in a brick wall. This construction procedure minimizes the possibility of leakage through the construction joints to the outlying environment (USEPA 1985).

Various forms of liquid asphalt may be sprayed onto a prepared soil surface to form a flexible membrane liner. As the asphalt solidifies in place, it forms a seamless liner. A protective soil cover is often placed on the asphalt liner for protection from mechanical damage or from wave action. The major drawbacks to sprayed-on asphalt liners are the potential for formation of small bubbles and pinholes during field installation and the incompatibility of asphalt with some wastes, particularly organic wastes.

Both asphalt cement and emulsified asphalt have been and continue to be used to form durable, functional, flexible membrane liners. These materials often are applied in two or more coats with a 1- to 2-ft overlap between sections; typical final liner thickness is about 0.25 in. Today there are a myriad of products that can be added to asphalt to enhance its properties for certain applications. For example, virgin rubber added to asphalt will provide greater resistance to flow, increase elasticity and toughness, decrease brittleness at low temperatures, and provide greater resistance to aging (USEPA 1988c, Asphalt Institute 1976).

When proper conditions of moisture and temperature exist, plant growth can cause severe disruption to asphaltic liners, so it is often prudent to spray the area to be lined with a sterilization product (e.g., sodium chlorate, borates, arsenates, or some urea compounds, but these can pose their own contamination problems). Proper placement of the asphaltic liner will minimize the potential for cracks through which plants may grow.

10.3.2 Portland cement concrete

Properly placed and cured portland cement concrete has a very low permeability on the order of 10^{-10} cm/s. However, as it hydrates, changes in moisture content, or changes in temperature, it changes in volume. To accommodate these volume changes, either joints must be provided periodically, or else the concrete will form its own joints by cracking. These joints or cracks are the major potential source of leakage. With proper design and construction, portland cement concrete can be used successfully to contain fluids. Special waterstops of neoprene, galvanized iron, and zinc are available to seal the joints effectively, and efficient slipform paving equipment is available and adaptable for lining many hydraulic structures. Detailed guidance on portland cement concrete materials and construction is provided by the American Concrete Institute (1995).

Portland cement concrete is a relatively expensive product, which limits its use as a liner in large impoundments. For liner applications, cement-stabilized soils are a less costly alternative to conventional portland cement concrete and are discussed in more detail in Sec. 6.4. The following fundamental requirements for soil cement liners are necessary if one is to obtain the permeability of 10^{-7} cm/s that is commonly required for waste containment: (1) the soil must have sufficiently low porosity to permit construction of a low-permeability liner, (2) the moisture content required to obtain maximum density should be used, (3) the minimum cement content required to meet specifications should be used, (4) the soil cement must be compacted to the design density, and (5) the constituents of the soil cement must be compatible with the waste to be contained.

10.3.3 Compacted clay

Compacted clay is used extensively in waste containment systems as both a primary liner and an underlying component of other primary liners. The small particle size, platy structure, and plasticity characteristics of natural clays often will provide a permeability of 10^{-7} cm/s or less in a properly compacted material. This makes clay a highly effective and economical liner material.

Several properties of compacted clay liners are critical to their successful long-term performance: low permeability, low diffusivity, ductility (resistant to cracking), internal and interface shear strength, chemical compatibility, chemical retardation, minimum of preferential flow paths, and good constructability. These properties are affected by a number of factors that Mitchell et al. (1990) have grouped into four categories: soil composition, placement and construction conditions, postconstruction changes, and chemical compatibility. The first two categories are discussed below; the latter two are mentioned briefly but are beyond the scope of this book.

Most natural soils are a composite of different material types (clays, silts, sands, and gravels) and plasticity characteristics depending on their geologic origins. They also may have internal structures that can greatly affect the permeability of the deposit and may cause it to have permeability values very different from those anticipated from material classification or laboratory testing results. These natural conditions often tend to increase permeability by several orders of magnitude (10, 100, or 1000 times) above what is expected. For instance, varved clays have high horizontal permeability, but low vertical permeability, and loess has high vertical permeability but low horizontal permeability. Residual soils contain layers, discontinuities, etc., of variable strength and permeability reflecting the original rock's internal structure and jointing system. Consequently, it is unwise to rely on native unprocessed soil for a primary barrier in waste disposal facilities, even if it is a clay soil that would be expected to have a low permeability. (Unprocessed or in situ clay soils can, however, be used effectively as a backup for engineered and constructed containment systems.) At two waste disposal sites (one in the northern United States and one in southern Canada), in situ glacial till was used as the primary clay liner. This glacial till was very clayey and was expected to have a low permeability. In actuality, it had extensive vertical fissures that allowed rapid leakage into and contamination of local aquifers.

The natural condition and variability of the soil deposit(s) should be investigated carefully before they are accepted for use as a liner material. Both vertical and horizontal variability must be quantified. Proper assessment of variations in material is critical, since changes in material characteristics may disqualify the material for liner use or at least require completely different construction specifications for successful use. For example, a superficial sampling plan for one construction project collected several surficial weathered samples and one sample from a 3-ft depth (C-horizon unweathered material) in an easily accessible valley. These samples were combined and mixed prior to laboratory testing. The resulting classification tests indicated that the material was a CL clay. All compaction and permeability testing in preparation for construction was performed on the CL soil. When construction began, field checks of density and permeability did not agree with prior testing. Further investigation showed that material for field construction was not obtained from the valley surface of weathered clay but by removing a hill of unweathered ML silt, a material with very different engineering properties. This example illustrates that it is imperative to properly identify and adequately sample the source of material for a clay liner if accurate engineering characteristics are to be obtained for these construction materials.

Proper selection of soils for use in a compacted clay liner is critical. The material chosen ideally should be a clay of medium plasticity that can, without great difficulty, be processed and compacted to meet the specified moisture, density, and permeability values (Mitchell 1990). For compaction and permeability considerations, soils used in liners should contain at least 20 percent fines, no more than 20 percent gravel, and no particles larger that 1 to 2 in (USEPA 1988a). It should be noted that a compacted clay liner used with a geomembrane in a composite liner should *not* contain any gravel or sharp particles that might puncture the geomembrane. Well-graded soils with a plasticity index (*PI*) greater than 10 but less than 30 to 40 will compact easily to high densities and will have "uniform distribution of small pores" (Mitchell 1990), resulting in a low permeability.

The permeability of a compacted clay is affected not only by material type but also by construction procedures. In typical earthwork construction, the entire procedure is monitored and controlled by a required minimum dry density and a range of moisture contents with the purpose of ensuring adequate strength and minimizing compressibility; permeability is not of paramount importance. However, it is generally recognized that (1) as compaction effort increases (more passes of a particular roller or use of a heavier roller), permeability decreases, and (2) compaction at or somewhat above the optimum moisture content, as determined by a Proctor compaction test, also decreases permeability. (Chapter 5 presents detailed discussions of compaction methods, attainable densities, and effects on engineering properties.)

In construction of compacted clay liners, permeability is of utmost importance and normally is used for acceptance testing in addition to the standard moisture and density values. When permeability contours are plotted over the compaction curve, as shown in Fig. 10.5, it

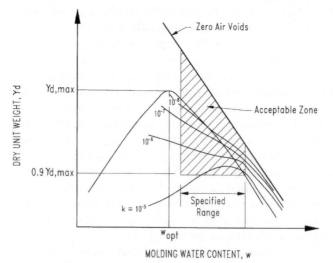

Figure 10.5 Traditional method for specifying acceptable ranges of moisture content and density for compacted clay liners.

becomes obvious that further restrictions should be imposed on compacted clay liner construction if adequately low permeabilities are to be obtained. In the example shown in Fig. 10.5, the lowest permeability values are obtained at a water content about 2 to 3 percent above the optimum water content.

Because of the location of compacted clay liners (under the waste fill and on side slopes), low permeability must not be obtained to the detriment of shear strength and compressibility. Thus a procedure has been developed for clay liner construction that optimizes the desirable properties of a compacted clay liner (Daniel and Benson 1990, Daniel 1993a). As described in the following paragraphs, the approach requires defining the moisture content and density requirements over a broad range of compactive energies and then overlaying these with permeability and other pertinent requirements.

First, one should develop compaction curves covering the range of compactive effort anticipated in the field (see Sec. 5.3.1 for a discussion of compactive efforts and Sec. 4.2.7 for compaction test methods). This usually will require compaction of five to six soil samples (at varying water contents) at each of three compactive efforts, as shown in Fig. 10.6a. Daniel (1993a) recommends using a "reduced" (15-blow Proctor), standard, and modified Proctor. Permeability tests should then be run on the compacted soils (see Fig. 10.6b); care must be taken to ensure that this critical test is run properly and that field conditions are simulated accurately. Guidance on details of permeability testing relevant to compacted clay liners is given by Daniel et al. (1984, 1985) and Carpenter and Stephenson (1986). General information on permeability testing is found in Sec. 4.2.10. The compaction curves should be replotted differentiating between the samples with acceptable permeability values (≤ required permeability) and samples that were too permeable. In Fig. 10.6c, solid symbols were used for samples with acceptable permeability values and open symbols for those which were too permeable. An "acceptable zone" to provide sufficiently low permeability is then sketched onto the figure to encompass the acceptable points. This zone is then modified as necessary to account for other needed properties, e.g., shear strength, limited desiccation, local construction practices, etc. Figure 10.7 shows how a final acceptable zone may be identified by overlaying different criteria. This acceptable zone is obviously much more restrictive and will result in placement of a higher-quality liner than the traditional acceptable zone shown in Fig. 10.5.

Long after construction is completed, changes may occur in the conditions to which a compacted clay liner is subjected. These changes may affect the clay liner and its performance adversely and therefore

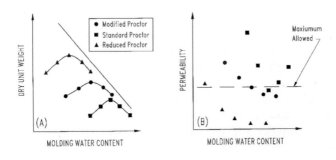

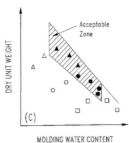

Figure 10.6 Determination of acceptable zone for clay liner moisture content and dry density. *(Source: "Water Content-Density Criteria for Compacted Clay Liners," D.E. Daniel and C.H. Benson, 1990,* Journal of Geotechnical Engineering, *reproduced by permission of American Society of Civil Engineers.)*

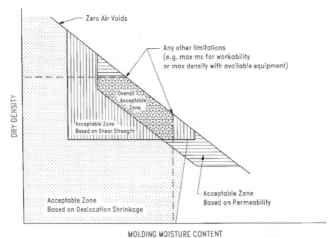

Figure 10.7 Final acceptable zone meeting multiple criteria *(based on original concepts of Daniel and Benson 1990 and Daniel 1993).*

should be considered during facility design. Such factors include changes in confining stress (both increases and decreases), particle migration possibly leading to piping, differential settlement with subsequent cracking of the clay layer, repeated freeze-thaw cycles leading to an increase of one to two orders of magnitude in permeability, biologic clogging in drainage layers with possible adverse effects on the clay liner, slope instability or downslope movement of the liner system, and thixotropic changes in the clay layer causing an increase in permeability.

Chemical compatibility of a clay liner with the fluids it is intended to contain is a critical consideration. The permeability of a soil may be changed by the type and chemistry of the pore fluid, or leachate, passing through the soil. For instance, if a laboratory permeability test on a clay is run using an organic fluid such as a petroleum product or one of many industrial solvents, the measured permeability often will be higher than if the test were run with water (Table 10.3). Thus the permeability of a clay liner may be higher for some waste products than it would be for water. Consequently, the characteristics of the fluids to be stored in a waste containment facility and their effect on the clay must be evaluated. The types of waste liquids are often grouped for analysis as acids and bases, neutral inorganic liquids, neutral organic liquids, and leachate. A summary of concerns about and methods for testing chemical compatibility are given by Mitchell and Madsen (1987), Bowders et al. (1986), and in Chap. 4 of U.S. EPA (1988a).

The practicality of construction also needs to be considered during design of a compacted clay liner. One very important aspect of construction is that there should be adequate width of features to allow construction equipment to operate efficiently. For example, if the trench for a drainage pipe is designed to the width of a piece of construction equipment (e.g., rollers in the United States commonly have drums 7 ft wide), this allows a grader to angle its blade to cut the trench efficiently. It then allows a roller to get in and compact the

TABLE 10.3 Differences in Permeability Values for Montmorillonite Permeated with Water and Naphtha

Clay	Water		Naphtha	
	Void ratio	Permeability, cm/s	Void ratio	Permeability, cm/s
Ca-montmorillonite	1.72	1.6×10^{-9}	1.52	6.4×10^{-5}
Na-montmorillonite	3.75	5.2×10^{-11}	1.31	3.8×10^{-5}

SOURCE: From Buchanan, 1964.

clay soil adequately to achieve the desired density and permeability. Thus installation is facilitated, and the final compacted fill around the pipe is of higher quality than if it were placed and compacted by hand.

Construction problems may occur when a clay liner is to be placed on the side slope of an embankment (Fig. 10.8). Often these liners will be specified to be 1 to 3 ft thick. It is problematic whether construction equipment can achieve the levels of compaction needed to achieve the low permeability required in clay liners by rolling up and down the slope. A sounder approach would be for the liner to be built in horizontal lifts of sufficient width to allow equipment to place them effectively. The side slopes can then either be left at this construction thickness or trimmed back to the design thickness. If the latter approach is used, it is critical that proper scarification and bonding be achieved between layers; otherwise, a thin zone of higher permeabili-

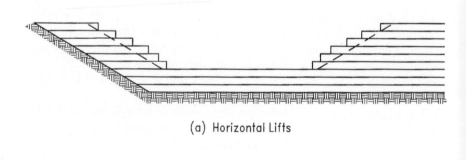

(a) Horizontal Lifts

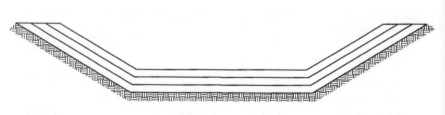

(b) Continuous Lifts

Figure 10.8 Lift placement for clay liners.

ty is constructed into the liner that will allow an easy, rapid path for escape of leachate. When horizontal lifts are constructed, they are sometimes gently sloped down into the containment facility in an attempt to minimize any escape of leachate.

When properly constructed, compacted clay liners will function as economical and reliable barriers to leachate migration. When problems occur with compacted clay liners, they are generally caused by inadequate field investigations prior to construction or to improper or inadequate construction and/or construction quality assurance. The need and procedures for good construction quality control and quality assurance for compacted clay liners are discussed by Daniel (1990).

10.3.4 Modified soils

When neither the in situ soils nor soils from local borrow areas are suitable for construction of a low-permeability liner, commercially produced bentonites or other clay minerals may be used as a soil additive to lower the permeability of the local soils and provide an economical liner material. Because of the expansive nature of bentonite (resulting mainly from its sodium-montmorillonite component), relatively small amounts can be added to a noncohesive soil to cause the composite to behave as a cohesive soil.

The use of a modified soil for liner construction requires careful consideration of the form of bentonite to use (granular or powdered), the mineralogy of the bentonite (percentages of sodium- and/or calcium-montmorillonite), and the rate of application of bentonite. Experience has indicated that powdered bentonite mixes more intimately and more uniformly with the native soils than does the granular form (USEPA 1938a). The mineralogy of bentonite affects its swell potential; as the swell potential increases, the amount of product required to produce the necessary reduction in liner permeability decreases. Since sodium-montmorillonite has a higher swell potential than does calcium-montmorillonite, use of bentonite with a higher percentage of sodium-montmorillonite usually will minimize the volume of bentonite needed. The bentonite application rate must be determined for each specific project and will depend on the existing soil conditions. For many granular materials, the addition of 3 to 8 percent bentonite will lower the permeability to a value below 10^{-7} cm/s (Kozicki and Heenan 1983).

The optimum application rate for bentonite and the optimum water content to achieve minimum permeability must be determined for each specific soil-bentonite admixture. This is done by conducting laboratory permeability tests on a series of specimens compacted at various moisture contents and containing different percentages of bentonite.

When bentonite or any other additive is to be used with a soil, it is imperative that the additive and the soil be thoroughly and homogeneously mixed. For field applications, there are two approaches that can be used to achieve the desired end. The most reliable method of mixing is in a pugmill, where the soil and bentonite are introduced in predetermined proportions; water can be added simultaneously with the bentonite or in a separate processing step. Alternatively, bentonite can be spread over a loose lift of soil, and a pulvamixer can be used to mix the materials. Water should be added in the appropriate proportion during this mixing process. This method can provide an acceptable final mix when done carefully, but it is not as controlled as the pugmill operation.

Obviously, the successful use of a bentonite-modified soil liner depends not only on the type, form, and application rate of bentonite but also on the characteristics of the soil to be modified, the quality of mixing, the adequacy of construction (lift thickness, moisture content, and the size, type, and operation of the roller), and the existence of a good quality control operation. Additional information on bentonite-modified soil is available in Alther (1983, 1987) and Chapuis (1990).

10.3.5 Geomembranes

Geomembranes are very low permeability synthetic materials used in geotechnical engineering construction to control liquid and vapor movement within earth materials. They provide an effective barrier to fluid migration. The materials most often used in geomembrane manufacture are classified as thermoplastic polymers. (This means that the polymer can be repeatedly heated to its softening point, shaped as required, and cooled to retain its new shape.) The types of thermoplastic geomembranes commonly used in the United States are

- High density polyethylene (HDPE)—smooth or textured
- Very low density polyethylene (VLDPE)—smooth or textured
- Chlorinated polyethylene (CPE)—nonreinforced or reinforced
- Chlorosulfonated polyethylene (CSPE)—usually reinforced
- Ethylene interpolymer alloy (EIA)—always reinforced
- Polyvinyl chloride (PVC)

The particular resin used in manufacture provides the name for each geomembrane. Geomembranes are also referred to as *membranes, flexible membrane liners* (FMLs), and *impermeable geosynthetic sheets.*

Geomembranes are manufactured in a variety of standard thicknesses, ranging from 1 to 3 mm (30 to 120 mils). In the United States, the thickness is usually reported in mils (one mil is one one-thousandth of an inch). Generally, at least an 80-mil geomembrane is needed for liners in landfills or similar structures where containment is important. Thinner geomembranes are easier to handle during installation because of their more flexible nature and lighter weight. However, significant problems (e.g., burn-through) are normally encountered in seaming the thinner ones; most of these problems are almost eliminated with thicker geomembranes. Currently, the 80-mil geomembrane is widely favored because of regulatory requirements, practicality of achieving a good-quality geomembrane with minimum defects, and/or concern for environmental protection.

Geomembrane selection for a particular installation should be based on comparison of the physical, mechanical, chemical, thermal, and biologic properties required for the project and those provided by the various products. The physical and mechanical properties are normally specified by the project designer and are available from the manufacturer, although they should be spot-checked as the geomembrane is received on the project site. Laboratory testing is necessary to determine the chemical compatibility of the geomembrane with the site-specific leachate; this is by far the major area of investigation required for geomembrane selection. (U.S. EPA method 9090 immerses samples of geomembrane in the leachate at temperatures of 20 and 55°C for 120 days.) Biologic and thermal properties come into consideration only on projects with unusual circumstances. Guidance on the various test methods is available from Koerner (1993b). Specifics of required testing should be checked with the U.S. EPA regional office and the appropriate state EPA office for each waste containment project.

Because geomembranes are manufactured in finite widths (not exceeding about 10 m), field seaming, or welding, is necessary to obtain a continuous geomembrane of sufficient width to cover an entire waste-containment site. It is imperative that these seams be fabricated properly and uniformly to maintain the low permeability afforded by the geomembrane. There are several methods of seaming (Fig. 10.9) that may be grouped into four categories: extrusion welding, thermal fusion, solvent-based systems, and adhesives. These categories are described in Table 10.4. Corrective patches over holes, defects, and destructive test sampling sites use these same techniques.

Construction with geosynthetics, and particularly geomembranes, is different from other geotechnical construction in several significant ways (see also the discussion in Sec. 10.3.7). Since the geomembrane is impermeable to liquids, it is also impermeable to wind. Therefore, any deployed but uncovered geomembrane must be weighted down to prevent blowing in the wind. Sandbags typically are used. Because a huge weight would be required to resist the force of high-velocity winds, the surface area of exposed geomembrane should be kept to a

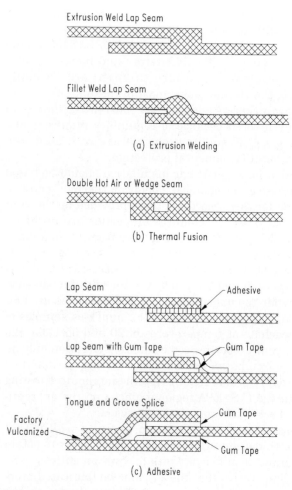

Extrusion Weld Lap Seam

Fillet Weld Lap Seam

(a) Extrusion Welding

Double Hot Air or Wedge Seam

(b) Thermal Fusion

Lap Seam — Adhesive

Lap Seam with Gum Tape — Gum Tape

Tongue and Groove Splice — Gum Tape

Factory Vulcanized

— Gum Tape

(c) Adhesive

Figure 10.9 Field seam types for geomembranes.

minimum. Because of the kind of material and its relative thinness, geomembranes must be protected from puncture after deployment and prior to placement of a protective soil cover. This protection entails hand removal of all rocks from the surface of the compacted soil that the geomembrane will contact, prohibiting use of construction equipment on the geomembrane, and wearing of flat-soled shoes without heels (i.e, athletic shoes, not work boots) by personnel who must walk on the geomembrane for testing and evaluation. To get a good seam, any surfaces to be welded must be cleaned first, since the presence of dust will prevent proper seaming.

Upon installation of the geomembrane, both laboratory and field testing of the seams themselves must be performed. Samples are cut

TABLE 10.4   Field Seaming Methods for Different Types of Geomembranes

Seaming method	Description	Suitable materials*
Extrusion welding	A ribbon of molten polymer is extruded over the edge (*filet weld*) or between geomembrane sheets (*flat weld*). This melts the adjacent surfaces, which are then fused together upon cooling.†	All polyethylenes (HDPE or VLDPE)
Thermal fusion	Adjacent surfaces are melted and then pressed together. Commercial equipment is available that uses a heated wedge (most common) or hot air to melt the materials. Also, ultrasonic energy can be used for melting rather than heat.	Thermoplastics (HDPE, CPE, PVC)
Solvent-based systems	A solvent is used with pressure to join adjacent surfaces. Heating may be used to accelerate the curing. The solvent may contain some of the geomembrane polymer already dissolved in the solvent liquid (*bodied solvent*) or an adhesive to improve the seam quality.‡	Must be compatible with solvent, primarily PVC, also CPE and CSPE
Contact adhesive	Solution is brushed onto surfaces to be joined, and pressure is applied to ensure good contact. Upon curing, the adhesive bonds the surfaces together.‡	Thermosets primarily, may be used with others

*See Table 9.4 for a description of the different types of polymer materials used in geomembranes.
†Clean surfaces, proper techniques, calibrated equipment, and trained operators are critical for good results.
‡Clean surfaces, correct application rates, appropriate delays before joining surfaces, timing and amount of pressure to be applied, and proper curing all strongly affect results.

from the seams at periodic intervals for destructive testing. The type and frequency of sampling is set by regulatory agencies, the design engineer, the owner, and/or the installer. Nondestructive field testing of seams is conducted between the destructive seam sampling locations. The methods of field testing are summarized in Table 10.5; further discussion of geomembrane seaming methods is given by

TABLE 10.5   Examples of Nondestructive Seam Tests

Method	General principles	Equipment cost	Speed	Reliability
Visual	Visual observation of defects	Low	Fast	Poor
Manual	Insertion of screwdriver or similar tool along edge of seam to try to locate unbonded areas	Low	Fast	Poor
Air lance	High-pressure air blast applied to edge of seam in an attempt to lift unbonded areas	Low	Fast	Poor
Pressurized double seam	A positive pressure is applied to the open chamber between double welds in the seam (see Fig. 10.9). Loss of pressure indicates a leak, but finding the location of the leak is difficult.	Low	Fast	Good
Vacuum chamber	A soap solution is applied to the seam, a vacuum is applied in a portable chamber covering the seam, and leaks are detected by bubbling of the soap solution. Widely used.	Low	Slow	Good
Electrical methods	Several variants exist. Most common approach leaves a thin metal conductor in the seam. A high voltage is applied to the conductor, and a probe moved along the seam can detect defects.	Low	Fast	Variable results reported
Ultrasonic methods	Several variants exist. A high-frequency signal is used with reflectance, impedance, or transmission principles to determine quality of seam.	Moderate	Moderate	Variable results reported, can be good
Electric current	A current is introduced into a liquid that covers the membrane. Surface potential is then measured on a predetermined grid to identify anomalies that indicate possible leaks.	High	Slow	Good

462

Koerner (1993b). A good installer construction quality control (CQC) program *and* an independent third party construction quality assurance (CQA) program are mandatory if seam and geomembrane defects are to be minimized and leakage controlled. Discussion of the importance and critical elements of a thorough CQA program is presented by Giroud and Peggs (1990).

Dr. Giroud summarized the importance of good design and quality assurance practice with geomembrane liners by observing that

> ...geomembranes are not always as successful as they should be. The main reason is that many people, when they use geomembranes, tend to neglect two important aspects, design and quality assurance, because they believe that geomembranes are absolutely impermeable in all circumstances. The fact that there is no such thing as an absolutely impermeable geomembrane cannot be emphasized enough....Of course, impermeability is a legitimate goal, especially for conservation of resources and protection of the environment, which are two of the most important challenges of our times, but this goal cannot be achieved by just putting a geomembrane on the ground—proper design and [construction] quality assurance are required [Giroud, 1984].

It has been over a decade since Dr. Giroud made these observations, and today's design and quality assurance practices routinely produce good geomembrane liners. However, this success has developed only as our design and quality assurance has improved over the years. Continuation of this success will require that good quality design and quality assurance practices remain the industry standard and not the exception.

10.3.6 Geosynthetic clay liners

In recent years, several composite products referred to as *geosynthetic clay liners* (GCLs) have been produced for use in place of compacted clay layers in waste containment liner and cover systems. The composite geosynthetic clay liners that are currently produced consist of an approximately 10-mm-thick (0.4-in) structure of (1) dry bentonite sandwiched between two nonwoven geotextiles that are needlepunched together, (2) dry bentonite mixed with an adhesive sandwiched between two geotextiles, or (3) dry bentonite mixed with an adhesive attached to a geomembrane by calendering. Each product uses approximately 5 kg/m^2 of bentonite, is produced in sheets about 4 to 5 m wide and 25 to 60 m long, and is delivered in rolls, as are other geosynthetic products. The GCLs are simply overlapped at their juncture; no seaming is required because the bentonite is generally considered to swell and seal the overlap when it hydrates. Two of the products recommend placement of additional bentonite between the sheets at overlaps to help in self-sealing upon hydration.

Installation of GCLs is relatively quick and easy. Each roll can be moved onto the construction site by lightweight equipment and easily unrolled into place. This product must be covered immediately upon installation, since the introduction of any water from precipitation, dew, etc., will cause the bentonite to begin swelling. If swelling occurs prior to placement of overlying (surcharge) layers, the bentonite will swell unevenly, it may not self-seal, it will become extremely slippery, and further construction cannot proceed.

The engineering properties of GCLs have been studied by various investigators in attempts to determine their properties and how they compare with compacted clay layers. Research reported by Daniel (1993a) indicates that the permeability of the GCLs currently produced falls in the range of 10^{-8} to 10^{-10} cm/s, with some of the variation caused by differing compressive stresses. However, if leachate reaches the bentonite before it is hydrated with water, this permeability may be much lower. In one study (Shan and Daniel 1991) it was found that "when the dry GCL was permeated directly with an organic chemical, the bentonite did not hydrate, did not swell, and did not attain a low hydraulic conductivity." Other tests have indicated that GCLs will likely maintain their low permeability through self-sealing despite punctures, desiccation, and freeze-thaw cycles (Shan and Daniel 1991).

Drained direct shear tests have been conducted on the three different types of GCLs described earlier (Daniel 1993a). The GCL with the needlepunched fibers connecting the surface geotextiles (type 1 described earlier) had an effective cohesion of approximately 4 psi (30 kPa) and an effective angle of internal friction of 26 degrees compared with an average effective cohesion of 1 psi (4 to 8 kPa) and an effective angle of internal friction of 8 or 9 degrees for the other two types of GCL. The needlepunched fibers appeared to provide significant reinforcement to the GCL.

An often ignored potential problem associated with the use of GCLs as the clay component of a composite liner is the presence of a geotextile between the bentonite and the overlying geomembrane. When the geomembrane has defects, e.g., punctures or holes, this geotextile will likely serve as a conduit for fluid flow between the geomembrane and the bentonite, thus nullifying the benefit of using a composite liner. (Sections 10.3.7 and 10.4.2 discuss composite liners and their purpose, design, and benefits in significantly reducing potential leakage.) Reportedly, at least one manufacturer is attempting to reduce the lateral transmissivity of the geotextile that will be in contact with the geomembrane (Daniel 1993a).

Table 10.6 compares several characteristics of conventional compacted clay liners with geosynthetic clay liners. The major advantages of the geosynthetic clay liner are probably speed and ease of construction and the high quality control possible in a manufacturing

TABLE 10.6 Comparison of Conventional Compacted Clay Liners and Geosynthetic Clay Liners

Characteristics	Compacted clay liner	Geosynthetic clay liner
Robustness		
Thickness	Thick (18–36 in)	Thin ($<\frac{1}{2}$ in)
Resistance to damage	Highly resistant	Vulnerable
Ease of repair	Difficult	Easy
Construction		
Location	Site specific	Factory manufacture
Difficulty	Hard to build	Easy to build
Equipment	Heavy equipment	Light equipment
Speed of construction	Slow	Fast
Protection needed	Prevent desiccation cracking	Cover immediately and protect from rain
Quality control	Difficult and variable	Easier
Containment effectiveness		
Past experience	Extensive	Limited
Leachate attenuation	High due to thickness	Limited
Desiccation cracking	Possible during and after construction	Only after initial wetting
Performance	Highly dependent on construction	Less dependent on construction quality
Structural stability	Low strength clays	Very low strength clays
Economics		
Cost	Site specific, dependent on availability of suitable clays	Off-the-shelf product
Landfill volume	Thick liner consumes valuable landfill volume	Maximizes available landfill volume

environment as opposed to field construction. The major disadvantages are probably a lack of experience with this relatively new product and its vulnerability to damage during construction. Conventional compacted clay liners and geosynthetic clay liners each have advantages and disadvantages, and specific site and project conditions will determine which is the most advantageous to use.

10.3.7 Composite liners

Some combination of several of the preceding types of liners is often used for solid waste disposal facilities. These combination, or composite, liners require careful design and construction to be most effective. The design of a liner system is more than just placing a geomembrane or compacted soil barrier between the waste and the surrounding environment. It is arranging and designing several system components so that the system as a whole will function to prevent or at least minimize any leakage into the surrounding environment, even though leakage may occur in one component.

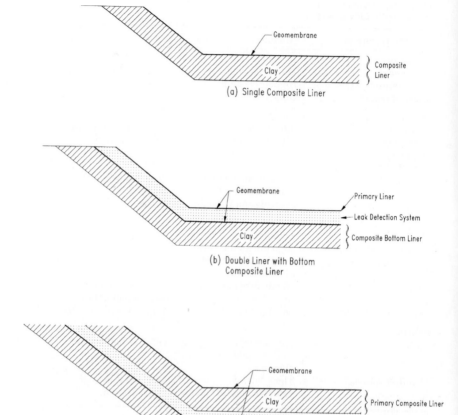

(a) Single Composite Liner

(b) Double Liner with Bottom
Composite Liner

(c) Double Composite Liner

Figure 10.10 Composite liner configurations including (a) single composite liner, (b) double liner with composite bottom liner, and (c) double composite liner.

The most commonly used composite liners consist of a low-permeability compacted clay or a geosynthetic clay liner overlain by a nearly impermeable (except for defects or holes) geomembrane. The configurations most often encountered (Fig. 10.10) are a single composite liner, a double liner with a composite bottom liner, or a double composite liner. By using a low-permeability clay layer under a geomembrane, any leachate that may escape through a hole in the membrane will then have to percolate through the thick, low-permeability backup liner. Looking from the other perspective, by placing a geomembrane above the low-permeability clay layer, the surface of the clay is protected from

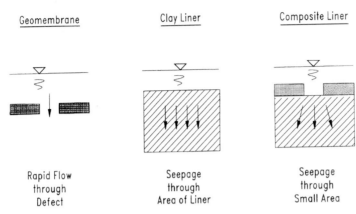

Geomembrane   Clay Liner   Composite Liner

Rapid Flow   Seepage   Seepage
through   through   through
Defect   Area of Liner   Small Area

Figure 10.11   Flow through different types of liners.

permeation by leachate except in a few locations where a hole exists in the membrane; i.e., the area of flow through the clay liner is greatly minimized. Figure 10.11 illustrates the different conditions of flow that will occur with various liner types. Calculation of flow rates through each liner type will show the significant decrease in effective permeability of a composite system over its individual components. These calculations are discussed in Sec. 10.4.2.

For a composite liner to function effectively, the geomembrane must be placed flatly against the clay layer so that there are no preferential flow paths for liquids between the geomembrane and the clay; this is often referred to as *intimate contact*. Intimate contact can be achieved by first finish-rolling the surface of a compacted clay liner with a steel-wheeled roller. If any ruts or other irregularities of the clay surface remain after finish rolling, these areas should be removed and recompacted to meet both specifications and smoothness requirements. The geomembrane is then placed flat against the smooth clay surface (Fig. 10.12). When geosynthetic clay liners are used, this philosophy leads to

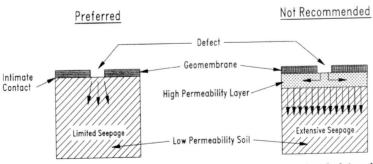

Preferred   Not Recommended

Intimate
Contact
Defect
Geomembrane
High Permeability Layer
Limited Seepage
Extensive Seepage
Low Permeability Soil

Figure 10.12   Intimate contact needed between geomembrane and underlying clay.

questions regarding their use below a geomembrane, since most GCLs have a geotextile above the bentonite, thus preventing the most intimate of contacts. This is of particular concern when thicker geotextiles are present. Work is ongoing to assess the performance of GCLs in composite liner systems and to redesign the GCLs to minimize lateral flow through the upper geotextiles.

Before placement of the geomembrane, the surface of a compacted clay liner must be checked carefully by personnel walk-over with close inspection, and any small rocks that could puncture the membrane must be removed. The geomembrane should then be placed on the clay surface in such a way as to minimize wrinkles and expediently covered with any overlying system components or a protective soil cover (see also the discussion in Sec. 10.3.5). This will minimize

- Membrane damage from blowing in winds
- Puncture by unauthorized trafficking (by vehicles, persons with heeled shoes/boots, wildlife, etc.)
- Degradation from exposure to sunlight
- Wrinkling of the geomembrane caused by expansion and contraction from cyclic heating and cooling
- Possible compacted clay liner cracking from extreme heating below a geomembrane (temperatures can reach 65°C under a black exposed membrane)
- Condensation of moisture under the geomembrane that can lead to reduced interface shear strength and/or ponding of water beneath the membrane at the toe of slopes

In geographic areas with large populations of deer, elk, and other sharp-hoofed wildlife, it may be necessary to erect fences or other obstacles around geosynthetic installation operations to prevent their intrusion (particularly at night) and subsequent damage to the geosynthetics.

There is a potential failure surface at the geomembrane and clay interface that must receive careful evaluation when assessing the slope stability of inclined composite liners. Mitchell et al. (1990) and Seed et al. (1990) provide a detailed examination of a waste landfill slope failure and the problems involved in evaluating the geomembrane and clay interface shear strength.

10.4   Evaluation of Liner System Effectiveness

Most surface impoundments and landfills must be "watertight" and thus must be lined and capped to prevent (or minimize) either inflow and/or outflow of liquids and potential contaminants. No absolutely impermeable materials exist, and if they did, construction defects would ensure that some leakage would still be possible. The design engineer can do many things to limit leakage and, if necessary, provide for protection of the surrounding environment; however, the finest design is no better than its construction. Consequently, the importance of thorough quality control inspection during construction cannot be overemphasized. Surely there can be fewer places where Murphy's law (i.e., "If anything can go wrong, it will") is so applicable as in the design and construction of waste containment systems.

Since the possibility of leakage always exists, the designer must provide systems for leachate detection, collection, and removal (often in duplicate). Monitoring during active operations and long-term monitoring after closure of a facility are also necessary to identify potential leakage problems and to allow corrective measures to be implemented expediently.

The geotechnical engineers Karl Terzaghi and Ralph Peck developed the concept of the observational method, wherein complex problems in soil mechanics could be solved by adjusting the project design during or even after construction as monitoring revealed new information about how the soils were really behaving. The key to their approach was picking an initial design concept that seemed reasonable for the project conditions and establishing a plan for monitoring the project's behavior to determine if the initial design was going to be adequate. If the project did not go as planned, then a contingency plan had to be ready to go into effect. Conceptually, this is very similar to what should happen at impoundment sites:

1. A containment plan must be developed consistent with the site conditions, the waste to be contained, and the threat to the surrounding environment.

2. A system of monitoring the site (appropriate to the threat it poses) should be developed so that if leakage does occur, it will be detected.

3. If leakage does occur, there should be a readily implementable plan of action to deal with it.

The plan to deal with possible future leakage should be part of the original landfill or impoundment design. Once leakage is detected, it is too late to start wondering what to do about it.

The design of a containment system entails selection of the type of liner to be used (which depends on the type of waste and compatibility considerations), initial material selection, and preliminary evaluation of the expected effectiveness of the liner system. If the evaluation indicates that the selected liner material or system will not function to the required level, the design must be modified to provide the necessary containment. Asphaltic materials, portland cement concrete, compacted clay, and modified soils are sometimes used in special applications. If these materials are selected for use, available literature and methods should be consulted to assess the potential performance of these liners, with special attention to the potential leakage rates through these materials.

The most commonly used lining (and capping) systems for landfills and surface impoundments are geomembranes, compacted clay, and/or composite geosynthetic liners. These are used in combination with various drainage layers that provide for leachate collection and/or leak detection. In the United States, single composite liners (geomembrane-clay), double liners with a composite bottom liner (geomembrane-geomembrane-clay), and double composite liners (geomembrane-clay-geomembrane-clay) are often used. The following subsections discuss various aspects of leachate collection and leak detection system design, including calculation of impingement rates, head on liners, and flow through defects in the liners. The leakage rate through a geomembrane liner may be affected by the following parameters:

- Type of geomembrane
- Thickness of geomembrane
- Size and shape of the flaw
- Characteristics of the subbase material
- Presence or absence of a geotextile between subbase and geomembrane
- Characteristics of the liquid to be retained

10.4.1 Leachate quantities

The greatest quantity of leachate will develop while the waste disposal site is open and actively receiving waste. After the site is closed with a cap in place, the inflow of water coming into contact with the waste should be greatly reduced (and may be negligible if the cap is properly designed). Leachate, however, continues to be generated long after landfill closure. During the first few postclosure years, the rate of leachate removal (through the leachate collection system) is almost 100 percent of that during construction. Approximately 2 to 5 years after closure, leachate generally levels off to a low-level constant leakage rate equal to the rate of leakage through the cap, or in a very tight, nonleaking closure, leachate production falls to zero.

Estimates of the quantity of leachate to be addressed in design are normally estimated in the United States with the U.S. EPA computer model "Hydrologic Evaluation of Landfill Performance (HELP)" (USEPA 1984a, 1984b). Manual methods and "rules of thumb" were used in the past, but today most regulatory agencies normally only accept the HELP model results. Quantities of leachate vary significantly from site to site and depend on such factors as waste type, method of landfill operation, construction materials and practices, precipitation (infiltration and overland flow onto the site), and local topography.

10.4.2   Leachate collection systems

The leachate collection system collects the leachate that percolates through the waste and removes it to a sump, where it can be extracted and treated. This system typically includes a granular drainage medium that carries the leachate to a pipe system, which then carries it to the sump. To minimize leakage through the liner, the leachate drainage system is commonly designed to keep the maximum head of leachate on the underlying liner to 0.3 m or less. This is normally calculated with Moore's equation (USEPA 1987):

$$h_{max} = L[\sqrt{(e/k) + \tan^2\beta} - \tan\beta]$$

where h_{max} = maximum head of leachate on the liner, m
L = horizontal length to drainage collector, m
e = leachate impingement rate, m/s (This is normally determined from the HELP model but can be crudely and conservatively estimated as 25 percent of the 100-year 7-day storm.)
k = coefficient of permeability of the drainage material, m/s
β = slope angle, degrees (usually at least 2 percent)

Liner leakage calculations.   Some basic equations for calculating leakage through liners have been developed and can be found in references such as USEPA (1987), Giroud and Bonaparte (1989a, 1989b), Giroud et al. (1991), and Brown et al. (1987). These equations are based on simplifying assumptions and limited testing, so they should be used with caution. However, they are useful for sizing collection and detection systems and for studying how different design factors affect potential leakage.

Permeation through the geomembrane.   The geomembrane material is essentially impermeable when no holes or punctures are present. The

TABLE 10.7   Moisture Vapor Permeability of Various Geomembrane Materials

Base polymer	Range of thickness, mm	Moisture vapor permeability (10^{-2} metric perm · cm*	
		Average	Range of values
Butyl rubber	0.85–1.18	0.081	0.016–0.170
Chlorinated polyethylene (CPE)	0.53–0.97	0.485	0.213–1.27
Chlorosulfonated polyethylene	0.74–1.07	0.460	0.234–0.845
Neoprene	0.51–1.59	0.354	0.147–0.517
Nitrile rubber	0.76	3.98	—
Polyethylene (low-density)	0.76	0.041	—
Polyethylene (high-density)	0.80–2.44	0.014	0.013–0.014
Polyvinyl chloride	0.28–0.79	1.30	0.77–1.45

*Permeability in metric perm · cm = permeance×thickness of geomembrane.
SOURCE: From Haxo, 1990.

TABLE 10.8   Permeability of Several Geomembranes to Water at Various Pressures*

Base polymer	Hydraulic pressure, kPa			
	100	400	700	1000
Butyl rubber	22	8.0	4.1	2.8
Chlorosulfonated polyethylene	300	120	77	60
Ethylene propylene rubber	115	57	34	25
Polyvinyl chloride	86	37	20	10

*Data reported in 10^{-15} m/s.
SOURCE: From Haxo, 1990.

equivalent permeability of HDPE or LDPE is 10^{-13} cm/s; it is 10^{-11} cm/s for PVC or nitrile rubber, and for neoprene the permeability is 10^{-12} cm/s. Of course, the permeability of a geomembrane depends not only on the geomembrane material type but also on the geomembrane thickness, the type of fluid or gas to which it is exposed, and the pressure in the fluid or gas. Some typical permeability values are given in Table 10.7 for moisture vapor permeability of different materials. Table 10.8 provides the permeability to water of four geomembranes at various pressures; the pressures were applied to the "upstream" side of the geomembrane while the water was on the "downstream" side of the membrane, thus "compressing" the geomembrane between the pressure and the incompressible water. Each of these tables shows that the permeability of a geomembrane is not a single value but will vary as a function of several factors.

Leakage through defects in the geomembrane. When a geomembrane is installed, typically it will have some defects (holes, punctures, improper seaming, etc.) through which leachate can escape. Various calculations can be made to estimate the quantity of leakage that will occur through these defects. For the case of a single liner, the geomembrane will normally be between two highly permeable layers. For calculation purposes, the defects are assumed to be either small (pinhole-sized) or larger. *Pinholes* are defined to have diameters less than the membrane thickness.

For small holes (pinholes), leakage calculations are based on Poiseuille's equation (Giroud and Bonaparte 1989a):

$$Q_p = \frac{\pi \rho g h^4}{128 \eta T_g}$$

For larger holes, the leakage calculations are based on Bernoulli's equation for free flow through an orifice (Giroud and Bonaparte 1989a):

$$Q_h = C_B a \sqrt{2gh_w}$$

where Q_p = leakage through a pinhole, m³/s
 Q_h = leakage through a hole, m³/s
 h_w = liquid depth on the membrane, m
 d = pinhole diameter, m
 a = area of the hole, m²
 T_g = thickness of the membrane, m
 g = acceleration of gravity, m/s²
 C_B = dimensionless coefficient related to shape of the edges of an aperture
 = 0.6 for sharp edges
 ρ = density of liquid, kg/m³
 = 1000 for water at 20°C
 η = dynamic viscosity, kg/(m · s)
 = 10^{-3} for water at 20°C

Leakage through composite liners. The existence of a low-permeability soil beneath the geomembrane in a composite liner significantly reduces the leakage through the composite liner compared with the geomembrane alone. Based on empirical tests and theoretical studies, the following empirical equations for leakage through round holes in composite liners have been developed (Giroud et al. 1989). Because these are empirical equations, the dimensions used in them must be as shown, and other units cannot be substituted for them.

For the head of liquid on the membrane less than the thickness of the low-permeability soil (conditions similar to a well-designed landfill with a leachate collection system that will maintain small heads on the liner) and good contact between the geomembrane and soil (geomembrane is largely free of wrinkles and the soil layer has been compacted and graded smooth), the following equation may be used:

$$Q_{sg} = 0.21a^{0.1}h_w^{0.9}k_s^{0.74}$$

For poor contact between the geomembrane and the soil, the equation becomes

$$Q_{sp} = 1.15a^{0.1}h_w^{0.9}k_s^{0.74}$$

where Q_{sg} = leakage due to a hole in the geomembrane portion of a composite liner subject to a small liquid head and with good contact between the geomembrane and soil, m³/s

Q_{sp} = leakage due to a hole in the geomembrane portion of a composite liner subject to a small liquid head and with poor contact between the geomembrane and soil, m³/s

a = area of the hole in the geomembrane, m²

h_w = head of liquid, m

k_s = permeability of the soil under the geomembrane, m/s

For leakage under larger heads such as may be found in impoundments, several intermediate calculations are needed. As before, these are empirical equations and units must be used as given below. The leakage can be calculated from the following equations (Giroud et al. 1991):

$$R_g = 0.26a^{0.05}h_w^{0.45}k_s^{-0.13}$$

$$R_p = 0.61a^{0.05}h_w^{0.45}k_s^{-0.13}$$

$$i_{avg} = 1 + \frac{h_w}{2t_s \ln(R_x/R_o)}$$

$$Q_{lg} = 0.21i_{avg}a^{0.1}h_w^{0.9}k_s^{0.74}$$

$$Q_{lp} = 1.15i_{avg}a^{0.1}h_w^{0.9}k_s^{0.74}$$

where Q_{lg} = leakage under large head, good contact, m³/s

Q_{lp} = leakage under large head, poor contact, m³/s

i_{avg} = average hydraulic gradient

R_p = radius of wetted area, poor contact, m

R_g = radius of wetted area, good contact, m
R_x = R_p or R_g as appropriate, m
R_o = radius of the hole in the geomembrane, m
t_s = thickness of the low-permeability soil below the geomembrane, m
a = area of the hole in the geomembrane, m^2
h_w = head of liquid, m
k_s = permeability of the soil under the geomembrane, m/s

Some common assumptions used with the various leakage equations are:

1. Hole sizes of 1 cm^2 may be used for calculations concerned with sizing lining system components (sump size, leakage detection thicknesses or permeability, etc.) for maximum flow.

2. Hole sizes of 3.1 mm^2 may be used for calculations under normal operating conditions.

3. One hole per 4000 m^2 is a reasonable estimate of defects when good, independent quality assurance testing of the geomembrane occurs.

4. If quality assurance is limited to periodic, visual spot checks by the engineer, defects may exceed 25 per hectare.

10.4.3 Leak detection system

In double-lined systems, the leak detection system collects any leakage through the primary liner and carries it away for collection and treatment. The impingement rate for this layer would be the leakage through the primary liner, and Moore's equation can be used to determine a maximum head, which then allows the preceding equations to be used to determine leakage rates through the underlying secondary liner. A few trial calculations will rapidly show that the permeability of sand is too low for this material to be effective in the leak detection system. Gravel or one of the synthetic drainage materials such as geonet normally will be needed in this layer. Note that not all water in the leak detection system is from leakage. Consolidation of clay (if included) in the overlying liner will expel consolidation water into the leak detection system. The quantity and rate of water from this consolidation can be calculated from classic soil mechanics approaches.

10.5 Essential Components of Waste Containment Systems

The essential components of a complete waste containment system include (from the top down) the cover system, the gas collection system, waste, and a liner system (single, double, or composite). In addition to these elements, there are other considerations that must play into design and construction if the waste containment facility is to perform satisfactorily. Additionally, monitoring of the performance of the containment system components and the surrounding environment is essential. The system components and considerations are discussed in the Secs. 10.5.1 through 10.5.4; monitoring is presented in Sec. 10.6.

10.5.1   Cover system

After a waste disposal facility has been filled, a multilayer final cover normally is constructed over the site to isolate the waste from the environment, minimize infiltration of surface water, and thus minimize liquid migration and leachate formation. In some cases, the cover also is intended to control release of gas from the waste. The cover system should function with a minimum of maintenance, promote surface drainage, accommodate settlement and subsidence, and have a permeability no greater than that of the liner system.

Table 10.9 shows the typical components of a cover system in descending order from top to bottom. Each of these elements performs a specific function. The vegetation prevents surface erosion but should be selected carefully to include only varieties that have shallow root systems that will not penetrate to the depth of the impermeable layer (geomembrane-clay). The topsoil–protective soil layer pro-

TABLE 10.9   Component Layers and Materials in a Cover System

		Materials		
Category	Components	Option 1	Option 2	Option 3
Surface protection	Vegetation	Shallow-rooted plants	Shallow-rooted plants	Shallow-rooted plants
	Top soil	Top soil	Top soil	Top soil
	Protective soil	Local soil	Local soil	Local soil
Drainage	Filter	Soil or geotextile	Thick geotextile	Geotextile
	Drainage	Sand or gravel	Thick geotextile	Geonet
Barrier	Geomembrane	Geomembrane	Geomembrane	Geomembrane
	Clay	CCL or GCL	CCL or GCL	CCL or GCL
Gas collection	Drainage (gas)	Sand or gravel with geotextile filter	Thick geotextile	Geonet drain with geotextile filter
Foundation	Foundation soil	Local soil	Local soil	Local soil

vides a growing medium for plants while providing sufficient thickness to prevent disturbance of underlying layers by roots, burrowing animals, freeze-thaw cycles, mechanical disturbance, and excessive wetting and drying. The filter layer protects the drainage layer from intrusion of the overlying protective soil. The drainage layer provides interception and removal of water infiltrating from the surface. The composite geomembrane-clay layer acts as a barrier to downward percolation of water into the waste. A gas collection layer intercepts gases generated during decomposition of waste and moves them out of the containment facility in a controlled (and sometimes useful) manner; inclusion of this layer prevents buildup of gases beneath the impermeable layer and subsequent rupture of the clay layer and development of large bubbles (whales) and possible rupture of the geomembrane. The soil foundation layer is placed directly on the waste to level and smooth the surface prior to cover construction. More specific guidance on cover system design is provided by Daniel and Koerner (1993) and USEPA (1989a, 1989b).

Typical layer thicknesses for covers in a waste containment system are surface protection layer ≥ 2 ft, drainage layer ≥ 1 ft, and for the barrier layer, a geomembrane ≥ 20 mil and a compacted clay layer ≥ 2 ft. These thicknesses obviously will be smaller if geosynthetic products such as geonet or geosynthetic clay layers are used instead of natural materials such as gravel or compacted clay.

It is important to remember that each waste disposal facility differs from all others. Therefore, it is necessary to design a cover system for each site individually, although the factors to be considered during cover design are the same for all sites. During design of the cover, the type of waste and its placement (technique and location) must be considered, since these factors significantly affect settlement (both total and differential) of the cover. The designer also should analyze the cover design for percolation of precipitation, slope stability, and erosion. Not only must these factors be incorporated into the final cover design, but they also affect the type of impermeable barrier used in the cover. To reduce erosion, drainage swales and concrete-lined ditches often are incorporated into the cover surface design. It is important that the geomembrane (if used) in the cover be secured properly to completely isolate the waste and to ensure minimal (or no) water migration into the waste. The geomembrane in the cover may be bonded to the geomembrane from the liner system (when one is used), or the cover liner must be anchored securely in an appropriate trench. The latter approach is generally preferred nowadays (see Sec. 10.5.5). Appropriate details should be provided by the designer.

When a cover system is designed and installed over a surface impoundment, the considerations and requirements for the cover's

TABLE 10.10   Landfill Cover Requirements

Cover material	Minimum thickness	Exposure time (days)
Daily	150 mm	0–7
Intermediate	300 mm	7–365
Final	1.5 m	≥365

functions and components are the same as those for a landfill, with the one addition that the cover alone must support construction equipment. Surface impoundments usually contain weak slurried materials. Consequently, it may be necessary to dewater or otherwise stabilize the upper portion of the waste to provide a working platform to allow equipment access to the interior of site (Rollings et al. 1988, Broms 1987).

When *cover* is mentioned with regard to design or construction of waste containment facilities, it normally refers to the final cover that is placed after the site is filled with waste. However, one should be aware that three types of surface covers are used during the active life of a waste disposal facility—daily, intermediate, and final covers. The type of cover used depends on the length of time the cover will be exposed to erosion by wind and water. Suggested minimum thickness and exposure time for the three cover types are shown in Table 10.10.

10.5.2   Waste characteristics

If waste is placed into a designed landfill, information should be available regarding the types and locations of waste placement, amount of compaction used during landfill construction, etc., which will be helpful in cover design. For uncontrolled (usually old) landfills or "dumps," such records probably will not be available. In the latter case, best estimates should be used. As a general rule of thumb, the amount of waste consolidation to be expected in a landfill is 5 to 10 percent, and the unit weight of waste for engineering calculations is often assumed to be 50 to 100 lb/ft^3. Data recently compiled by Oweis (1993) shows solid municipal waste varying in density from 30 to 90 lb/ft^3, incinerator residue varying from 46 to 106 lb/ft^3, and organic waste varying from 15 to 33 lb/ft^3. More information on settlement of waste and ways to calculate settlement can be found in Rao et al. (1977) and Sowers (1973).

Shear strengths of a small number of wastes have been measured, and some results are shown in Table 10.11. However, there is very little field verification of the reported laboratory data.

TABLE 10.11   Shear Strengths of Municipal Wastes

Waste type	Apparent cohesion, c_a	Apparent friction angle ϕ_a, degrees
Shredded refuse	23 kPa (480 lb/ft²)	24
Old refuse	16 kPa (334 lb/ft²)	33
Artificial refuse	0	27–41
Fresh artificial refuse	0	36

SOURCE: From Landva and Clark, 1990. Copyright ASTM. Reprinted with permission.

10.5.3   Liner system

The components of a landfill or surface impoundment lining system are intended to isolate the waste from the underlying environment and to prevent contaminant migration in a downward or lateral direction. The exact components will depend on the local regulations, type of waste to be contained, geology of the site, groundwater location and quality, and other site-specific factors (see Secs. 10.2 and 10.3).

The most commonly used liner materials are compacted clay, geomembranes, geosynthetic clay liners, and composite liners. Less commonly used liner materials are modified soils, asphalt, and portland cement. All these liner materials are discussed in Sec. 10.3. Because of current laws and regulations, composite liners are the most prevalent liner type used in the United States today.

The configuration of a liner system incorporating a composite liner may take one of several forms. The waste containment facility may utilize a single composite liner. It may use a composite liner as the secondary liner in a double liner system, or a double composite liner may be used. Each of these options is illustrated in Fig. 10.10. If either of the double liner systems is used for a landfill, a leachate collection system would be located above the upper geomembrane (i.e., between the waste and the primary liner), and a leak detection system would be placed between the primary liner and the geomembrane of the secondary composite liner. If used for a liquid surface impoundment, the leachate collection system would be omitted. Terminology typically used for the various component layers in a double composite liner is (from top to bottom)

- Waste
- Filter layer
- Primary leachate collection (and recovery) system (or leachate collection system, LCS)
- Primary geomembrane (or flexible membrane liner, FML)

TABLE 10.12   Major Elements of Common Liner Systems

| | Types of liners | | | |
| | Landfill | | | Surface impoundment |
Element	Single composite	Double with composite secondary	Double composite	Double with composite secondary
Protective cover	*	*	*	*
Filter	*	*	*	*
Leachate collection system (LCS)†	Gravel, coarse sand, or geonet	Gravel, coarse sand, or geonet	Gravel, coarse sand, or geonet	—
Primary liner†	FML	FML over CCL or FML over GCL	FML over CCL or FML over GCL	FML
Leak detection system (LDS)†	Gravel, coarse sand, or geonet	Gravel, coarse sand, or geonet	Gravel, coarse sand, or geonet	Gravel, coarse sand, or geonet
Secondary liner	FML over CCL	FML over CCL	FML over CCL	FML over CCL

NOTE: FML = flexible membrane liner (geomembrane); CCL = compacted clay liner; GCL = geosynthetic clay liner.
*Use throughout as needed.
†When multiple entries appear, select one.

- Primary clay layer (compacted or geosynthetic)
- Secondary leachate collection (and recovery) system (or leak detection system, LDS)
- Secondary geomembrane (or flexible membrane liner, FML)
- Compacted clay layer
- Native soil (unsaturated zone)

The materials that are most often used in various liner configurations are shown in Table 10.12.

Between the waste and the primary liner is the leachate collection system (LCS), which collects leachate as it percolates out of the waste. The LCS moves leachate through the network of permeable material and pipes to its exit from the landfill. The LCS is not included in the liner system for a liquid impoundment; instead, the primary liner (uppermost liner) is designed for the large head to be imposed by the ponded liquid. The primary liner in a waste containment facility may consist of a geosynthetic membrane or a composite liner of different materials. Beneath the primary liner is a leak detection system (LDS), which collects any leachate that leaks through the primary liner. This leak detection layer may be a highly permeable granular material (gravel or coarse sand) or a geosynthetic material such as geonet that is designed to collect and transport liquids. Depending on the particular design, a geotextile filter layer may be needed between the drainage layer in the LCS or the LDS and any adjacent soil layer. Beneath the leak detection layer is a secondary liner, which should be a composite liner.

Geonet or gravel are desirable in both the leachate collection and leak detection systems, since they are much more permeable than any sand. Sands are much less effective because they do not drain as rapidly as we often believe. It is essential that both the LCS and the LDS drain quickly to permit rapid removal of leachate, thus preventing a buildup of head on the underlying impermeable liner. When the head on a liner is kept to a minimum, the leakage through any defects in that liner also will be minimized. For comparison purposes, the permeability/transmissivity for several soils and geonet are given in Table 10.13. Sand is a poor choice for use in leak detection systems because the presence of leaks could not be detected in a timely manner with this relatively slow-draining medium. Geonet and gravel are better choices where high capacity or rapid drainage is needed in waste containment facilities.

Considerable head will exist on the primary liner in a surface impoundment. Because of the strong impact this will have on quantities and rates of leakage, these facilities are usually pretested for leaks by filling with water before commencement of actual waste containment operations.

10.5.4 Gas collection system

After coverage or closure of a waste facility, degradation of waste materials proceeds very quickly from aerobic to anaerobic decomposition. (Liquid wastes in surface impoundments, except at the very sur-

TABLE 10.13 Typical Ranges of Permeability/Transmissivity for Various Soils and Geonet

Material	Permeability (cm/s)
Geonet	5*
Clean gravel	1×10^{-5} to 1.0
Clean sand or sand + gravel mixtures	1.0 to 1×10^{-3}
Fine sands and silts	1×10^{-2} to 1×10^{-6}
Silty clay and clay	1×10^{-5} to 1×10^{-9}

*Transmissivity in cm²/s (see ASTM D 35).

face, are usually subject to anaerobic conditions before closure.) As wastes decompose in this anaerobic environment, various gases are produced and should be controlled (Shafer et al. 1984). Almost 98 percent of the gases formed are either carbon dioxide (CO_2) or methane (CH_4). Because carbon dioxide is heavier than air, it will move downward and be removed with the leachate. The methane (composing approximately 50 percent of landfill gases), however, is lighter than air and will move upward through the waste (Barlaz and Ham 1993).

Thus potentially large volumes of methane gas will collect beneath the impermeable liner of the closure system. If a gas collection and removal system is not incorporated into the waste site design, this collection of gas can cause disruption of the closure system (e.g., "whales," i.e., bubbles in the geomembrane, and blowouts or ruptures of the geomembrane and/or clay layer). Figure 10.13 shows small whales in a geomembrane under a surface water detention pond at a hazardous waste facility in western Canada.

A complete gas collection and removal system should include

- A drainage layer consisting of a granular soil layer or nonwoven geotextile placed below the low-permeability liner component(s), i.e., below the FML and compacted soil

- A gas collection pipe within the granular soil layer or connected pneumatically with the geotextile

Figure 10.13 Bubble or whale in geomembrane under surface water detention basin.

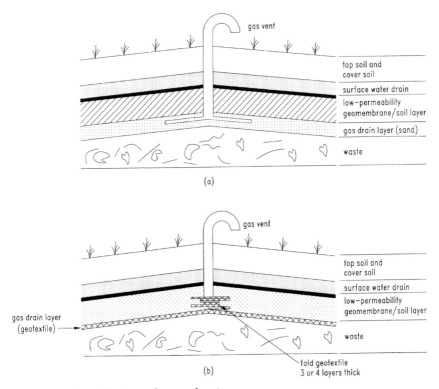

Figure 10.14　Gas collection and removal system.

- A riser pipe that extends through the compacted clay and geomembrane
- (Optional) feeder lines from riser pipes to gas use area

Figure 10.14 shows a typical gas collection and removal system. Alternatively, various types of gas collection wells can be installed vertically (or at other angles) into the waste to collect gases from deep within the facility. Whether deep wells or the gas collection system described above are used, careful connection must be made where the riser penetrates the membrane.

Since the uncontrolled release of gases is often unacceptable from an environmental and/or human neighbor perspective, the gas riser pipes or wells may be equipped to flare and burn the escaping gases to prevent noxious odors. This is especially useful in areas with close human population centers. Alternatively, the risers or wells may be connected to a gas collection (suction) system that transports the waste gases to an on-site or off-site recovery center.

10.5.5 Other considerations in waste containment systems

Global stability of the waste containment facility must be considered. Standard slope stability analyses should be performed to determine the resistance of the facility to both rotational and translational (wedge-shaped) failures. Analyses should be conducted for deep, toe, and side-slope failures in addition to the typical soil-geosynthetic interface stability testing and calculations performed for waste containment facilities (Oweis 1993, LaGrega et al. 1994).

In many recently designed waste facilities, multiple layers of various geosynthetics are used in the side slopes of the containment system. (As many as 8 to 12 geosynthetic layers are often used.) In cases where the interface stability is questionable, textured geomembranes often are specified to increase the friction between the adjacent layers. When used on slopes, the internal stability of hydrated geosynthetic clay liners also must be addressed.

Laboratory testing of the various geosynthetics that may be used in a waste containment facility should be conducted in a manner to accurately assess real field conditions. This is a relatively new arena of testing, and procedures for correctly predicting field behavior are still being developed and refined. Subjects to be addressed include performance of drainage geosynthetics under pressure (i.e., reduced transmissivity from compression and from intrusion of other materials into the flow channels), interface friction (both geosynthetic-to-geosynthetic and geosynthetic-to-soil) at various angles and pressures, long-term durability of geosynthetics under field conditions, and chemical compatibility of geosynthetics with actual leachate.

The interaction of filters and drains and their adequate long-term performance are critical to successful operation of waste containment facilities. Filters must be selected carefully to provide the necessary protection of drainage layers and adjacent soil layers; this is true whether soil or geosynthetic filters are used (Cedegren 1989). Also, because of the nature of many wastes and their leachates, the potential for biologic clogging of both filters and drains should be assessed, and methods for remediating any biologic clogging should be incorporated into the facility design (Koerner 1993a).

Consideration should be given to the method of completing the encapsulation of the wastes, e.g., not only the design of the liner and capping systems but also the juncture of these systems. Previously, the philosophy had been to physically connect the geomembrane in the cap to the geomembrane in the liner so that the waste was entirely and continuously enclosed by an impermeable material. These geomembranes usually were connected by extrusion welds, as were the primary and secondary geomembrane liners (USEPA 1989a). The problem with this system was that any settlement or movement of the waste containment system would cause tension in the geomembrane likely followed by rupture of the membrane or tearing of the weld. Current practice recommends that the various geomembranes be anchored in separate trenches, with the cover geomembrane preferably being anchored beyond those of the liner system (USEPA 1989a, Koerner 1993a), as shown in Fig. 10.15. This method provides positive closure with a more forgiving nature. If movement causes pullout of the geomembrane from the trench, repair and reanchorage are relatively easy.

Cleanout of leachate collection pipes is normally required, from both operational and regulatory perspectives, throughout the life of a waste containment facility. Therefore, it is essential that the leachate collection and removal system be designed to provide access points for cleanout equipment.

It is imperative that both the leachate collection and recovery systems and the surface water collection and removal systems be designed for rapid and safe removal of the maximum design flows to which they will be subjected. This requires use of uniform slopes, proper sizing and spacing of collector pipes or drainage swales, piping with adequate load-carrying capability, and erosion-resistant surfacing in surface drainage swales (USEPA 1985, Wagner et al. 1986).

After the waste containment facility is closed to active operations, it must be maintained to ensure its continued proper functioning throughout the closure and postclosure care period. The care required during this period should include monitoring of leachate levels in the waste system, monitoring of groundwater, monitoring of air emissions, and

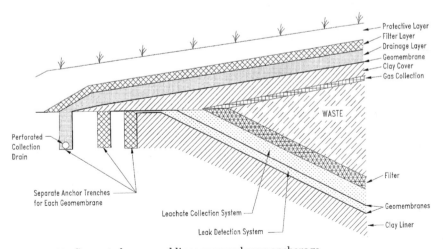

Figure 10.15 Suggested cover and liner geomembrane anchorage.

maintenance of the cover system to prevent surface disruption, erosion, gully formation, and potential moisture intrusion into the waste.

10.6 Performance Monitoring of Containment Systems

Monitoring is an essential part of the long-term operation and care of a waste disposal facility, just as it is with most environmental projects. The complete monitoring of facility performance requires visual monitoring of the surface of the facility, groundwater monitoring, gas monitoring, and monitoring of the various collection systems (leachate and gas).

10.6.1 Visual monitoring

Visual monitoring of the facility surface is intended to identify potential problems with differential settlement (particularly of the cap), slope stability, and erosion of the surface soils. If any of these potential problems occur, there is a real possibility that water can enter the waste, forming additional leachate, or waste may be exposed, causing additional contamination, nuisance vector problems (e.g., rats, flies, and mosquitos), and odor problems. Visual surface monitoring can identify these problems early, and corrective actions can be taken before the situation becomes critical.

10.6.2 Groundwater monitoring

Groundwater monitoring is essential around and possibly under any waste disposal facility. This monitoring effort determines whether there is a developing problem with leachate leakage and groundwater contamination. The number and location of monitoring wells needed is site-specific, depending on site geology, waste type, waste disposal facility type, groundwater quality, density and proximity of population, etc. The frequency of sampling and testing of monitoring well water is usually determined by prevailing regulations governing the site. Groundwater monitoring is normally conducted quarterly if contaminants have been detected previously or twice per year otherwise for the waste facility's active life and postclosure care period. If groundwater contamination is detected, it is necessary to initiate remedial actions promptly; thus it is imperative to have a remedial action plan in place throughout the life of a waste disposal facility.

10.6.3 Gas monitoring

Since gas generation at waste containment facilities is common, it is essential that a long-term monitoring program be conducted to detect and prevent any explosion hazards. Methane and carbon dioxide gases are generated wherever biodegradable organic matter is buried, i.e., in an anaerobic condition (see Sec. 10.5.4). Thus gas generation is inevitable at nearly all municipal waste landfills and at many surface impoundments; in these situations, the U.S. EPA recommends installation of a gas collection system. Where no decaying matter is buried, gas will probably not be a problem, and no collection system is needed. Since it is not known how much, if any, gas may be formed in later years at hazardous waste facilities, the U.S. EPA recommends installing gas collection systems at these sites.

Monitoring for gas at a waste facility should include air monitoring both on and near the facility, especially in low areas or other places where gas may collect; monitoring of buildings within 300 m of the waste facility; and monitoring of the quantity and composition of gas collected in any gas collection system. The frequency and intensity of gas monitoring will depend on the local requirements as well as the quantities and types of waste at the site (Shafer et al. 1984).

10.6.4 Monitoring of collection systems

Monitoring of the leachate collection system, the leak detection system, and the gas collection system is essential and must be continued throughout the active life and the postclosure period of the waste disposal facility. The frequency of sampling and the specific testing to be conducted on the collected samples depends on local regulations and is usually itemized in the operating permit(s). Not only should the monitoring results be collected as required, but they also should be analyzed when collected, and if problems are detected, the appropriate remedial action plan should be implemented immediately.

10.7 Problems and Remedies Associated with Containment Systems

To properly plan a waste containment facility, the length of time the facility is expected to operate should be determined or estimated. Some recommended time frames for different kinds of facilities are as follows (USEPA 1989a):

Heap leach pads	1–5 years
Waste piles	5–10 years
Surface impoundments	5–25 years
Solid waste landfills	30–100 years
Radioactive waste landfills	100–1000 + years

It should be noted that none of these time frames are regulated in the United States. The only time frame that the U.S. EPA mandates is a 30-year postclosure care period for all hazardous waste landfill facilities.

Most long-term problems associated with waste disposal facilities are associated with liners, collection systems (both leachate and gas), and caps/covers. Table 10.14 summarizes areas of concern in various waste disposal facility materials and components. Factors that should be considered in judging the impact of any specific waste containment facility include geology, stratigraphy, seismicity, groundwater location and quality, population density, facility size, leachate quantity and quality, and nontechnical factors (political, social, and economic).

10.7.1   Flexible membrane liners and other geosynthetics

The major long-term considerations in any geosynthetically lined facility are the integrity and durability of the flexible membrane liners (Koerner et al. 1990).

Potential problems associated with flexible membrane liner design and installation include

- Excessive numbers of protrusions (usually on concrete structures) or penetrations (such as pipes) that make lining system installation difficult and increase the number of locations where stress concentrations are likely to develop in the liner or where liner movements are likely to be restrained

- Improper connections between penetrations (such as pipes or concrete structures) and the lining system that may cause leakage

- Complex shape of sump structures that make it difficult to properly seam geomembranes in this area

- Improper interfaces between geosynthetics and concrete or other materials, i.e., sharp corners or voids, which can generate excessive concentrated stresses in geosynthetics

- Shape or location of concrete structure conducive to differential settlements that can cause excessive stresses in the lining system

- Anchor trenches that are improperly shaped, too shallow to contain geosynthetics, or contain incomplete geomembrane seams that can initiate tears in the geomembrane liners or introduce water into the subgrade

- Protective cover specified on slopes where it can cause instability because of insufficient friction angles between geosynthetics (Such problems can be addressed by redesigning the geosynthetics, slopes, or protective cover, improving the material's characteristics by, for example, using textured geomembranes for higher friction.)

TABLE 10.14 Long-Term Concerns in Waste-Disposal Facilities

	Mechanism					
Landfill component	Movement of subsoil	Subsidence of waste	Aging	Degradation	Clogging	Disturbance
Cap						
FML	No	Yes	Yes	Yes	No	Yes
Clay	No	Yes	Yes	No	No	Yes
Cap-SWCR						
Natural	No	Yes	Yes	No	Yes	Yes
Synthetic	No	Yes	Yes	Yes	Yes	Yes
Primary FML	Yes	No	Yes	Yes	No	No
Leachate collection system						
Natural	No	No	Yes	No	Yes	No
Synthetic	No	No	Yes	Yes	Yes	No
Secondary FML						
FML	Yes	No	Yes	Yes	No	No
Clay	Yes	No	No	No	No	No
Leak detection system						
Natural	No	No	No	No	Yes	No
Synthetic	No	No	Yes	Yes	Yes	No

SOURCE: From USEPA, 1989a.

- Improper geomembrane layout with horizontal seams on slopes or at the toe of slopes (Such seams are likely to be subjected to excessive stresses.)

- Use of geonets that do not have the proper compressive stiffness to withstand overburden stresses

- Geonet layout that does not account for increasing flow quantities from top of slope to drainage outlet, thus creating bottlenecks in the drainage system

- Improper connections between geonets and other drainage media, thus generating a risk of clogging

- Use of geonet in contact with an improperly selected geotextile that partially blocks its channels and therefore significantly reduces its flow capacity

- Geotextiles intended as filters that do not meet filter criteria for the application(s)

- Use of geotextile filters in locations inappropriate for a filter and where they may generate clogging

- Drawings that do not clearly indicate the location of the extremities of geotextiles (This might cause the installer to extend geotextiles to locations where they can be harmful, e.g., by causing slope instability or clogging.)

- Improper connection between landfill cover geosynthetics and bottom liner geosynthetics

Other long-term concerns regarding flexible membrane liners include degradation (by oxidation, UV exposure, high-energy radiation, chemical degradation, biologic degradation, thermal processes, ozone, extraction via diffusion, and delamination) and stress-induced mechanisms (freeze-thaw, abrasion, creep, and stress cracking).

10.7.2 Compacted clay liners

Clay has been used as a liner material for many years. Intensive study and evaluation of the effects of leachate on clay liners have been conducted over the past 10 or more years. Indications are that clay liners generally perform satisfactorily, although there are some areas of concern. While clays are not subject to stress cracking or degradation as are geomembranes, they can have problems with moisture content (variation, desiccation, etc.), clods, high concentrations of organic solvents, severe volume changes, and repetitive freeze-thaw cycles. Detrimental effects of moisture content and clods can be controlled through meticulous construction and good construction quality control/assurance. Most potential volume changes can be controlled by proper material selection to minimize shrink/swell potential and adequate compaction. Freeze-thaw problems usually can be alleviated by proper design (to provide adequate depth of cover to prevent frost penetration into critical soil layers).

Other potential clay liner design and installation problems are listed in Table 10.15. Possible preventative measures are also listed.

10.7.3 Leachate collection and removal systems and cap drainage systems

The leachate collection and removal system includes all materials involved in the primary leachate collection system and the leak detection system. The cap drainage system includes all materials in the final waste disposal facility cap. Leachate collection and cap drainage systems can fail or clog through a variety of physical, chemical, biologic, and biochemical mechanisms. Bass (1985) provides a detailed examination of causes for failure within these drainage systems.

Mechanisms that can lead to failure include pipe breakage, pipe separation, pipe deterioration, differential settlement of system components, other nonclogging problems (including tank failure and exceeding design capacity), chemical deterioration of system components, sedimentation, biologic growth, chemical precipitation, and biochemical precipitation. All these mechanisms are known to occur at municipal solid waste facilities; all but biologic precipitation are known or strongly suspected to be active in hazardous waste facilities.

Problems with physical mechanisms (i.e., pipe problems, settlement, design capacity, etc.) normally can be eliminated through proper waste facility design and leachate collection and cap drainage system sizing and material selection.

Chemical deterioration of system components can be avoided or minimized by using materials chemically resistant to the leachates they are handling. Gravel and sand (except some limestones) are generally quite resistant to chemical attack. Flexible membrane liners are normally tested (in the United States) for chemical resistivity by the U.S. EPA Method 9090 test; no equivalent test protocols have been established for geotextiles and geonets but the U.S. EPA recommends the tests listed in Table 10.16.

Clogging is the primary cause of concern for the long-term performance of leachate collection and cap drainage systems, since this is the most common failure mechanism. *Clogging* is defined as the physical buildup of material in the collection pipe, drainage layer, or filter layer to the extent that leachate flow is significantly restricted. Clogging may be caused by the buildup of soil, biologic organisms, chemical and/or bio-

TABLE 10.15   Potential Problems and Solutions in Clay Liner Design and Construction

Problems	Cause	Preventative measures
Sidewall slump and collapse	Improper characterization of soil strength profile that results in improper side wall design	Properly characterize subsurface conditions
	High inward hydraulic pressure on side walls (sites below water table)	Design gentler side wall slope (depending on shear strength of foundation soil)
		Reduce hydraulic head by:
		■ Installing slurry wall around site perimeter to cut off groundwater
		■ Trenching and pumping around site to cut off groundwater
		■ Pumping from wells to lower groundwater table
Bottom heave or rupture	High inward hydraulic pressure on bottom (sites below water table)	Control depth of excavation to lower potential head
		Reduce hydraulic head by slurry wall, pumping, or other technique
		Fill landfill before heaving occurs
Accumulation of water in landfill during construction	Rainfall	Cover site with inflatable dome (reduces leachate treatment requirements for continuous operation facilities)
		Seal roll liner at end of construction day to ensure proper runoff of precipitation into sump
	Seepage into site (sites below water table)	Reduce hydraulic head in surrounding soil
		For all infiltration:
		■ Operate LCS to remove water
		■ Design LCS or LDS to handle extra water input
Drying and cracking of clay liner, greatly increasing permeability	Desiccation	Do not construct during extremely hot, dry periods
		Wet down liner during dry periods
		Cover liner with plastic sheet or soil layer if liner construction is interrupted
		Do not leave liner exposed prior to waste emplacement or LCS installation

TABLE 10.15   Potential Problems and Solutions in Clay Liner Design and
Construction (*Continued*)

Problems	Cause	Preventative measures
Loss in liner density (increased permeability)	Freeze-thaw	Do not construct during winter in cold climates
		Cover liner with soil blanket or other insulation material when cold weather is expected
Reduction in clay workability	Low temperatures	Increase compactive effort
		Cease construction until spring
Erosion of upper liner after construction (liquid impoundments)	Wave erosion	Place rip rap on lagoon side walls extending from below liquid level into freeboard area
Visible partings between liner lifts and increased permeability parallel to lifts	Liner lifts not properly tied together	Scarify or disk lower lift prior to installing next lift
		Ensure moisture content of last lift and lift being installed are the same
Pockets of high-permeabiliy material (e.g., sand and gravel) in liner material	Heterogeneous clay liner material	Closely inspect liner material at borrow site or as it is being installed and reject coarse-grained material
Leakage around designed liner penetrations	Improper sealing around penetrating objects	Avoid liner penetrations in design
		Seal properly around penetrating objects
Leachate collection system clogging	Sediment entering into systems	Cover system with geotextile or graded soil layer
Leachate collection system damage during waste emplacement	Inadequate management of personnel	Initiate personnel training program
		Cover system with geotextile or soil cover prior to waste placement
		Design system to minimize protrusions (manholes, etc.) in waste emplacement areas
Areas of high permeability	Substandard compaction	Conduct proper CQC and CQA including:
		■ Monitoring for material variability
		■ Observation of compaction operations
		■ Run moisture/density/ permeability tests

SOURCE: From USEPA, 1988a.

TABLE 10.16   Suggested Test Methods for Assessing Chemical Resistance of
Geosynthetics Used in Leachate Collection Systems

Test type	Geotextile	Geonet	Geocomposite
Thickness	Yes	Yes	Yes
Mass/unit area	Yes	Yes	Yes
Grab tensile	Yes	No	No
Wide width tensile	Yes	Yes	No
Puncture (pin)	No	No	No
Puncture (CBR)	Yes	Yes	Yes
Trapezoidal tear	Yes	No	No
Burst	Yes	No	No

SOURCE: USEPA, 1989a.

chemical precipitates, or combinations of these. Particulate clogging can occur in a number of locations. First, the sand filter itself can clog the drainage gravel. Second, the solid material within the leachate can clog the drainage gravel or geonet. Third (and most likely) the solid suspended material in the leachate can clog the sand filter or geotextile filter. The potential for particulate clogging can vary significantly depending on the particular site and waste involved; studies of leachate at 18 landfills have shown a variation in total solids from 0 to 59,200 mg/liter. Biologic clogging can result from many sources, including slime and sheath formation, biomass formation, ochering, sulfide deposition, and carbonate deposition. The order in which system components are likely to clog is (from most likely to least likely) sand filters and geotextile filters, gravel, geonets, and geocomposites. The most effective method for relieving particulate and/or biologic clogging is to create a high-pressure water flush to clean out the filter and/or the drain. In cases of high biologic growth, a biocide also may need to be used.

The final factor to be considered in leachate collection systems is extrusion and intrusion of materials into the leak detection system. In a composite primary liner system, clay can readily extrude through a geotextile into a geonet if the geotextile has continuous open spaces, i.e., a percentage open area (POA) greater than 1 percent. Therefore, relatively heavy nonwoven geotextiles are recommended. Elasticity and creep can cause geotextiles to intrude into geonets from composite primary liners as well. An FML above or below a geonet also can intrude into the geonet due to elasticity and creep. Laboratory evaluations that simulate field conditions can be used to alert designers to these potential problems.

10.7.4   Gas collection systems

Gas collection systems can have many of the same problems associated with them as do leachate collection and cap drainage systems—

physical deficiencies, failures, and infiltration and physical, chemical, or biologic clogging. (See the discussion above as it relates to granular venting media.) Infiltration of adjacent soil into the gravel vents and subsequent clogging can occur with water drainage into the vents. This can be avoided or minimized by use of appropriately graded filter materials. Too many fines (>5 percent passing the No. 100 sieve) in the backfill around vents can inhibit gas flow to the vents. The presence of soil and vegetation on the exposed surfaces of gravel vents and trenches can hinder gas venting; these materials should be kept off of vent surfaces. Since gas vents often must pass through the waste facility's cap, it is imperative that any connections between vent pipes and geosynthetics, especially geomembranes, be water- and gas-tight. When designing the gas venting system, the possible downdrag forces on the vents and vent pipes should be considered to prevent possible rupture of these structures during any settlement of the surrounding waste. If proper venting of gas from a waste facility is not provided, there can be problems with explosion hazards, vegetation distress, odors, property-value deterioration, physical disruption of the cover, and toxic vapors.

10.7.5 Cap/closure systems

Closure systems of waste containment facilities may suffer degradation or disruption and failure from a number of sources including erosion (from wind and water), lack of vegetation, sunlight, disturbance by burrowing animals (or by people), and surface settlement. The effects of wind, rain, hail, snow, and freeze-thaw can be detrimental to both the geosynthetics and the soil cover systems; these environmental factors should be considered and the cap design appropriately modified to eliminate the risk from each factor for a particular site. Sufficient depth of vegetative soil cover will prevent damage from most of these sources. Selection of an appropriate cover slope (steep enough to shed water but flat enough to minimize or avoid erosion) also will help to prevent future problems. Healthy vegetation growing over the cap will minimize the potential for surface and internal (piping) erosion of the cover soil.

The detrimental effects of animals and sunlight (UV and ozone) can be minimized by using adequate depths of soil cover over the other cap components. Soil depths over FMLs used in covers range from 3 to 6 ft (1 to 2 m) in thickness. Human intrusion usually can be prevented by posting signs and erecting fences.

The successful performance of the capping system is critical to the overall performance of the waste disposal facility. If the cap is disrupted or breached, precipitation will be allowed to enter, thus generating leachate on a continuing basis. This could add significantly to any potential disaster. If a breach of the cap were to occur many years after closure, it is likely that no maintenance forces or monies would be available. Thus the initial design must be developed carefully with long-term considerations in mind.

10.8 References

Alther, G. R. 1983. "The Methylene Blue Test for Bentonite Liner Quality Control," *Geotechnical Testing Journal,* 6 (3): 128–132.
Alther, G. R. 1987. "The Qualifications of Bentonite as a Soil Sealant," *Engineering Geology,* 23: 177–191.
American Concrete Institute. 1995 (updated annually). *ACI Manual of Concrete Practice,* 5 parts, Detroit, Mich.
Asphalt Institute. 1976. *Asphalt in Hydraulics,* Manual Series no. 12, Lexington, Ky.
Barlaz, M. A., and R. K. Ham. 1993. "Leachate and Gas Generation," in *Geotechnical Practice for Waste Disposal,* D. E. Daniel, ed., Chapman and Hall, London.
Bass, J. 1985. "Avoiding Failure of Landfill Drainage Systems," Arthur D. Little, Inc., Cambridge, Mass. (prepared for U.S. EPA).
Bowders, J. J., D. E. Daniel, G. P. Broderick, and H. M. Liljestrand. 1986. "Methods for Testing the Compatibility of Clay Liners with Landfill Leachate," *Hazardous and Industrial Solid Waste Testing, 4th Symposium,* STP886, American Society for Testing and Materials, Philadelphia.
Broms, B. 1987. "Stabilization of Very Soft Clay Using Geofabric," *Geotextiles and Geomembranes,* 5: 17–28.
Brown, K. W., J. C. Thomas, R. L. Lytton, P. Jayawickrama, and S. C. Bahrt. 1987. "Quantification of Leak Rates Through Holes in Landfill Liners," EPA/600/2-87/062, Environmental Protection Agency, Washington.
Buchanan, P. N. 1964. "Effect of Temperature and Adsorbed Water on Permeability and Consolidation Characteristics of Sodium- and Calcium-Montmorillonite," Ph.D. dissertation, Texas A&M University, College Station, Texas.
Carpenter, G. W., and R. J. Stephenson. 1986. "Permeability Testing in the Triaxial Cell," *Geotechnical Testing Journal,* 9 (1), ASTM, Philadelphia, pp. 3–9.
Carrier, W. D., ed. 1994. *Proceedings of the First International Congress on Environmental Geotechnics,* sponsored by International Society for Soil Mechanics and Foundation Engineering and Canadian Geotechnical Society, Edmonton, Alberta, Canada.
Cedegren, H. R. 1989. *Seepage, Drainage, and Flownets,* Wiley, New York.
Chapuis, R. P. 1990. "Soil-Bentonite Liners: Predicting Permeability from Laboratory Tests," *Canadian Geotechnical Journal,* 27 (1): 47–57.
Daniel, D. E. 1993a. "Clay Liners," in *Geotechnical Practice for Waste Disposal,* D. E. Daniel, ed., Chapman and Hall, London.
Daniel, D. E., ed. 1993b. *Geotechnical Practice for Waste Disposal,* Chapman and Hall, London.
Daniel, D. E. 1990. "Summary Review of Construction Quality Control for Compacted Soil Liners," in *Waste Containment Systems: Construction, Regulation, and Performance,* special technical publication no. 26, American Society of Civil Engineers, New York.
Daniel, D. G., S. J. Trautwein, S. S., Boynton, and D. E. Foreman. 1984. "Permeability Testing with Flexible-Wall Permeameters," *Geotechnical Testing Journal,* 7 (3), ASTM, Philadelphia, pp. 113–122.
Daniel, D. E., S. J. Trautwein, and D. McMurtry. 1985. "A Case History of Leakage from a Surface Impoundment," in *Seepage and Leakage from Dams and Impoundments,* R. L. Volpe and W. E. Kelly, ed., American Society of Civil Engineers., New York, pp. 220–235.
Daniel, D. E., and C. H. Benson. 1990. "Water Content-Density Criteria for Compacted Clay Liners," *Journal of Geotechnical Engineering,* 116 (12): 1811–1830.
Daniel, D. E., and R. M. Koerner. 1993. "Cover Systems," in *Geotechnical Practice for Waste Disposal,* D. E. Daniel, ed., Chapman and Hall, London.

Department of the Army. 1986. "Laboratory Soils Testing," EM 1110-2-1906, Washington.

Giroud, J. P. 1984. "Opening Address," *Proceedings of the International Conference on Geomembranes,* Denver, Colo., Industrial Fabrics Association International, St. Paul, Minn.

Giroud, J., and R. Bonaparte, 1989a. "Leakage Through Liners Constructed with Geomembranes: I. Geomembrane Liners," *Geotextiles and Geomembranes,* 8: 27–67.

Giroud, J., and R. Bonaparte, 1989b. "Leakage Through Liners Constructed with Geomembranes: II. Composite Liners," *Geotextiles and Geomembranes,* 8: 71–111.

Giroud, J. P., and I. D. Peggs. 1990. "Geomembrane Construction Quality Assurance," in *Waste Containment Systems: Construction, Regulation, and Performance,* special technical publication no. 26, American Society of Civil Engineers, New York.

Giroud, J., K. Badu-Tweneboah, and R. Bonaparte. 1991. "Rate of Leakage Through a Composite Liner Due to Geomembrane Defects," *Geotextiles and Geomembranes.* 10: 71–111.

Giroud, J., A. Khatami, and K. Badu-Tweneboah. 1989. "Evaluation of the Rate of Leakage through Composite Liners," *Geotextiles and Geomembranes,* 8 (4): 337–340.

Haxo, H. E. 1990. "Determining the Transport Through Geomembranes of Various Permeants in Different Applications," in *Geosynthetic Testing for Waste Containment Applications,* R. M. Koerner, ed., ASTM special technical publication 1081, American Society for Testing and Materials, Philadelphia.

Haxo, H. E., R. S. Haxo, N. A. Nelson, R. M. White, and S. Dakessian. 1985. "Liner Materials Exposed to Hazardous and Toxic Wastes," EPA-600/2-84-169, U.S. Environmental Protection Agency, Cincinnati, Ohio.

Hickey, M. E. 1971. "Asphaltic Concrete Canal Lining and Dam Facing," report no. REC-ERC-71-37, Bureau of Reclamation, U.S. Department of Interior, Denver, Colo.

Hinkle, R. D. 1976. "Impermeable Asphalt Concrete Pond Liner," *Civil Engineering,* August 1976: 46 (8): 56–59.

Hodge, V., N. DeSalvo, S. Mahmud, T. Margolis, and M. Evans. 1986. "Guidance Manual on Overtopping Control Techniques for Hazardous Waste Impoundments," EPA/600/2-86/012, U.S. Environmental Protection Agency, Washington.

Koerner, R. M. 1993a. "Collection and Removal Systems," in *Geotechnical Practice for Waste Disposal,* D. E. Daniel, ed., Chapman and Hall, London.

Koerner, R. M. 1993b. "Geomembrane Liners," in *Geotechnical Practice for Waste Disposal,* D. E. Daniel, ed., Chapman and Hall, London.

Koerner, R. M., Y. H. Halse, and A. E. Lord. 1990. "Long-Term Durability and Aging of Geomembranes," in *Waste Containment Systems: Construction, Regulation, and Performance,* special technical publication no. 26, American Society of Civil Engineers, New York.

Kozichi, P., and D. M. Heenan. 1983. "Use of Bentonite as a Soil Sealant for Construction of Underseal Sewage Lagoon Extension, Glenboro, Manitoba," Short Course on Waste Stabilization Ponds, Winnipeg, Manitoba, Canada.

LaGrega, M. D., P. L. Buckingham, and J. C. Evans. 1994. *Hazardous Waste Management,* McGraw-Hill, New York.

Landva, A. O., and J. I. Clark. 1990. *Geotechnics of Waste Fill: Theory and Practice,* A. O. Landva and G. D. Knowles, eds., STP1070, American Society for Testing and Materials, Philadelphia.

Mitchell, J. K., and F. T. Madsen. 1987. "Chemical Effects on Clay Hydraulic Conductivity," in *Geotechnical Practice for Waste Disposal '87,* R. D. Woods, ed., American Society of Civil Engineers, New York.

Mitchell, J., R. B. Seed, and H. B. Seed. 1990. "Kettleman Hills Waste Landfill Slope Failure: I. Liner-System Properties," *Journal of Geotechnical Engineering,* 116 (4): 647–668.

Oweis, I. S. 1993. "Stability of Landfills," in *Geotechnical Practice for Waste Disposal,* D. E. Daniel, ed., Chapman and Hall, London.

Quarles, J. 1982. *Federal Regulation of Hazardous Waste: A Guide to RCRA,* Environmental Law Institute, Washington.

Rao, S., L. K. Moulton, and R. K. Seals. 1977. "Settlement of Refuse Landfills," *Proceedings of a Conference on Geotechnical Practice of Disposal of Solid Waste Materials,* ASCE, Ann Arbor, Mich., pp. 574–598.

Rollings, M. P. 1994. "Geotechnical Considerations in Dredged Material Management," *Proceedings of the First International Congress on Environmental Geotechnics,* Edmonton, Alberta, Canada, International Society for Soil Mechanics and Foundation Engineering, Edmonton, Alberta, Canada.

Rollings, M. P. 1991. "Strategies for Environmental Engineering: Solutions for Progress," seminar papers, Longman Professional, South Melbourne, Australia.

Rollings, R. S., M. E. Poindexter, and K. G. Sharp, 1988. "Heavy Load Pavements on Soft Soils," part 5, *Proceedings of the 14th Australian Road Research Board Conference,* Australian Road Research Board, South Vermont, Canberra, Australia.

Seed, R., J. K. Mitchell, and H. B. Seed. 1990. "Kettleman Hills Waste Landfill Slope Failure: II. Stability Analyses," *Journal of Geotechnical Engineering,* 116 (4): 669–690.

Shafer, R. A., A. Renta-Babb, J. T. Bandy, E. D. Smith, and P. Malone, 1984. "Landfill Gas Control at Military Installations," technical report N-173, U.S. Army Corps of Engineers Construction Engineering Research Laboratory, Champaign, Ill.

Shan, H. Y., and D. E. Daniel. 1991. "Results of Laboratory Tests on a Geotextile/Bentonite Liner Material," *Geosynthetics '91,* Industrial Fabrics Association International, St. Paul, Minn., pp. 517–535.

Sowers, G. 1973. "Settlement of Waste Disposal Fills," *Proceedings of the Eighth International Conference on Soil Mechanics and Foundation Engineering,* vol. 2, Moscow, International Society for Soil Mechanics and Foundation Engineering.

USEPA. 1984a. "The Hydrologic Evaluation of Landfill Performance (HELP) Model," vol. I. "User's Guide for Version I," EPA/530/SW-84/009, Washington.

USEPA. 1984b. "The Hydrologic Evaluation of Landfill Performance (HELP) Model," vol. II: "Documentation of Version I," EPA/530/SW-84/010, Washington.

USEPA. 1985. "Covers for Uncontrolled Hazardous Waste Sites," EPA/540/2-85/002, U.S. EPA, Cincinnati, Ohio.

USEPA. 1987. "Background Document: Proposed Liner and Leak Detection Rule," EPA/530/SW-87/015, Washington.

USEPA. 1988a. "Design, Construction, and Evaluation of Clay Liners for Waste Management Facilities," EPA/530/SW-86/007F, U.S. EPA, Cincinnati, Ohio.

USEPA. 1988b. "Guide to Technical Resources for the Design of Land Disposal Facilities," EPA/625/6-88/018, U. S. EPA, Cincinnati, Ohio.

USEPA. 1988c. "Lining of Waste Containment and Other Impoundment Facilities," EPA/600/2-88/052, U.S. EPA, Cincinnati, Ohio.

USEPA. 1989a. "Requirements for Hazardous Waste Landfill Design, Construction, and Closure," seminar publication EPA/625/4-89/022, U.S. EPA, Cincinnati, Ohio.

USEPA. 1989b. "Technical Guidance Document: Final Covers on Hazardous Waste Landfills and Surface Impoundments," EPA/530/SW-89-047, U.S. EPA, Cincinnati, Ohio.

USEPA. 1991. "Technical Guidance Document, Inspection Techniques for the Fabrication of Geomembrane Field Seams," EPA/530/SW-91/051, Cincinnati, Ohio.

Wagner, K., K. Boyer, R. Claff, et al. 1986. *Remedial Action Technology for Waste Disposal Sites,* 2d ed., Noyes Data Corporation, Park Ridge, N.J.

Conversion Guide to SI Units

From	Multiply by*	Converts to†
Area, square mile	4046.9	square meters
Area, square yard	0.8361	square meters
Area, square foot	0.0929	square meters
Area, square inch	0.0006451	square meters
Bending moment, lb-force-foot	1.3558	newton-meter
Density, pounds/cubic yard	0.5932	kilograms/cubic meter
Density, pounds/cubic foot	16.0185	kilograms/cubic meter
Force, kips	4.4482	kilonewton
Force, pounds	4.4482	newtons
Length, miles	1609.344	meter
Length, yards	0.9144	meter
Length, feet	0.3048	meter
Length, inches	0.0254	meter
Force/length, pounds/foot	15.5939	newtons/meter
Force/length, pounds/inch	17.5127	newtons/meter
Mass, ton	907.184	kilogram
Mass, pound	0.4536	kilogram
Mass, ounce	28.35	gram
Pressure or stress, pound/square foot	47.8803	pascal
Pressure or stress, pound/square inch	6.8947	kilopascal
Temperature, °F	0.555556	°C
	$(t_F{}^O - 32)/1.8 = t_C{}^O$	°C
Volume, cubic yard	0.7646	cubic meters
Volume, cubic foot	0.02831	cubic meters
Volume, cubic inch	1.6387×10^{-5}	cubic meters

*The precision of a measurement converted to other units can never be greater than that of the original. To go from SI units to U.S. customary units, divide by the given constant. ASTM E 380 provides guidance on use of the SI system.

†The common SI prefixes are

mega	M	1,000,000.
kilo	k	1,000.
centi	c	0.01
milli	m	0.001
micro	μ	0.000001

B

American Society for Testing and Materials (ASTM) Test Methods Commonly Used in Construction Sampling and Testing

Following are some laboratory tests and specifications that are commonly encountered in geotechnical work. The list is not all-inclusive but should cover the bulk of the requirements normally found in work with soils, aggregates, and stabilization of these materials. ASTM publishes a complete set of over 8500 current standards annually. These are available as individual volumes covering specific areas (e.g., Volume 4.08 covers soil and rock, dimension stone, and geosynthetics), as copies of individual standards, or on CD ROM. ASTM may be contacted at 1916 Race Street, Philadelphia, PA 19103-1187; telephone: (215) 299-5400; fax: (215) 977-9679, for assistance in obtaining these materials.

ASTM standard	Test name
C 29	Test Method for Unit Weight and Voids in Aggregate
C 33	Specification for Concrete Aggregates
C 42	Methods of Obtaining and Testing Drilled Cores and Sawed Beams of Concrete
C 70	Test Method for Surface Moisture in Fine Aggregate
C 88	Test Method for Soundness of Aggregates by Use of Sodium Sulfate or Magnesium Sulfate
C 110	Methods for Physical Testing of Quicklime, Hydrated Lime, and Limestone
C 117	Test Method for Materials Finer than 75-μm (No. 200) Sieve in Mineral Aggregates by Washing

ASTM standard	Test name
C 127	Test Method for Specific Gravity and Absorption of Coarse Aggregate
C 128	Test Method for Specific Gravity and Absorption of Fine Aggregate
C 131	Test Method for Resistance to Degradation of Small-Size Coarse Aggregate by Abrasion and Impact in the Los Angeles Machine
C 136	Method for Sieve Analysis of Fine and Coarse Aggregate
C 150	Specification for Portland Cement
C 227	Test Method for Potential Alkali Reactivity of Cement-Aggregate Combinations (Mortar-Bar Method)
C 289	Test Method for Potential Reactivity of Aggregates (Chemical Method)
C 295	Recommended Practice for Petrographic Examination of Aggregates for Concrete
C 342	Test Method for Potential Volume Change of Cement-Aggregate Combinations
C 535	Test Method for Resistance to Degradation of Large-Size Coarse Aggregate by Abrasion and Impact in the Los Angeles Machine
C 566	Test Method for Moisture Content of Aggregate by Drying
C 586	Test Method for Potential Alkali Reactivity of Carbonate Rocks for Concrete Aggregates (Rock Cylinder Method)
C 593	Specifications for Fly Ash and Other Pozzolans for Use with Lime
C 595	Specification for Blended Hydraulic Cements
C 618	Specification for Fly Ash and Raw or Calcined Natural Pozzolan for Use as a Mineral Admixture in Portland Cement Concrete
C 666	Test Method for Resistance of Concrete to Rapid Freezing and Thawing
C 702	Recommended Practice for Reducing Field Samples of Aggregate to Testing Size
C 977	Specification for Quicklime and Hydrated Lime for Soil Stabilization
C 989	Specification for Ground Iron Blast-Furnace Slag for Use in Concrete and Mortars
C 1105	Test Method for Length Change of Concrete Due to Alkali-Carbonate Rock Reaction
D 5	Test Method for Penetration of Bituminous Materials
D 75	Methods for Sampling Aggregates
D 420	Recommended Practice for Investigating and Sampling Soil and Rock for Engineering Purposes
D 421	Method for Dry Preparation of Soil Samples for Particle-Size Analysis and Determination of Soil Constants
D 422	Method for Particle-Size Analysis of Soils
D 448	Specification for Standard Sizes of Coarse Aggregate for Road and Bridge Construction
D 558	Test Method for Moisture-Density Relations of Soil-Cement Mixtures
D 559	Methods for Wetting-and-Drying Tests of Compacted Soil-Cement Mixtures
D 560	Methods for Freezing and Thawing Tests of Compacted Soil-Cement Mixtures
D 698	Test Methods for Moisture-Density Relations of Soils and Soil-Aggregate Mixtures Using 5.5-lb (2.49-kg) Rammer and 12-in (305-mm) Drop

ASTM standard	Test name
D 854	Test Method for Specific Gravity of Soils
D 946	Specification for Penetration-Graded Asphalt Cement for Use in Pavement Construction
D 977	Specification for Emulsified Asphalt
D 1195	Test Method for Repetitive Static Plate Load Tests of Soils and Flexible Pavement Components, for Use in Evaluation and Design of Airport and Highway Pavements
D 1241	Specification for Materials for Soil-Aggregate Subbase, Base, and Surface Courses
D 1556	Test Method for Density of Soil in Place by the Sand-Cone Method
D 1557	Test Methods for Moisture-Density Relations of Soils and Soil-Aggregate Mixtures Using 10-lb (4.54-kg) Rammer and 18-in (457-mm) Drop
D 1559	Test Method for Resistance to Plastic Flow of Bituminous Mixtures Using Marshall Apparatus
D 1560	Test Methods for Resistance to Deformation and Cohesion of Bituminous Mixtures by Means of Hveem Apparatus
D 1586	Method for Penetration Test and Split-Barrel Sampling of Soils
D 1587	Practice for Thin-Walled Tube Sampling of Soils
D 1632	Methods of Making and Curing Soil-Cement Compression and Flexure Test Specimens in the Laboratory
D 1633	Test Methods for Compressive Strength of Molded Soil-Cement Cylinders
D 1883	Test Method for CBR (California Bearing Ratio) of Laboratory Compacted Soils
D 2026	Specification for Cutback Asphalt (Slow-Curing Type)
D 2027	Specification for Cutback Asphalt (Medium-Curing Type)
D 2028	Specification for Cutback Asphalt (Rapid-Curing Type)
D 2166	Test Method for Compressive Strength of Cohesive Soil
D 2167	Test Method for Density and Unit Weight of Soil in Place by the Rubber Unconfined Balloon Method
D 2170	Test Method for Kinematic Viscosity of Asphalts
D 2171	Test Method for Viscosity of Asphalts by Vacuum Capillary Viscometer
D 2216	Method for Laboratory Determination of Water (Moisture) Content of Soil, Rock, and Soil-Aggregate Mixtures
D 2217	Method for Wet Preparation of Soil Samples for Particle-Size Analysis and Determination of Soil Constants
D 2397	Specification for Cationic Emulsified Asphalt
D 2419	Test Method for Sand Equivalent Value of Soils and Fine Aggregate
D 2434	Test Method for Permeability of Granular Soils (Constant Head)
D 2435	Test Method for One-Dimensional Consolidation Properties of Soils
D 2487	Test Method for Classification of Soils for Engineering Purposes
D 2488	Recommended Practice for Description and Identification of Soils (Visual-Manual Procedure)
D 2573	Test Method for Field Vane Shear Test in Cohesive Soil
D 2844	Test Method for Resistance R-Value and Expansion Pressure of Compacted Soils
D 2850	Test Method for Unconsolidated, Undrained Compressive Strength of Cohesive Soils in Triaxial Compression

ASTM standard	Test name
D 2872	Test Method for the Effect of Heat and Air on a Moving Film of Asphalt (Rolling Thin Film Oven Test)
D 2901	Test Method for Cement Content of Freshly Mixed Soil-Cement
D 2922	Test Methods for Density of Soil and Soil-Aggregate in Place by Nuclear Methods (Shallow Depth)
D 2937	Test Method for Density of Soil in Place by the Drive-Cylinder Method
D 2940	Specification for Graded Aggregate Material for Bases or Subbases for Highways or Airports
D 3017	Test Method for Water Content of Soil and Rock in Place by Nuclear Methods (Shallow Depth)
D 3080	Method for Direct Shear Test of Soils Under Consolidated Drained Conditions
D 3155	Test Method for Lime Content of Uncured Soil-Lime Mixtures
D 3282	Recommended Practice for Classification of Soils and Soil-Aggregate Mixtures for Highway Construction Purposes
D 3381	Specification for Viscosity-Graded Asphalt Cement for Use in Pavement Construction
D 3385	Test Method for Infiltration Rate of Soils in Field Using Double-Ring Infiltrometers
D 3398	Test Method for Index of Aggregate Particle Shape and Texture
D 3628	Recommended Practice for Selection and Use of Emulsified Asphalt
D 3744	Test Method for Aggregate Durability Index
D 4123	Method for Indirect Tension Test for Resilient Modulus of Bituminous Mixtures
D 4220	Practices for Preserving and Transporting Soil Samples
D 4223	Practice for Preparation of Test Specimens of Asphalt Stabilized Soils
D 4253	Test Methods for Maximum Index Density of Soils Using a Vibratory Table
D 4254	Test Methods for Minimum Index Density of Soils and Calculation of Relative Density
D 4318	Test Method for Liquid Limit, Plastic Limit, and Plasticity Index of Soils
D 4429	Test Method for Bearing Ratio of Soils in Place
D 4564	Test Method for Density of Soil in Place by the Sleeve Method
D 4609	Standard Guide for Screening Chemicals for Soil Stabilization
D 4643	Test Method for Determination of Water (Moisture) Content of Soil by the Microwave Oven Method
D 4718	Practice for the Correction of Unit Weight and Water Content for Soils Containing Oversized Particles
D 4791	Test Method for Flat or Elongated Particles in Coarse Aggregate
D 4792	Test Method for Potential Expansion of Aggregates from Hydration Reactions
E 11	Specifications for Wire-Cloth Sieves for Testing Purposes
E 380	Use of the International System of Units (SI) (the Modernized Metric System)

Index

ABOUT THE AUTHORS

MARIAN P. ROLLINGS has over 20 years of experience working with geotechnical materials. Her varied experience encompasses a broad range of geotechnical topics including site improvement, behavior of and construction with soft soils, laboratory testing, field monitoring, wetlands, and liners. She obtained her B.S. and M.S. degrees from the University of Tennessee and her Ph.D. from Texas A & M University.

RAYMOND S. ROLLINGS, JR. has spent most of his last 23 years working worldwide on a variety of geotechnical problems that have been primarily associated with heavy-duty pavements. He is currently with the USAE Waterways Experiment Station where he is engaged in a number of research programs and trouble-shooting activities. He obtained his B.S. degree from the U.S. Military Academy, his M.S. from the University of Illinois, and his Ph.D. from the University of Maryland.